THE LINEAR IC HANDBOOK

THE

LINEAR IC HANDBOOK

MICHAEL S. MORLEY

TAB Professional and Reference Books

Division of TAB BOOKS Inc.
P.O. Box 40, Blue Ridge Summit, PA 17214

FIRST EDITION

FIRST PRINTING

Printed in the United States of America

Library of Congress Cataloging in Publication Data

Morley, M. S.
The linear IC handbook.

Includes index.
1. Linear integrated circuits. I. Title.
TK7874.M5338 1986 621.381′73 86-5857
ISBN 0-8306-0472-3

Cover photograph courtesy of National Semiconductor.

Contents

Introduction

THE PURPOSE OF THIS BOOK IS TO PROVIDE BOTH TUTORIAL information and technical data to aid electronics circuit designers in the selection of standard linear integrated circuits. This book was written for engineers, technicians, and hobbyists—in short, anyone who designs electronics circuits for any purpose.

During the course of a design assignment, the circuit designer must answer the following questions:

(1) What are the requirements?
(2) What are the alternative circuit design approaches?
(3) What components are available to do the job?
(4) How do I apply these components to my design to assure high yield and reliability?

The number one job is to understand the requirements. If this is done poorly, then the design will have to be redone (assuming there is some means in place for determining whether your design meets the requirements, i.e., design reviews and product assurance reviews). After you understand the design requirements, then you will answer questions 2-4 for each "paper design" you think of. The answer to these questions usually drives the design to one or two good design alternatives. The rest of the time, or a big percentage of it, is spent doing component trade studies. For that, you need specific technical information. You are looking for the "right" IC for the job—the least expensive IC that will do the job with some (but not too much) design margin. What do you do? You wade through a dozen or more IC books, looking up this or that IC, comparing the IC specifications with your requirements. And you look up the prices, or call the local sales representative to get a budgetary price.

The purpose of this book is to reduce the amount of time that you spend looking for a particular linear IC. Suppose, for example, that you need a low-noise op amp. The selection guides in Table 4-22 through 4-24 list low-noise op amps for the commercial, industrial, and military temperature ranges by part number, manufacturer, input noise voltage, input noise current, and price. From there you can go to Appendix A and get some more data—bandwidth, slew rate, input offset voltage, power supply current, and other data. Beyond this you will have to go to the manufacturer's data sheet or databook.

There are no pinouts in this book. No curves. It is not the purpose of this book to replace the databooks of a dozen or more linear IC manufacturers. (1) This book isn't big enough. (2) It's not, and never will be, an official databook. (3) Although I have tried to carefully copy (and in some cases, interpret) the manufacturers' data, I am sure that in a book this size that there are at least a few errors. The point is this: use this book to quickly locate an IC, but do not base any designs on the data in this book alone.

Also, there is price information in this book, but you should not consider this book an official price book for two reasons: (1) The price data in this book is out of date. It was out of date when I got it (fall of 1984). In fact, the only up-to-date prices are those you get when you formally ask a supplier for a quotation. (2) The price of any particular IC in this book is the lowest price for that IC. For example, if there is more than one package available for a given part type, I have listed the unit price for the lowest cost package version, probably plastic.

Then why has the price been given? Because IC cost is a big factor in circuit design. The prices in this book are old, but they are important because they tell you the *relative* price you must pay to get a certain performance. The name of the game is to get the most performance for your money. Never select a more expensive part than you need to do the job (with some performance margin for worst case conditions). (For me, learning the prices of the ICs listed in this book was one of the more interesting aspects of writing this book. I quickly found out which manufacturers generally produce the more expensive parts. And not all of these parts had the best performance!)

The structure of this book is as follows:

Part I is about linear IC fundamentals: fabrication (Chapter 1), components (Chapter 2), and design techniques (Chapter 3). This is background information for linear IC users, not a course for linear IC designers. Advanced IC design techniques are beyond the scope of this book.

Part II provides a basic tutorial section and a set of selection guides for seven major linear IC types: op amps (Chapter 4), comparators (Chapter 5), voltage references (Chapter 6), voltage regulators (Chapter 7), D/A converters (Chapter 8), A/D converters (Chapter 9), and sample-and-hold amplifiers (Chapter 10). The largest chapter is on op amps for two reasons: (1) There are more op amp types available than any other linear IC. (2) The design techniques used in op amps are fundamental to all other linear ICs. A thorough understanding of op amp design techniques will help you understand other linear IC types more quickly. The tutorials in the other chapters assume you understand the op amp material. This is true in general: the later chapters are based on the ideas presented in earlier chapters.

Part III is devoted to other linear ICs: special-purpose amplifiers (Chapter 11), analog math blocks (Chapter 12), timers and oscillators (Chapter 13), and transistor, amplifier, and diode arrays (Chapter 14). Each of these chapters presents tutorial and selection guides for four or five types of linear ICs that have similar or related functions. None of these categories have a large number of available part types. Nevertheless, the linear IC types presented in these chapters perform standard functions and have broad applications. Every circuit designer should be aware of these parts, how they work, and about how much they cost.

The Appendices are a set of specification summary tables by IC type and then by manufacturer. The data in the Appendices was compiled first, before the text sections were written and before the selection guide tables were created. I used a database program to enter the raw data into my computer and then used a sorting/report program to create the selection guides given in Part II.

The Appendices are purposely incomplete. For example, Texas Instruments makes the following op amps: μA709, LM108, SE5534, MC1558, and the OP-07. None of these op amps are included in the Texas Instruments section of Appendix A. However, you can find the μA709 under Fairchild, the LM108 under National, the SE5534 under Signetics, the MC1558 under Motorola, and the OP-07 under Precision Monolithics Incorporated (PMI) because these manufacturers are the original sources for these ICs. (In general, the prefix tells you the original source—μA for Fairchild, LM for National, and so forth.) Therefore, any part that has exactly the same prefix as the original part is not listed in Appendix A. Since the specifications are the same, the only thing lost is the price data, which should be approximately the same.

This does not mean that all alternate sourced parts are not included. For example, National makes the LM709, an alternate source to the Fairchild μA709. Both parts are in the Appendix. The rule I used for including parts in the Appendix (and also the selection guides) is this: If manufacturer A made a part with the same prefix (and number) as manufacturer B (the original manufacturer), then the part made by manufacturer A was not included. If manufacturer A made a part with the same number but a different prefix than manufacturer B, then both parts were included. These conditions correspond to two cases. In the first case (same prefix), the IC appears to be a straight copy, probably a mask exchange and licensing agreement. In the second case, it appears that a new IC has been designed to previously published specifications. The schematic is different, and probably there is no licensing agreement. Also, the redesigned part may not have exactly the same specifications. It's your responsibility (not your purchasing agent's) to make sure that alternate sourced ICs will meet your application's requirements.

Chapter 1

Linear IC Fabrication

A PHOTOGRAPH OF A SIMPLE LINEAR INTEGRATED CIRCUIT is shown in Fig. 1-1. Understanding how this particular chip and linear ICs in general are made is the purpose of this chapter. This is important because the performance of linear ICs is directly related to how the circuit components are fabricated.

MASK GENERATION

The chip shown in Fig. 1-1 was made using a set of masks, analogous in some ways to the artwork used in the generation of a multilayered printed circuit, or PC, board. In the simplest linear IC process, there are at least seven masks. These masks are listed in Table 1-1, on page 16. Figures 1-2 through 1-8 show simplified artwork drawings for the seven masks used to fabricate the chip shown in Fig. 1-1. The process steps associated with these masks will be explained later in this chapter.

As in the case of a PC board, the artwork for these masks is drawn at a much larger scale than the final product. IC mask artwork, in fact, is generally drawn several hundred times (as high as 1000 times) actual size.

The artwork generally is drawn on gridded mylar and then digitized. Digitization is the process whereby the artwork is entered into a computer. To the computer, each layer of the artwork is a

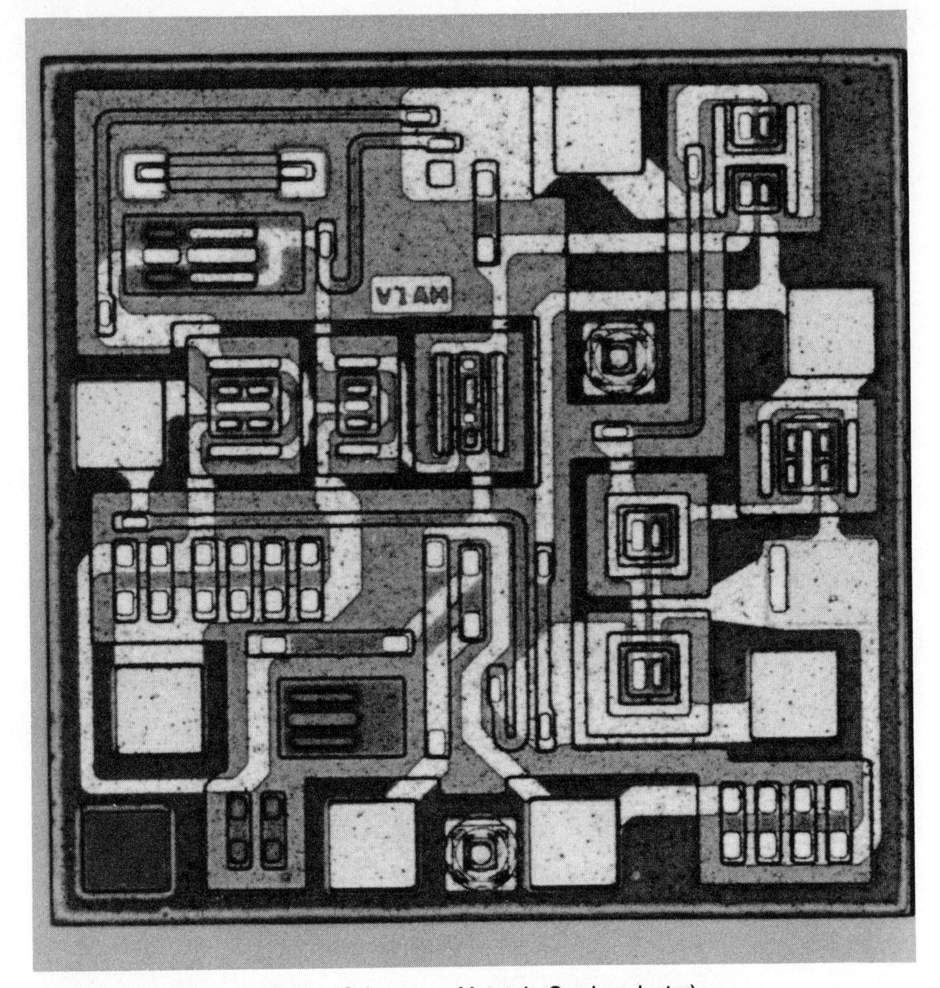

Fig. 1-1. Photograph of a linear IC (courtesy Motorola Semiconductor).

set of geometries defined by sets of coordinates connected by line segments. With the advent of computer-aided design (CAD) workstations, many IC designs today are drawn directly on the terminal screen, thus eliminating the tedious process of drawing and digitizing the artwork.

The physical design, or layout, of the IC is governed by a set of mask design rules, such as minimum rectangle size, minimum width, minimum spacing to another geometry of the same layer or other layer. Some of these rules are the result of the limitation of the photolithography technology used in mask generation. Other rules are defined in order to assure certain electrical performance criteria. For example, base-to-collector breakdown voltage for an NPN transistor is directly related to the spacing between the isolation diffusion (mask 2) and the base diffusion (mask 3).

Next, a 10× (10 times actual size) plate, called a reticle, for each layer is generated from the layout database. The reticle is used to photographically step and repeat a 1× image of a single layer in rows and columns on a square glass plate large enough to cover a 3-inch, 4-inch, or possibly even a 5-inch diameter wafer. (Wafer diameter size historically has been increasing with improved processing methods.)

The plates made from the reticles are called masters. From these are generated a set of masks, called working plates, which are used to fabricate the IC. A simplified drawing of a finished wafer is shown in Fig. 1-9. This figure shows only the chip boundaries and is not to scale. The darkened areas indicate the location of test patterns, which are used to monitor the electrical performance of test transistors and resistors during and after wafer fabrication.

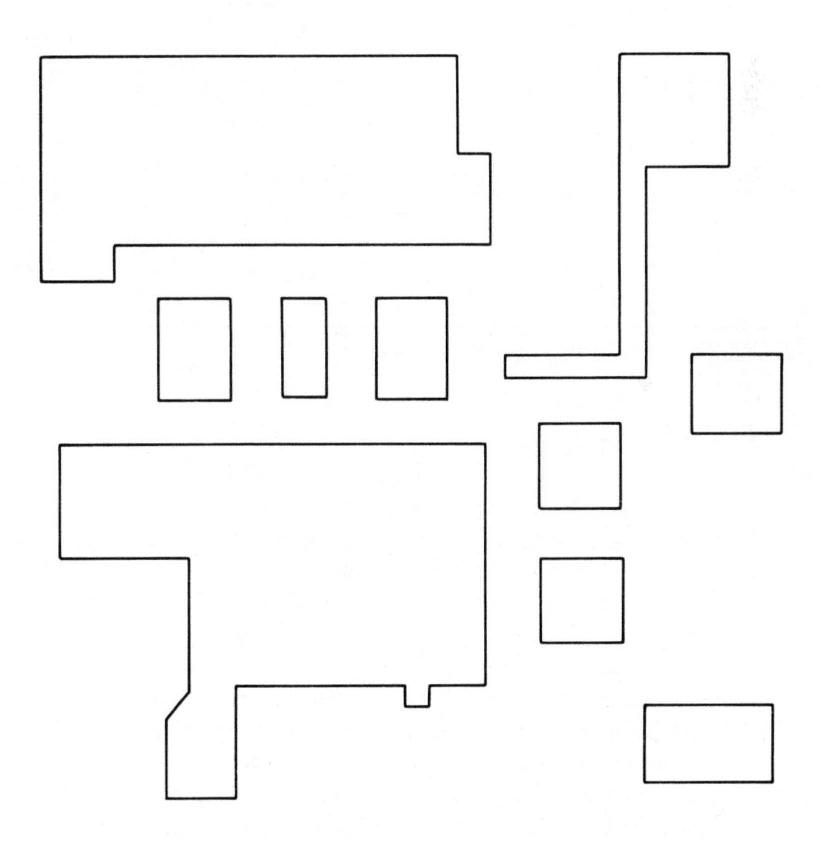

Fig. 1-2. Line drawing of buried layer diffusion artwork (mask 1).

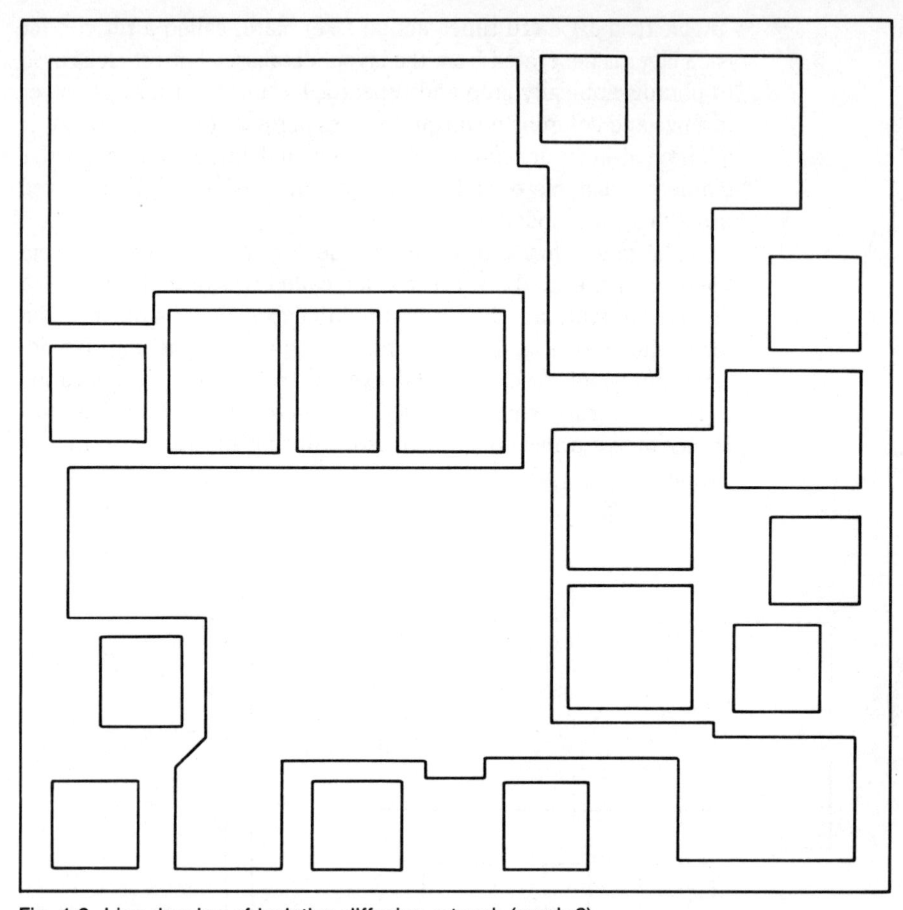

Fig. 1-3. Line drawing of isolation diffusion artwork (mask 2).

WAFER FABRICATION

A small portion of an IC chip, or die, is shown in Fig. 1-10. This figure shows the top view and side view of a hypothetical, small NPN transistor. The side view is the view seen if you could make a horizontal slice through the middle of the top view of the transistor. Dashed lines are used to indicate that only a portion of the chip is shown.

The top view is approximately to scale with the collector region measuring 95 microns long by 67 microns wide, or 0.00374 inches by 0.00264 inches (25.4 microns equals 0.001 inch, or 1 mil). The side view is not to scale. The diffusions and layers are deeper or thicker (as in the case of the epitaxial layer) than the scale of the top view would indicate. This liberty has been taken in order to be able to clearly show the various transistor regions.

Now, let's see how each of these regions is formed.

Buried Layer Diffusion

The formation of the buried layer is shown in Fig. 1-11. The starting material is a P-type silicon wafer approximately 15 to 20 mils thick (Fig. 1-11A). P-type silicon is made by introducing atoms with three electrons in their outer shells. Because silicon has four electrons in its outer shell, vacancies, or holes, are formed when the covalent bonds are formed (two electrons per covalent bond, eight electrons to completely fill the outer shell). Because holes have a positive charge, the material is called P-type. (N-type material is made by introducing atoms with five electrons in their outer shells. When the covalent bonds are formed, one free electron is generated for each dopant atom. Since electrons are negatively charged, the material is called N-type.)

Step 1. A thin layer of silicon dioxide is grown on the surface (Fig. 1-11B). The silicon dioxide is a form of glass and acts as a

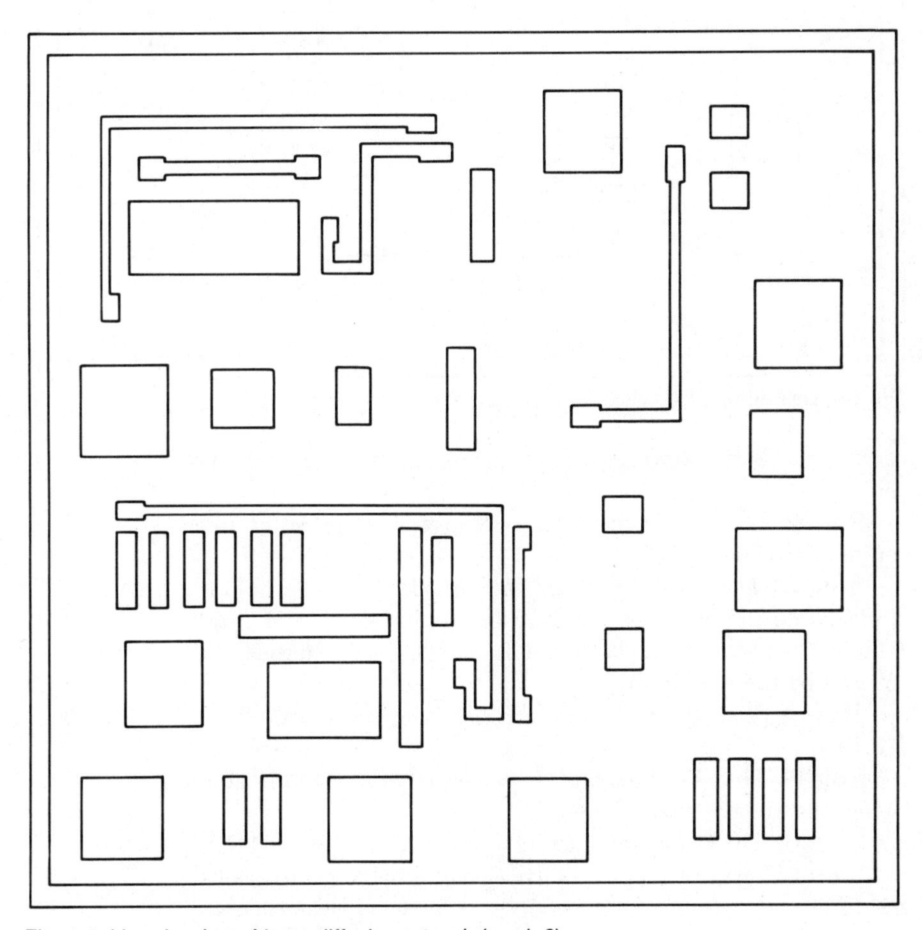

Fig. 1-4. Line drawing of base diffusion artwork (mask 3).

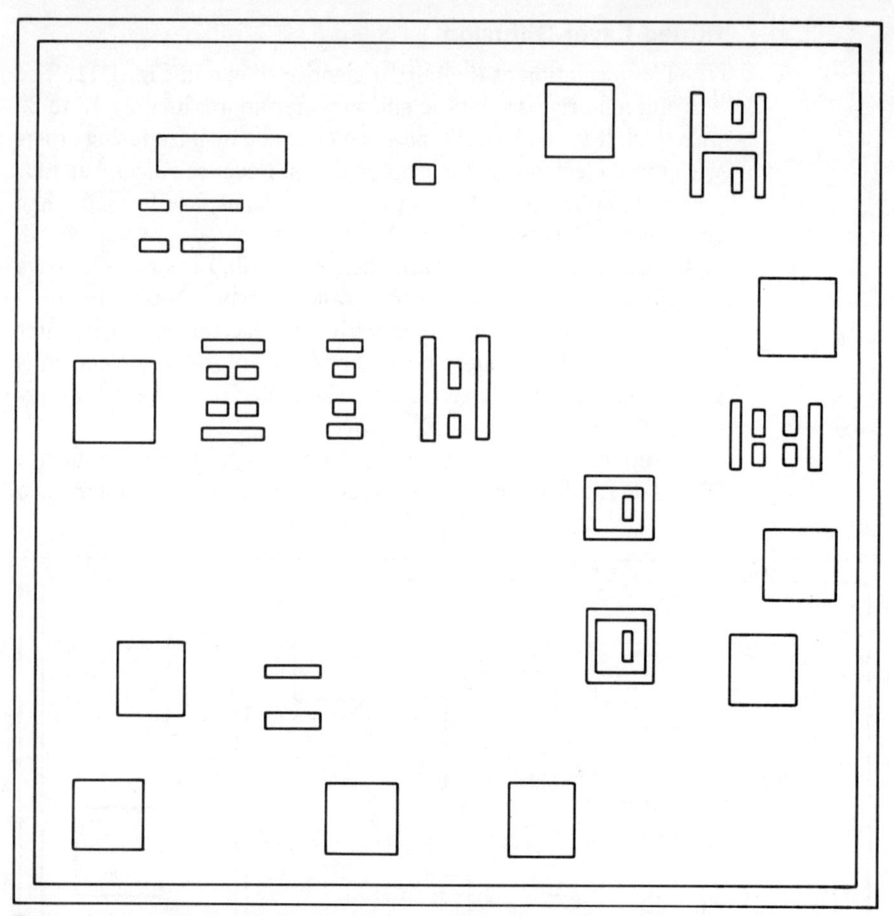

Fig. 1-5. Line drawing of emitter/collector diffusion artwork (mask 4).

barrier, preventing unwanted impurities from diffusing into the silicon.

Step 2. A layer of photoresist is deposited on top of the silicon dioxide (Fig. 1-11C).

Step 3. The glass mask with the buried layer pattern on it is placed on the wafer and is exposed to ultraviolet light (Fig. 1-11D). The photoresist is hardened in all areas except those areas covered by the mask pattern.

Step 4. The unexposed photoresist is washed away (Fig. 1-11E).

Step 5. The silicon dioxide not covered by the photoresist is chemically etched away (Fig. 1-11F).

Step 6. The remaining photoresist is removed (Fig. 1-11G).

Step 7. Impurity atoms with five electrons in their outer shells are diffused into the silicon through the oxide windows formed in steps 1-6 (Fig. 1-11H). The silicon just below the exposed surface becomes N-type. The buried layer is called "N+" because the number of free electrons is higher in this region than in the so-called epitaxial N-type layer formed in the next step. The depth and impurity concentration of the buried layer diffusion is a function of oven temperature and exposure time. The purpose of the buried layer is to reduce the collector resistance of the NPN transistors.

Epitaxial Layer

The oxide remaining from the formation of the buried layer is removed and N-type silicon is grown on the surface of the wafer (Fig. 1-12). This is the epitaxial layer. The thickness and impurity concentration of the epitaxial layer directly affect transistor breakdown voltage and collector resistance. For a medium voltage linear IC process, the epitaxial layer thickness may be in the range

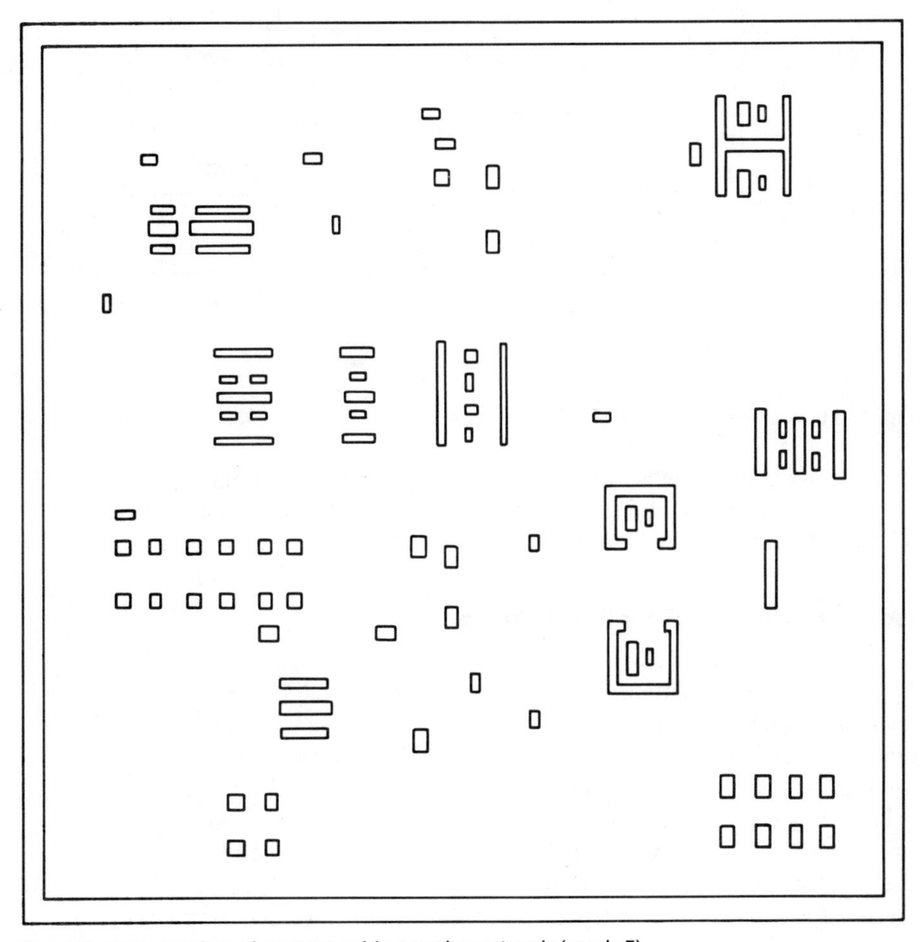

Fig. 1-6. Line drawing of contact oxide opening artwork (mask 5).

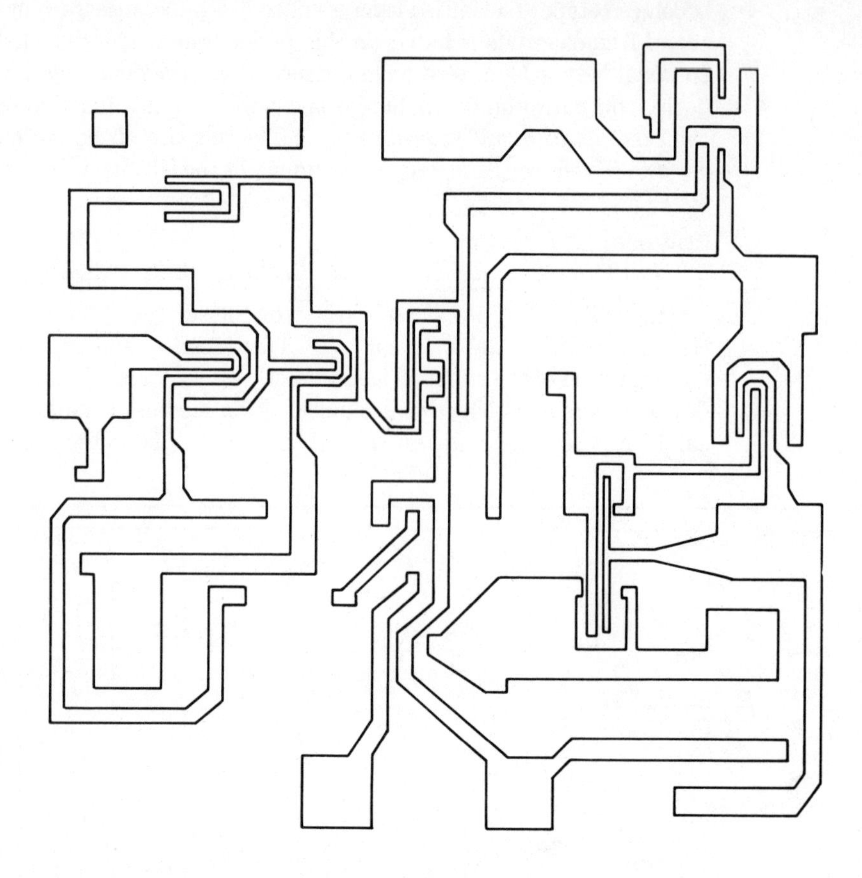

Fig. 1-7. Line drawing of metalization artwork (mask 6).

of 10-15 microns. This is very thin when compared to the thickness of the substrate, which, as stated earlier, is in the range of 15 to 20 mils (381 to 508 microns).

Isolation, Base, and Emitter Diffusions

The same process steps shown in Fig. 1-11A through 1-11H is repeated for the isolation, base, and emitter diffusions. The P+ isolation diffusion (Fig. 1-13), as its name implies, isolates one epitaxial island, or tub, from another. The isolation diffusion extends all the way to the P-substrate to insure isolation which is achieved by connecting the substrate to the most negative potential (ground or V−), thus backbiasing the PN junction formed by the P-substrate and the N-epitaxial layer.

An epitaxial island formed by the isolation diffusion may be the collector of an NPN transistor (as shown in Fig. 1-10), the base of PNP transistor, or an isolation region for P-type diffused resistors. (Top and side view drawings for PNP transistors and resistors will be shown and explained in the next chapter.)

The P-base diffusion (Fig. 1-14) is so-named because it forms the base region of a NPN transistor. The diffusion is also used for P-type diffused resistors, both the emitter and collector of the lateral PNP, and the emitter of the substrate PNP (the collector of the substrate PNP is the substrate, hence the name "substrate PNP"). The base diffusion depth is about 3 microns. Note that during the isolation and base diffusions the buried layer "up-diffuses" slightly into the N-epitaxy. In fact, the buried layer will continue to up-diffuse into the N-epitaxy during the remainder of the wafer processing.

Next, the NPN emitters and N+ collectors are diffused simultaneously (Fig. 1-15). The N+ collector diffusion is necessary be-

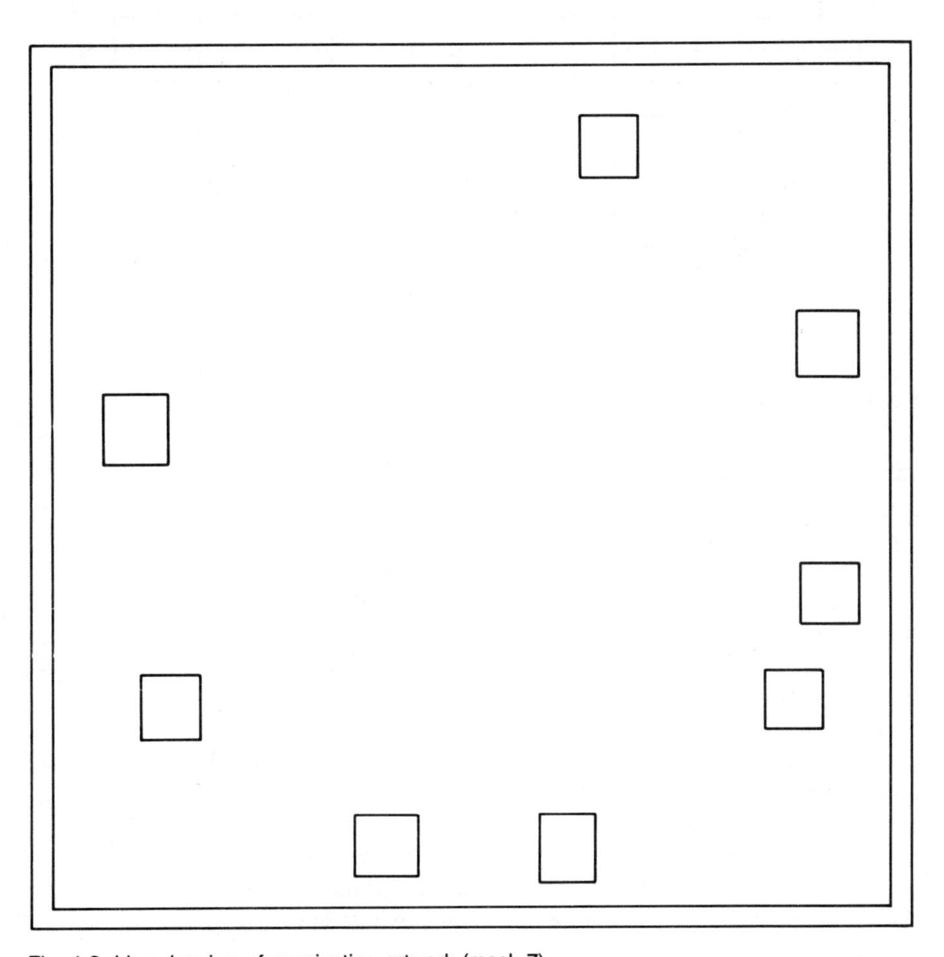

Fig. 1-8. Line drawing of passivation artwork (mask 7).

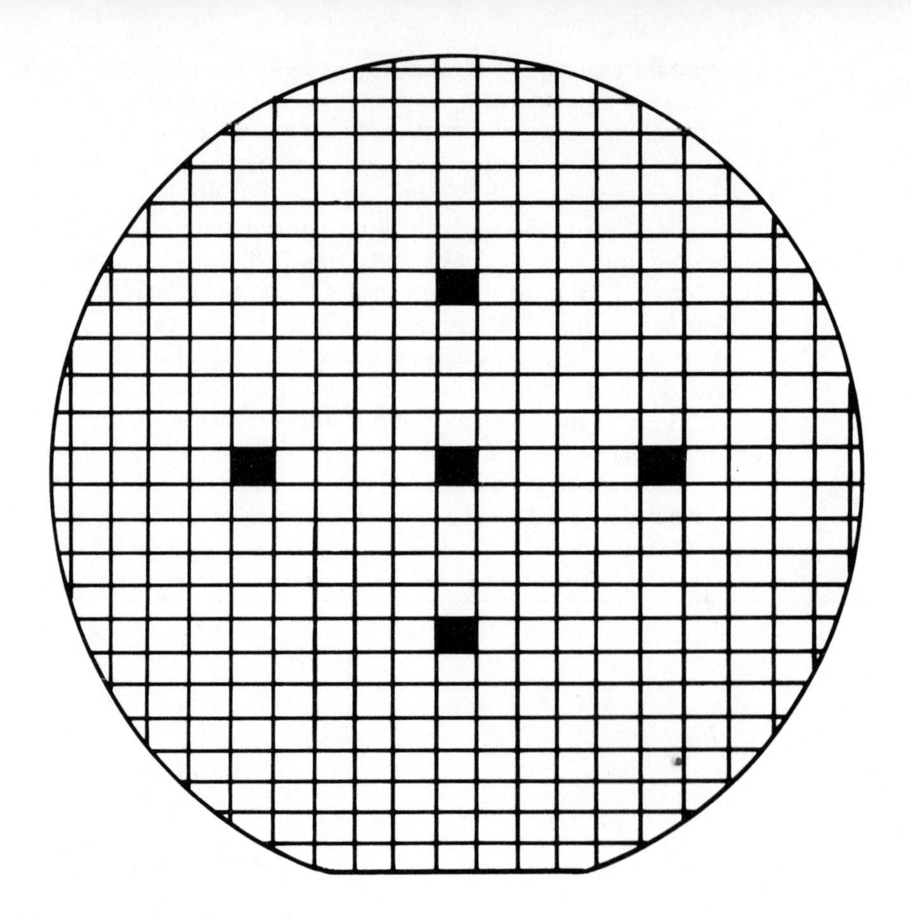

Fig. 1-9. Simplified drawing of a finished wafer.

cause direct contact to the epitaxy with aluminum metalization is poor at best. The N+ diffusion solves this problem and also reduces the overall collector bulk resistance. The total collector resistance begins at the base-collector junction immediately below the emitter, extends down through the high-resistance epitaxy, down into the low-resistance buried layer, laterally through the buried layer, up through the epitaxy, and, finally, up through the low-resistance N+ collector diffusion. For a small geometry, this collector resistance is on the order of several hundred ohms total.

The emitter diffusion is approximately 2-microns deep, which means that the base region underneath the emitter is about 1-micron thick. Controlling the emitter depth is critical to device performance. The thinner the base region the higher the current gain (beta), which is generally desirable, but also the thinner the base the lower the collector to emitter breakdown, which is generally not desirable.

Contact Openings, Metalization, and Passivation

At this point all devices on the wafer have been fabricated. To interconnect these devices it is necessary to oxidize the wafer once more and selectively etch openings for contacts (Fig. 1-16) in the same manner as was done for the previous diffusions. Only this time, instead of a diffusion, a thin layer (1 micron) of aluminum is deposited over the entire wafer and then etched (Fig. 1-17).

Finally, the wafer is passivated with a layer of oxide or other material to protect the finished chip. Bonding pad locations (see Fig. 1-8) and the area between chips (called the scribe channel) is etched. Bonding pads approximately 5 mils on a side are required in order to interconnect the die to the outside world. Gold or aluminum 1-mil wire is used to interconnect the die to the lead frame

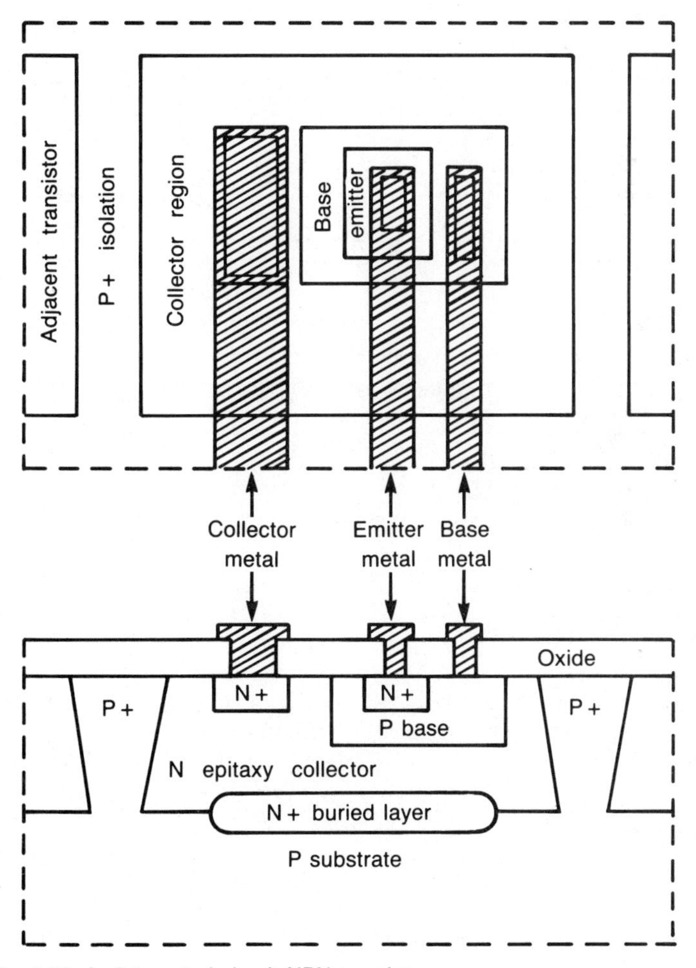

Fig. 1-10. An integrated circuit NPN transistor.

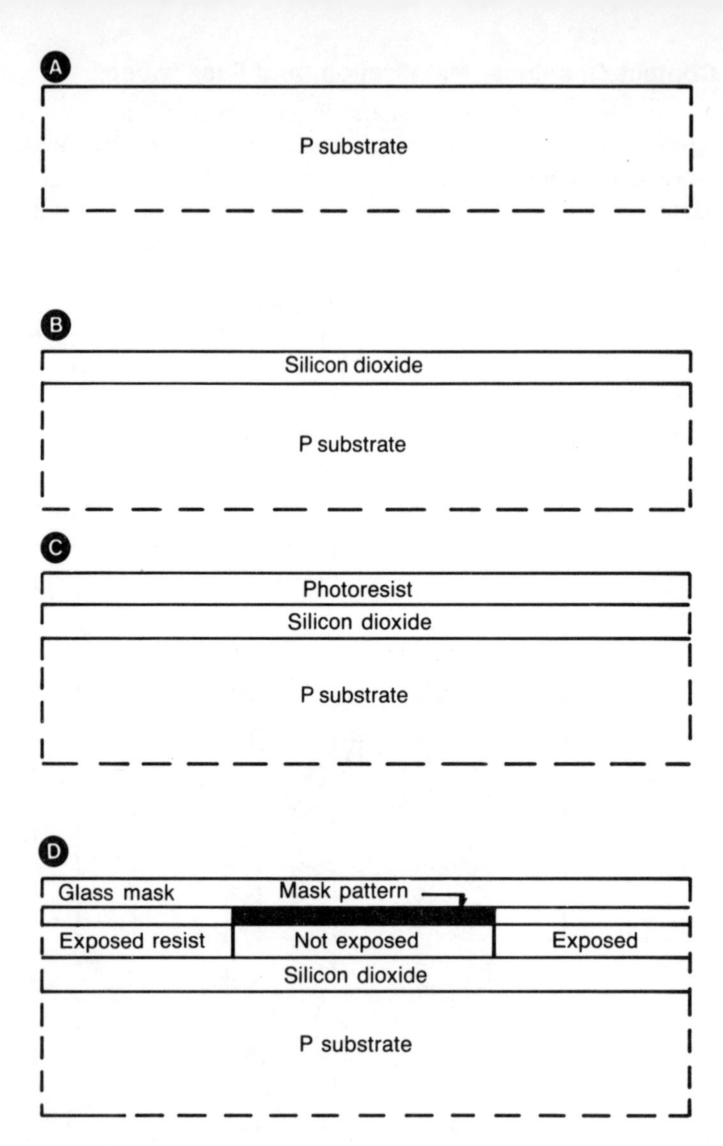

Fig. 1-11. Formation of the buried layer.

of dual-in-line packages (DIPs) or to the posts of a can-type package.

The scribe channel is usually etched in order to remove all oxide in the channel. This makes it easier to "saw" the wafer into dice: the wafer is actually scribed and gently stressed to break up the wafer along the chip boundaries.

WAFER TEST

Prior to the scribe and break operation, each die on the wafer is tested. This is accomplished by bringing a set of probes (with

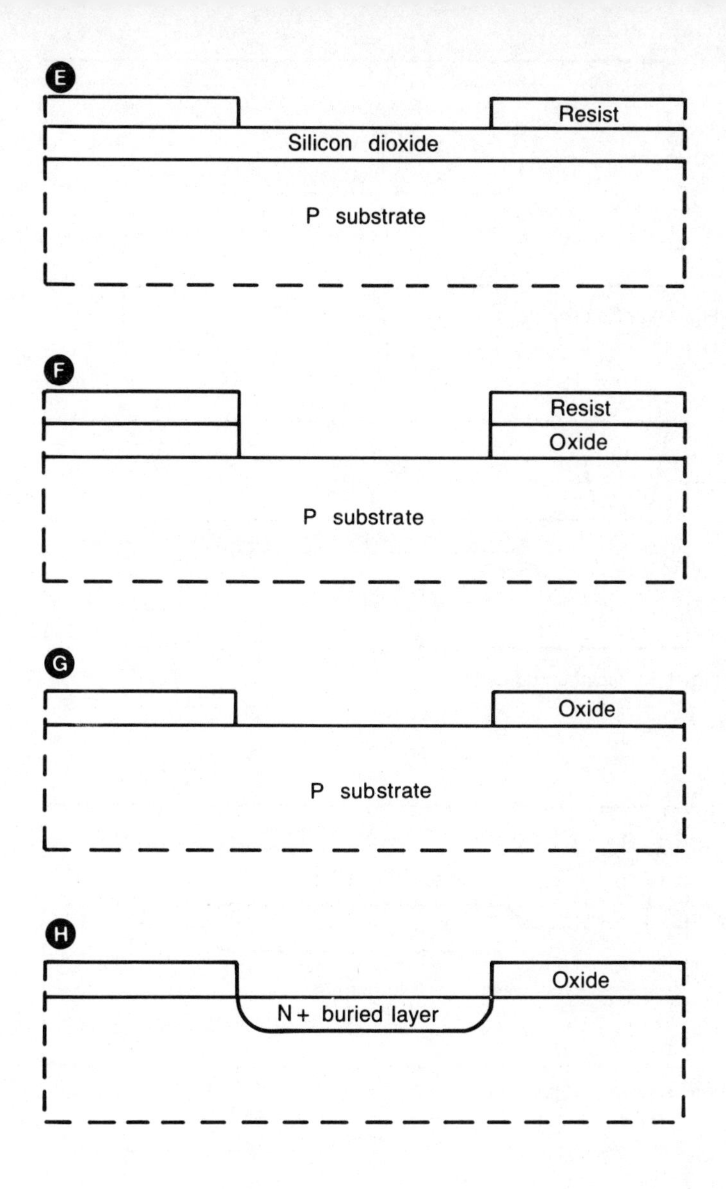

1-mil diameter tips) into contact with the bonding pads on each die and performing a set of electrical tests. The probes are connected to a special interface card (probe adapter) which is connected to automatic test equipment. Usually only dc tests are performed at this level. The probe station is equipped to mark all bad die with an ink blob. All inked die are removed after the scribe and break operation. Good die are placed in waffle packs (little plastic trays that look like one-half of a waffle iron with about 100 shallow receptacles for die) and sent to the assembly area.

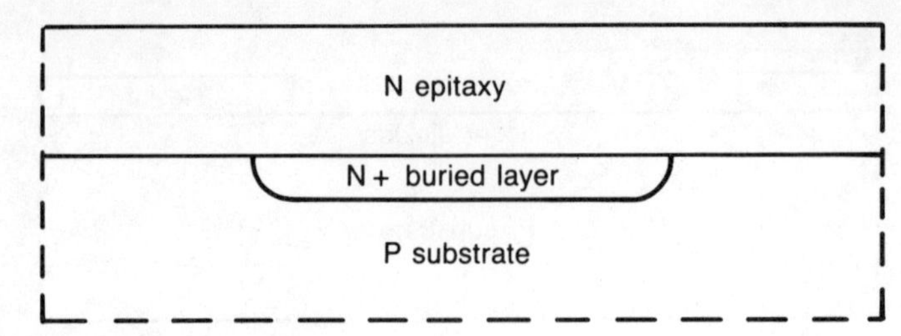

Fig. 1-12. Epitaxial layer.

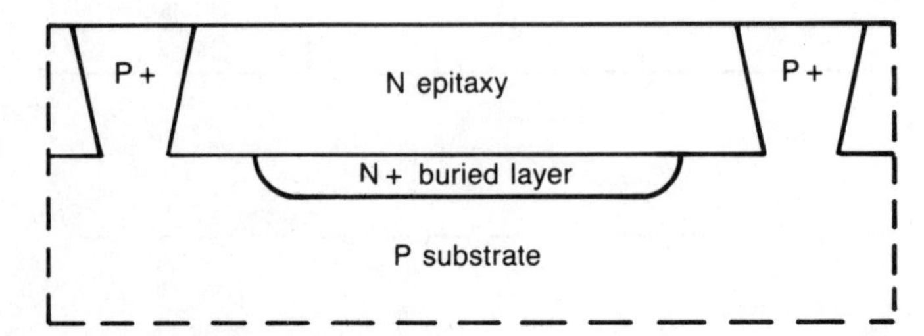

Fig. 1-13. Isolation diffusion.

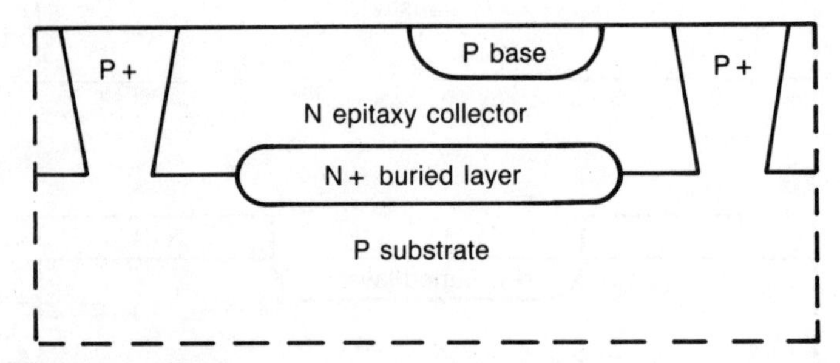

Fig. 1-14. Base diffusion.

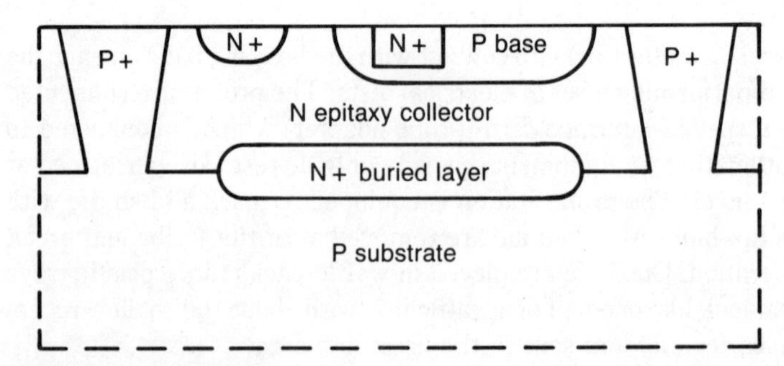

Fig. 1-15. Emitter diffusion.

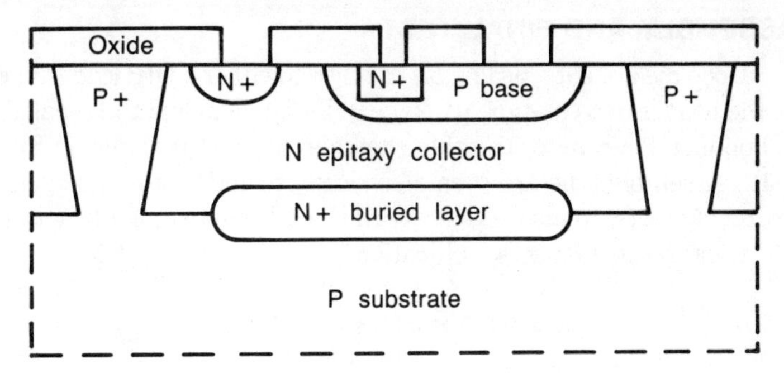

Fig. 1-16. Contact openings.

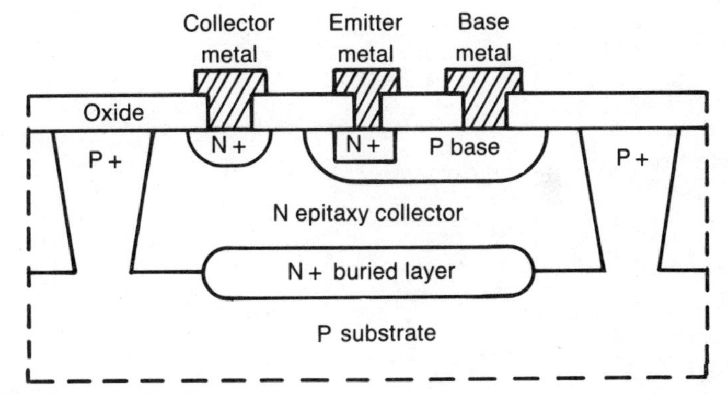

Fig. 1-17. Metalization.

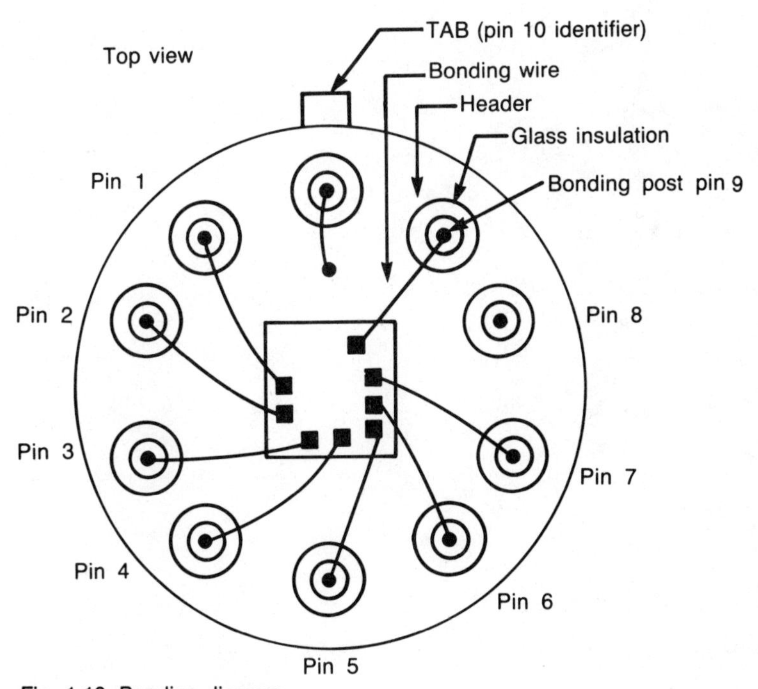

Fig. 1-18. Bonding diagram.

ASSEMBLY AND FINAL TEST

Good die are eutectically die attached to either a DIP lead frame or the header of a can-type package, wire-bonded, and then sealed. A bonding diagram for the chip shown in Fig. 1-1 is shown in Fig. 1-18. Assembled die are then electrically tested (both dc, ac, and switching tests) to insure that the finished ICs meet all advertised electrical performance specifications.

Table 1-1. Basic Linear IC Mask Set.

MASK	PROCESS STEP
1	N+ buried layer diffusion
2	P+ isolation diffusion
3	P base diffusion
4	N+ emitter and collector diffusion
5	Contact oxide openings
6	Metalization
7	Passivation

Chapter 2

Linear IC Components

LINEAR ICS ARE DESIGNED USING FIVE BASIC COMPONENTS:

- ☐ NPN transistors
- ☐ PNP transistors
- ☐ Diodes
- ☐ Resistors
- ☐ Capacitors

The purpose of this chapter is to explain the basic performance characteristics and limitations of these components.

NPN TRANSISTORS

The NPN transistor is, perhaps, the most important linear IC component. In fact, the basic linear IC process discussed in Chapter 1 was designed to optimize the performance characteristics of the NPN transistor more than any other IC component. Recall that the base and emitter masks are named after the corresponding regions of the NPN transistor.

For convenience, the top and side (cross-section) views of the NPN transistor previously shown in Fig. 1-10 are shown again in Fig. 2-1. When I drew this transistor I used the following design rules:

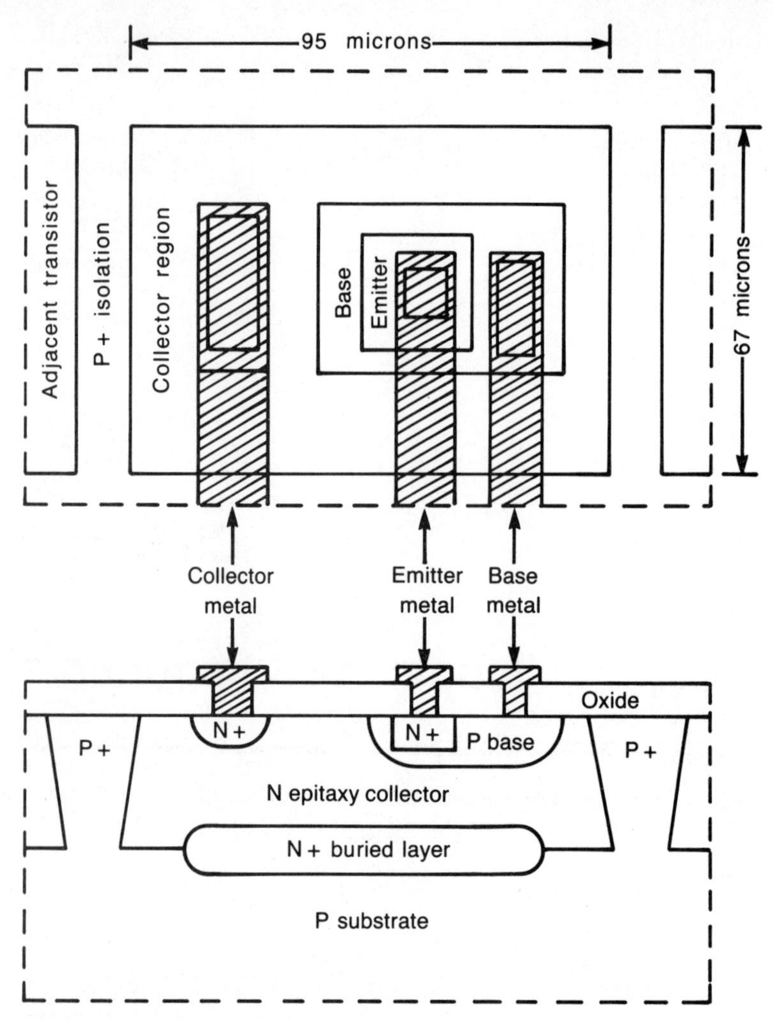

Fig. 2-1. An integrated circuit NPN transistor.

- ☐ Minimum size contact: 5 by 7 microns
- ☐ Minimum spacing, emitter contact inside emitter: 5 microns
- ☐ Minimum spacing, emitter inside base: 5 microns
- ☐ Minimum base-to-collector N+: 10 microns
- ☐ Minimum collector N+ width: 10 microns
- ☐ Minimum spacing, collector contact inside collector N+: 2 microns
- ☐ Minimum spacing, base to isolation: 20 microns
- ☐ Minimum metal overlap on contacts: 2 microns

The result is a device that measures 95 microns by 67 microns. These particular design rules were chosen for illustrative purposes and do not represent state-of-the-art linear IC process technology design rules (for which most of these rules could be cut in half). I have deliberately listed these rules in the order that I used them to design the minimum geometry NPN transistor, designing the transistor from the inside out, starting with the minimum allowable contact opening followed by the minimum spacing that this contact must be inside the emitter, and so forth.

Linear IC designers always use the minimum size NPN transistor (and other minimum geometry components) where possible in order to keep the chip size as small as possible. This is because the number of good die per wafer is a function of die size and percentage of active area, or area occupied by actual circuit components such as diodes, transistors and resistors, and not area occupied by metal, unused epitaxial area, isolation regions, or bonding pads. Silicon wafers inherently have a certain number of defects per square inch. Therefore, the smaller the die and the fewer the number of active components, the fewer die that will fail due to the lattice, or crystalline, defects associated with silicon.

Typical device characteristics for a small NPN are listed in Table 2-1, on page 31.

Dc Current Gain

In any circuit in which the transistor is biased in the linear region, beta (β) is defined as the ratio of the collector current to the base current and alpha (α) is defined as the ratio of the collector current to the emitter current:

$$\text{BETA} = \beta = I_C/I_B = \alpha/(1-\alpha)$$
$$\text{ALPHA} = \alpha = I_C/I_E = \beta/(\beta+1)$$

(The linear region is between cut-off, where the collector current is approximately equal to zero, and saturation, where an increase in base current causes no additional collector current to flow. In the linear region, by definition, ac signals are amplified without distortion.)

The transistor may be operated in either of two modes—normal and inverted—as shown in Fig. 2-2. Transistors are rarely intentionally operated in the inverted mode. This mode is discussed in this section mainly to introduce the term α_I, which will be used in the discussion on transistor saturation currents. In the inverted mode, the emitter is used as the collector and vice versa. Subscripts are used to differentiate between modes when the current gain parameters are discussed: N for normal, I for inverted.

β_N is typically around 150 (75 minimum, 300 maximum) for a small NPN transistor operating at a collector current of 1 mA. β_I is approximately 0.33. The corresponding values for α_N and α_I are 0.99338 and 0.248, respectively.

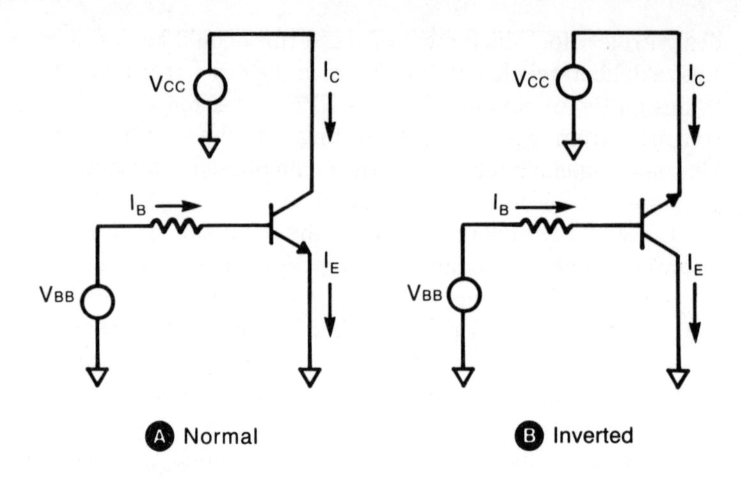

Fig. 2-2. Transistor operating modes.

Dc current gain is primarily a function of the base width, or the base thickness between the bottom of the emitter and the N+ buried layer. Recall from basic electronics that, with the base-emitter forward biased, electrons will flow from the emitter to the base. However, when the base-collector is reverse-biased a so-called "depletion region" is formed at the base-collector junction. If any electrons come too close to this region, they will be attracted to the positive collector voltage. If the base region is made sufficiently narrow, most of the electrons traveling from the emitter to the base will get collected by the collector.

Due to secondary effects, beta rolls off (from some peak or mid-range value) at both high and low currents. See Fig. 2-3 for a beta curve for a small NPN transistor. You can see from this graph that this transistor is useful up to about 10 mA. In order to increase the current range the size of the emitter must be increased. Since transistor action occurs primarily on the side of the emitter near the base contact, the emitter is made long and narrow rather than made into a large square or near-square rectangle. Also, multiple emitter stripes are preferred to one long emitter. (A problem called "emitter debiasing" can occur when emitter length is around 5 mils.) An 8× geometry (the minimum geometry is referred to as a 1× device) is shown in Fig. 2-4. Assuming the same design rules that were used for the 1× device, the 8× device is 135 by 124 microns. The peak beta for the 8× device should be approximately the same as the 1× device, only at a collector current of 8 mA. The 8× device should be useful up to about 80-100 mA.

Perhaps the most useful (from a circuit design standpoint) rule of thumb about transistor beta is its temperature characteristic: *Beta typically increases 0.5 percent per degree Centigrade.*.

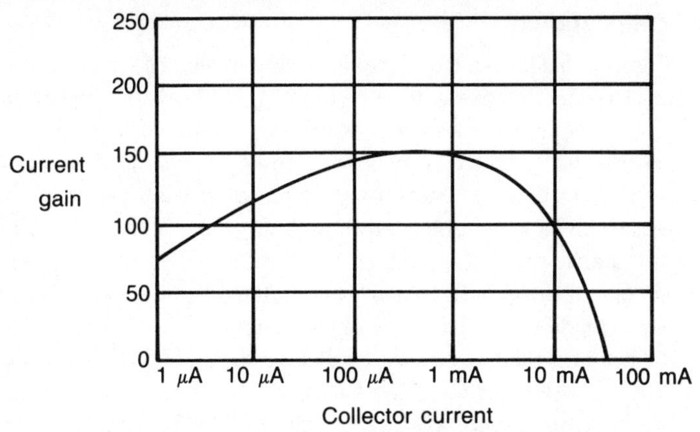

Fig. 2-3. Current gain versus collector current for a small NPN transistor.

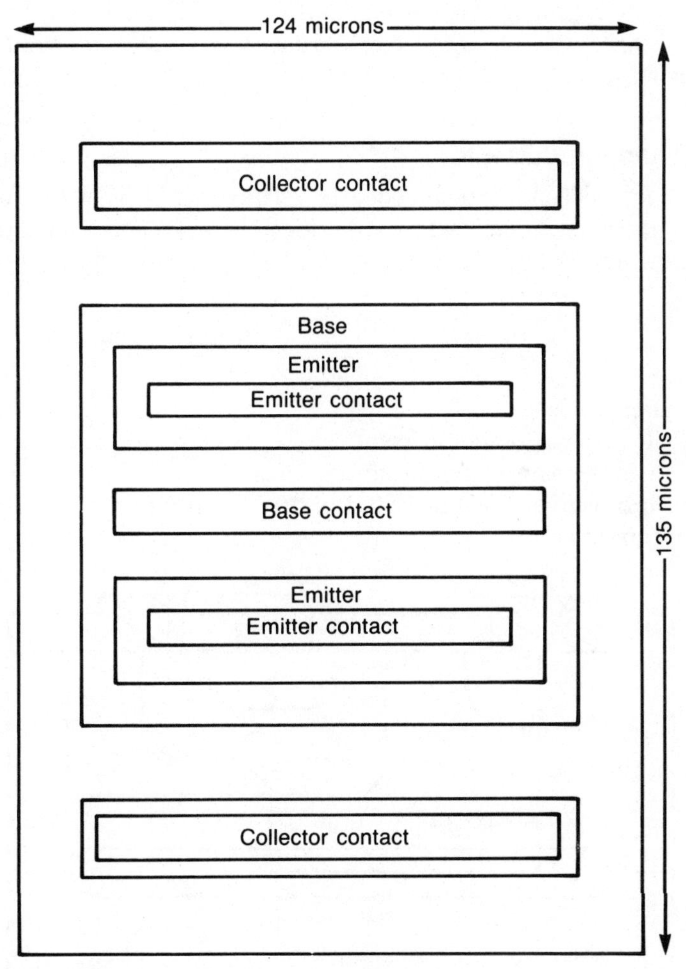

Fig. 2-4. Medium size NPN transistor geometry.

Bulk Resistance

Figure 2-5 shows a transistor and some resistors superimposed on the cross-section drawing. The intent is to show that integrated circuit transistors have a series resistance in each of the leads. This is also true for discrete transistors. However, the so-called bulk resistances for discrete transistors are much smaller primarily because the devices are physically much larger, effectively equivalent to many small transistors in parallel. Also, the discrete transistor collector bulk resistor is smaller because the collector contact is made on the backside of the transistor die, thus eliminating the need for the buried layer, collector N+, and topside contact.

You can see from Table 2-1 that the series base, collector, and emitter bulk resistances are on the order of 200, 100, and 10 ohms, respectively, for the 1× NPN. The 8× device has bulk resistances that are approximately one-eighth these values, or 25-ohms base resistance, 12.5-ohms collector resistance, and 1.25-ohms emitter resistance.

Saturation Currents

In a silicon junction diode, the relationship between the forward current, I, and the forward voltage, V, is given by the following equation:

$$I = I_S (e^{qV/kT} - 1) \qquad \text{[the diode equation]}$$

where:

I_S is the saturation current;
e is the exponential function;
q is charge, equal to 1.6021×10^{-19} coulomb;

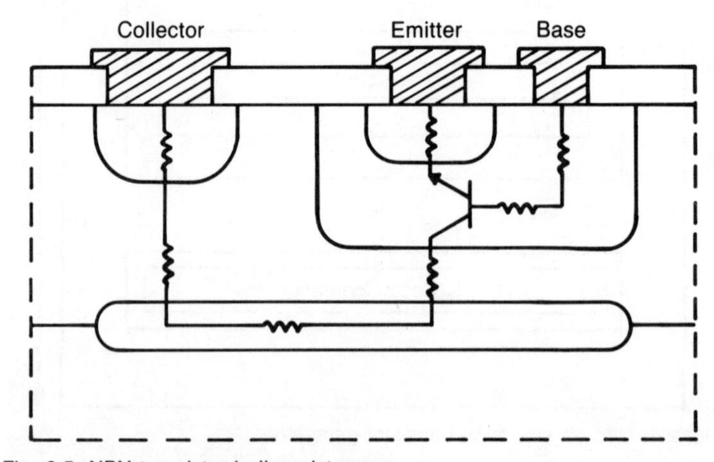

Fig. 2-5. NPN transistor bulk resistances.

k is Boltzmann's constant, equal to 1.3806×10^{-23} joules per degrees Kelvin (Kelvin = Centigrade + 273); and

T is temperature in degrees Kelvin.

At room temperature (27 degrees C, 300 Kelvin), the expression kT/q (sometimes referred to as the thermal voltage and designated by the symbol V_T) is approximately equal to 26 mV. Also, for forward voltages greater than 0.5 volts, the "−1" term in the above equation can be dropped:

$$I = I_S e^{V/0.026}$$

This equation can now be easily solved for the forward voltage, V:

$$V = 0.026 \ln (I/I_S)$$

where:

ln is the natural log function.

The saturation current is a function of doping level, junction area, and temperature. Although the value of the saturation current could be computed if these parameters were known, it is an easy task to compute I_S from the diode I-V curve using the above equations.

Because transistors have two junctions, there are two saturation currents, one for the base-emitter junction and one for the base-collector junction. The two saturation currents are related to each other through the transistor alphas:

$$\alpha_N I_{ES} = \alpha_I I_{CS}$$

where:

I_{ES} is the base-emitter saturation current; and

I_{CS} is the base-collector saturation current.

This equation is called the law of reciprocity, and is used by some computer circuit analysis programs to calculate I_{CS} from the user supplied values for I_{ES}, β_N, and β_I.

The equations for the corresponding junction voltages at room temperature are:

$$V_{BE} = 0.026 \ln (I_E/I_{ES}) \qquad \text{[collector open]}$$
$$V_{BC} = 0.026 \ln (I_C/I_{CS}) \qquad \text{[emitter open]}$$

Using these equations for the values given in Table 2-1 and a forward current of 10 μA results in a V_{BE} of 598 mV and a V_{BC} of 562 mV. A typical V_{BE} curve is shown in Fig. 2-6. The current axis is logarithmic and the voltage axis is linear, resulting in a straight-line plot except at high currents. This is because of the IR drop across the base bulk resistor. A similar curve could be drawn for the base-collector voltage.

There are three important rules of thumb associated with PN junctions:

1. The change in forward voltage drop equals 60 mV per decade of current. For example, V_{BE} in Fig. 2-6 is approximately 540 mV at 1 μA and 600 mV at 10 μA. In other words, the forward drop increased 60 mV when the forward current was made 10 times higher.
2. The saturation current doubles every 10 degrees Centigrade.
3. As a result of rule 2 the forward voltage decreases by approximately 2 mV per degrees Centigrade.

Another rule of thumb similar to rule 1 is that the junction voltage increases 18 mV when the current is doubled. All of these rules can be mathematically derived from the basic diode equation and verified by simple laboratory test. Rule 1 is a special case of the following equation:

$$V_{BE2} - V_{BE1} = \Delta V_{BE} = 0.026 \ln (I2/I1)$$

where:

V_{BE1} is the base-emitter voltage at I1; and
V_{BE2} is the base-emitter voltage at I2.

I strongly suggest that you memorize the diode equation in its various forms (if you haven't already) or be able to algebraically manipulate the basic equation quickly. I have used these equations at least as often as Ohm's Law in the design of electronic circuits.

One more comment about saturation currents: Since the saturation currents are proportional to junction area, the base-emitter saturation current for the 8× transistor is eight times that of the 1×. However, the base-collector saturation current is only about five times the saturation current of the 1× because the base-collector area is only five times the base-collector of the 1×, not eight times as is the case for the emitter.

Junction Capacitance

The base-emitter, base-collector, and collector-substrate junction capacitances for the small NPN are 0.65 pF (picofarads, 10^{-12} Farads), 0.65 pF, and 2.5 pF, respectively, at zero volts reverse bias. For reverse bias greater than 0 V, junction capacitance may be computed from the following equation:

$$C = \frac{C_0}{(1 - V/0.75)^{0.33}}$$

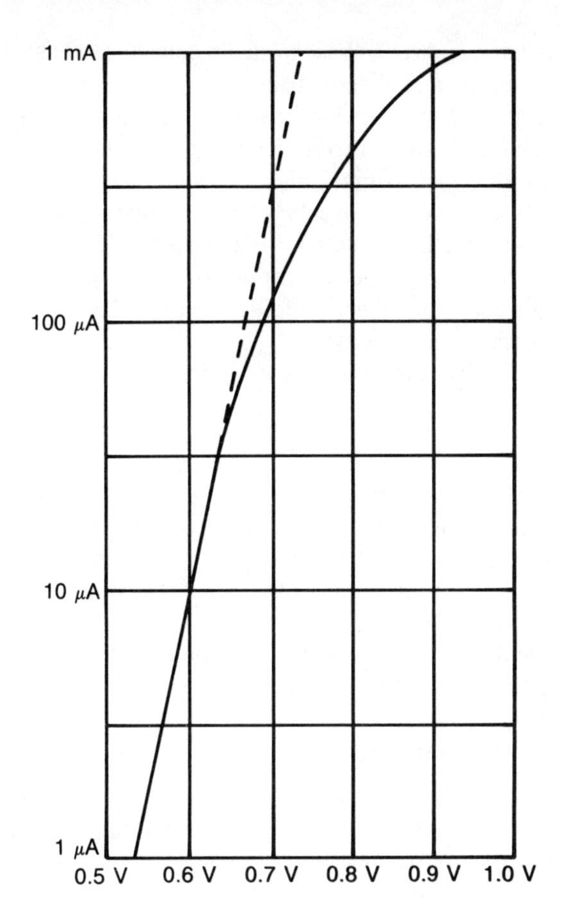

Fig. 2-6. V_{BE} versus I_E for a small NPN transistor.

where:

C_0 is the junction capacitance at 0 V; and
V is the reverse voltage in negative volts.

Note that the integrated circuit NPN has one junction capacitance—the collector-substrate capacitance—that the discrete NPN does not. It is essential that substrate diode (the anode is the P-substrate, the cathode is the N-epitaxial collector) be reversed biased in order to maintain isolation between transistors. This is accomplished by tying the substrate to the most negative potential.

Junction capacitance is a function of area. The area factors for the 8× NPN are 8× for the base-emitter, 5× for the base-collector, and 2.5× for the collector-substrate, resulting in 5.2 pF for C_{BE}, 3.25 pF for C_{BC}, and 6.25 pF for C_{CS}.

Time Constants

Current carriers (either holes or electrons) do not move instantaneously from emitter to base (or from collector to base in the inverted mode). As a result, the transistor current gain is less when the input current, whether base current (for common-emitter configuration) or emitter current (common-base configuration), is changing quickly with time. This is reflected in discrete transistor data sheets as a parameter called gain-bandwidth product, or f_T.

Gain-bandwidth product (f_T) is the frequency at which beta equals 1 and is usually specified at one or more values of collector current at a constant collector-emitter voltage. Table 2-1 specifies the base-emitter time constant T_E instead of f_T. f_T may be calculated from T_E by using the following equation:

$$f_T = 1/(2\pi T_E)$$

For a T_E equal to 0.5 ns, f_T is about 300 MHz.

The base-collector time constant, T_C, is about 50 ns for linear IC transistors. T_C is only important when the transistor is saturated, something that is generally avoided in linear IC design.

As a general rule, for a given process, junction time constants are independent of transistor size, i.e., the 8× device has the same time constants as the 1× device.

Matching Characteristics

You might think that all NPN transistors of the same size would all have the same characteristics. Unfortunately, this is not true. Mask patterns are never exactly the same even though they are drawn to be the same size, i.e., a 5-by-7 micron emitter is probably never exactly 5-by-7 anywhere on the chip. Also, diffusion depth will vary slightly from one side of the die to the other.

As a result, the transistors are not perfectly matched. IC designers are most concerned about two parameters in particular: beta, and V_{BE}. Beta match is typically about plus or minus 10 percent, 30 percent maximum. V_{BE} match is typically about 1 mV, 5 mV maximum. Identical transistors with larger emitters (4×, for example) laid side-by-side in the same direction may achieve a factor of 2 or 3 improvement over these numbers.

PNP TRANSISTORS

A layout and cross-section of a so-called lateral PNP is shown in Fig. 2-7. Typical characteristics are listed in Table 2-2, on page 31. Note that the emitter metal extends out beyond the actual emitter contact and slightly overlaps the PNP collector. This extra metal improves the PNP beta. A positive voltage with respect to the base causes transistor action to occur near the surface. Thus fewer current carriers go down to the N+ buried layer and out the base contact.

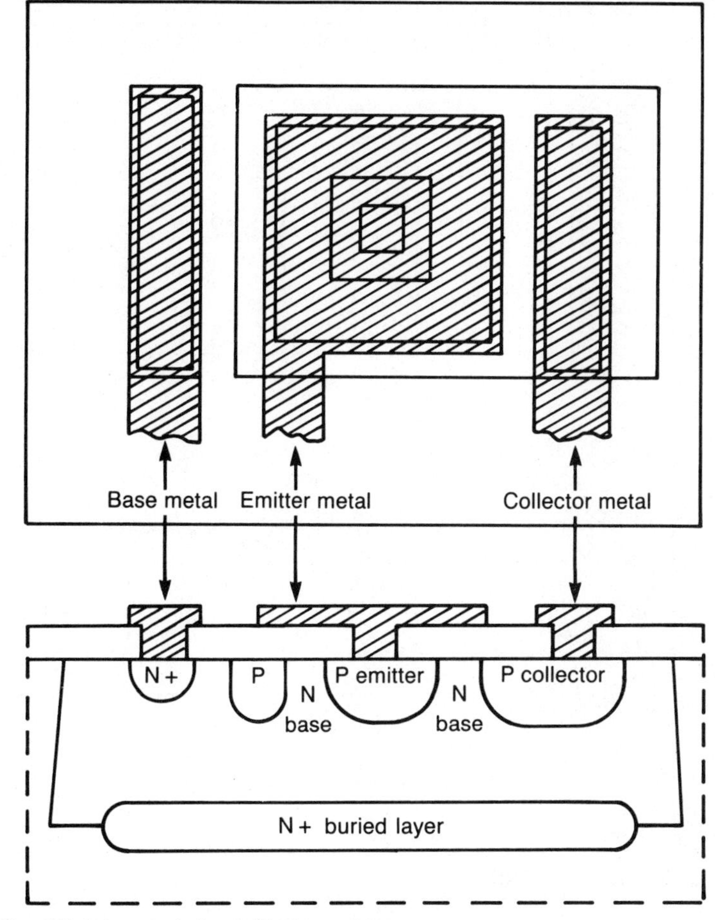

Fig. 2-7. Integrated circuit PNP transistor.

You can see from Table 2-2 that the base-emitter time constant is a factor of 100 greater than that of the small NPN. Accordingly, f_T is a factor of 100 smaller, or about 3 MHz for the PNP. If a particular circuit design required a beta of 10, this PNP would be useless above 300 kHz. As a result, linear IC designers avoid using a PNP in the signal path of wideband designs and use it mainly as a current source transistor.

Another type of integrated circuit PNP is shown in Fig. 2-8. This is the substrate PNP, so-called because the substrate is the collector. It is also called a vertical PNP because the transistor action is vertical (the transistor action is lateral in the lateral PNP). This PNP has the advantage that its beta is slightly higher (about 70) than the lateral PNP and has an f_T of about 30 MHz. The disadvantage of the substrate PNP is that its collector will always be tied to the most negative potential and, therefore, can only be used as an emitter-follower.

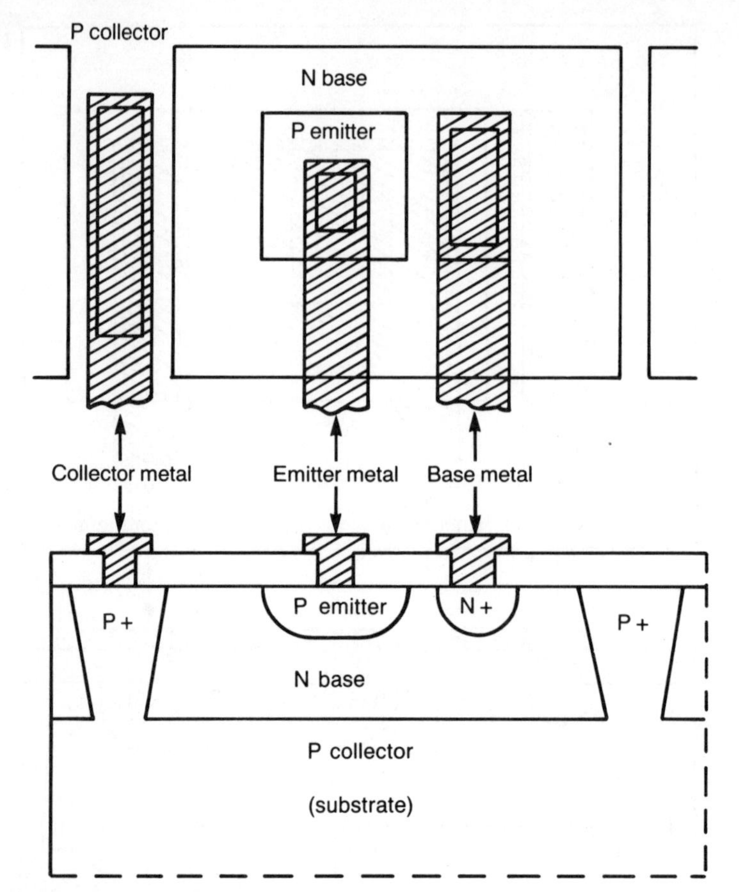

Fig. 2-8. Substrate PNP.

DIODES

In integrated circuits, diodes are made from transistors. Normally, the base-emitter junction of either the NPN or the PNP is used as a diode. The base-collector junction is shorted.

Zener diodes are created by using the NPN base-emitter diode in the reverse direction. The reverse breakdown of the NPN base-emitter is approximately 6.5 volts, plus or minus 0.5 volts. The temperature coefficient is +2 mV/°C.

RESISTORS

The layout and cross-section of a typical base diffusion resistor is shown in Fig. 2-9. The equation for resistance is:

$$R = \frac{\varrho}{t} \frac{l}{w}$$

where:

ϱ is the resistivity of the material in ohms-centimeters;
t is the thickness of the material in centimeters;
l is the length; and
w is the width.

The units of length and width are arbitrary as long as they are both the same. The linear distance between resistor contacts is l and w is the width of the resistor. The length of the resistor in squares is generally expressed as l/w. The diffusion depth for diffused resistors is t. For a known depth, ϱ/t is called the sheet resistance and is expressed in ohms-per-unit-length per-unit-width, or ohms-per-square. Therefore:

$$R = \varrho_S \, (l/w)$$

where:

ϱ_S is the sheet resistance in ohms-per-square; and
l/w is the number of squares.

Typical base diffusion sheet resistance is around 150-200 ohms-per-square. The tolerance on base diffusion resistors is plus or minus 30 percent, considerably higher than the low-cost 5-percent resistors used in discrete circuit design. Matching is typically 3 to 5 percent. As a result, linear IC designers must use circuit techniques that do not rely on the absolute value of resistors. In critical designs where resistor matching is important (such as

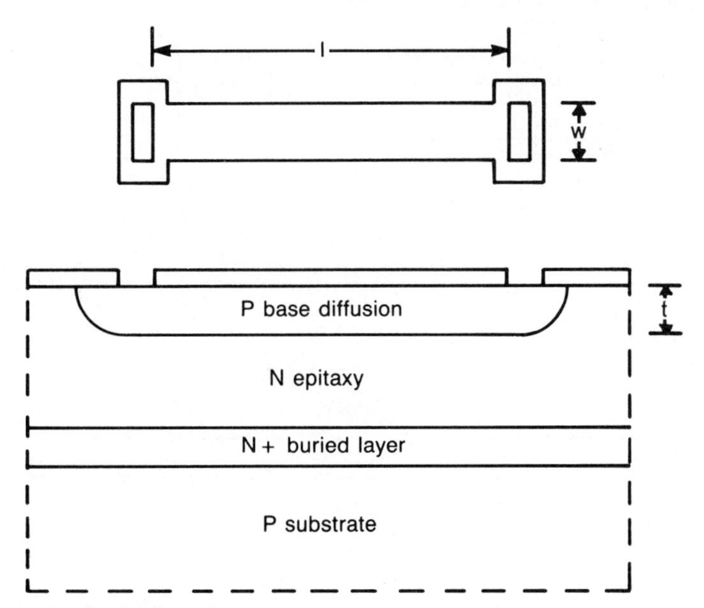

Fig. 2-9. Base diffusion resistor.

digital-to-analog converters), it is necessary to use ion-implanted resistors or thin-film resistors. Ion-implanted resistors can achieve 5 percent tolerance and 0.5 percent matching. Thin-film resistors are deposited on the surface of the chip and etched; can be laser trimmed to 0.1 percent tolerance; and have excellent temperature stability. Base-diffused resistors have a temperature coefficient of +0.15%/°C. The matching temperature coefficient is plus or minus 0.01%/°C.

Base diffusion resistors are diffused into an epitaxial island or tub. One or more resistors are placed in the tub and an N+ tub contact (same as the N+ collector contact of an NPN transistor) is connected to the positive supply voltage. Thus, isolation between resistors is achieved in a manner analogous to the isolation between transistors.

Another type of IC resistor is shown in Fig. 2-10. This resistor is called a pinched resistor. Basically, it is the same as the base-diffusion resistor, except that N+ has been diffused into the body of the resistor, effectively reducing the thickness of the base diffusion and significantly increasing the resistance between the contacts. The sheet resistance under the N+ is on the order of 5,000 to 10,000 ohms-per-square. The resistance tolerance of the pinched resistor is −50, +100 percent.

CAPACITORS

Small valued capacitors (less than 50 pF) are formed from the base-emitter and base-collector junctions of NPN transistors. Base-

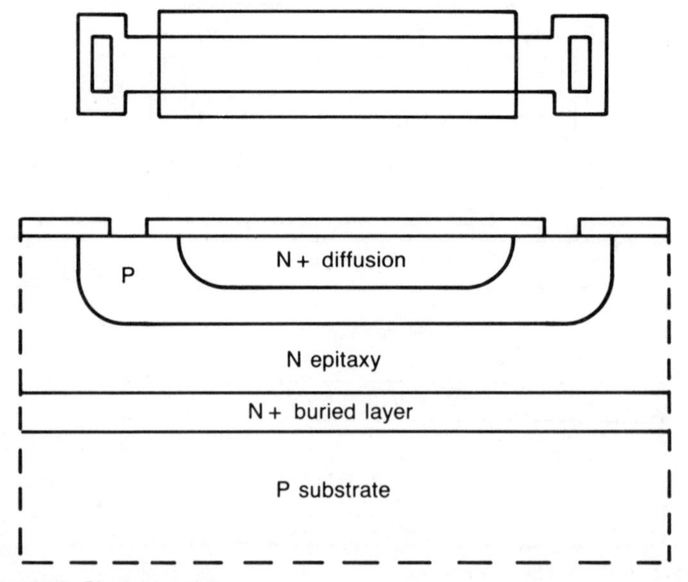

Fig. 2-10. Pinched resistor.

emitter capacitance is approximately 1.6 pF/mil². Base-collector capacitance is approximately 0.4 pF/mil². These are the zero-volt bias conversion factors. When the reverse voltage is greater than zero, the junction capacitance is reduced. See the junction capacitance equations in the section on NPN transistors.

Table 2-1. Small NPN Transistor Characteristics.

PARAMETER	SYMBOL	TYPICAL VALUE
DC CURRENT GAIN @ IC = 1 mA	β	150
BASE BULK RESISTANCE	R_B	200 OHMS
COLLECTOR BULK RESISTANCE	R_C	100 OHMS
EMITTER BULK RESISTANCE	R_E	10 OHMS
BASE-EMITTER SATURATION CURRENT	I_{SE}	1×10^{-15} AMPS
BASE-EMITTER JUNCTION CAPACITANCE	C_{BE}	0.65 pF
BASE-EMITTER TIME CONSTANT	T_E	0.5 nS
BASE-COLLECTOR SATURATION CURRENT	I_{SC}	4×10^{-15} AMPS
BASE-COLLECTOR JUNCTION CAPACITANCE	C_{BC}	0.65 pF
BASE-COLLECTOR TIME CONSTANT	T_C	50 nS
COLLECTOR-SUBSTRATE CAPACITANCE	C_{CS}	2.5 pF

Table 2-2. Small PNP Transistor Characteristics.

PARAMETER	SYMBOL	TYPICAL VALUE
DC CURRENT GAIN @ IC = 100 uA	β	50
BASE BULK RESISTANCE	R_B	750 OHMS
COLLECTOR BULK RESISTANCE	R_C	100 OHMS
EMITTER BULK RESISTANCE	R_E	10 OHMS
BASE-EMITTER SATURATION CURRENT	I_{SE}	1×10^{-15} AMPS
BASE-EMITTER JUNCTION CAPACITANCE	C_{BE}	0.5 pF
BASE-EMITTER TIME CONSTANT	T_E	50 nS
BASE-COLLECTOR SATURATION CURRENT	I_{SC}	1.5×10^{-15} AMPS
BASE-COLLECTOR JUNCTION CAPACITANCE	C_{BC}	1.5 pF
BASE-COLLECTOR TIME CONSTANT	T_C	50 nS
BASE-SUBSTRATE JUNCTION CAPACITANCE	C_{CS}	4 pF

Chapter 3

Linear IC Design Techniques

BEFORE THE LINEAR IC DESIGNER BEGINS TO DESIGN A NEW chip, he must know the basic process steps and masks used to fabricate the chip, and he must know what components are available to him and the characteristics and limitations of these components. Then he must design the smallest possible chip that meets the design requirements. Because resistors take up a lot of chip area ("silicon real estate") and because resistors have a relatively poor tolerance, he uses the fewest possible resistors. Also, he uses NPN transistors when possible and avoids using PNPs in the signal path of wideband designs.

In the face of these constraints, linear IC designers have produced many hundreds of low-cost, fairly high-performance circuits. How do they do it? Although linear IC designers may custom design each and every transistor geometry (something that circuit designers using discrete components cannot do) they, like any other specialized group of engineers, have their "bag of tricks." The purpose of this chapter is to explain the basic design techniques used by linear IC designers. As a user of their products, you will gain an appreciation for what they have done and also have considerably more insight than you might have had previously in the inherent performance limitations of linear ICs.

In this chapter, the following linear IC subcircuits will be presented and explained:

- ☐ The current source
- ☐ The differential pair
- ☐ The active load
- ☐ The output stage

Also, the design techniques used to create a complete circuit using these subcircuits will be presented.

THE CURRENT SOURCE

Figure 3-1 shows two basic current sources, one made with NPNs and the other with PNPs. Strictly speaking, the NPN version is a current "sink" because current flows into the NPN Q2. The PNP Q2 "sources" current. Nevertheless, both are referred to as current sources. The load shown in the figure represents a circuit or subcircuit that requires a constant current.

The current sources shown in the figure are also called "current mirrors." Look at the NPN version for a moment. When power is applied, current flows through the resistor R and into Q1, which is wired as a diode:

$$I_1 = (V+ - V_{BE1})/R$$

Assume for the moment that the beta of Q1 is infinite. Then the Q1 emitter current is equal to I_1. Next, the voltage generated at the base of Q1 is applied to the base of Q2. If Q1 and Q2 were exactly the same geometry (in particular, had exactly the same size emitter), then the emitter current of Q2 would exactly equal that of Q1. Hence, the current in Q1 is said to be "mirrored" in Q2.

Using this basic technique, the current flowing in Q2 can be scaled to be a fraction or a multiple of the Q1 current by making

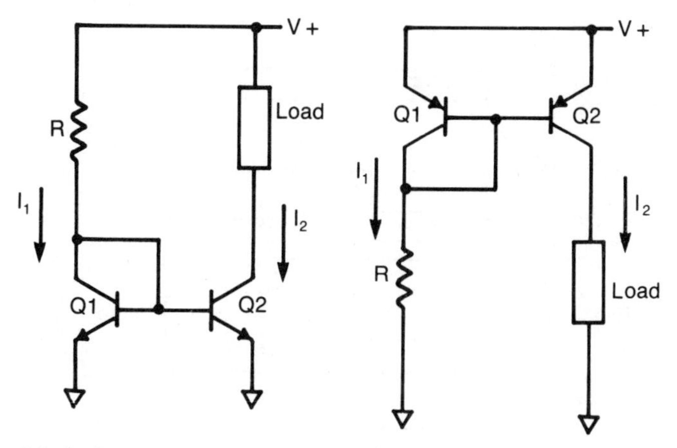

Fig. 3-1. Basic current sources.

the emitter size of Q2 smaller or larger than that of Q1. Thus if Q1 is a 1× device and Q2 is a 4× device then I_2 is 4 times I_1. In general,

$$I_2 = k\,I_1$$

where:

k is the ratio of Q2 emitter area to Q1 emitter area.

The above discussion assumes infinite betas and perfectly matched transistors (or perfectly ratioed). Emitter area mismatch for the same size devices is typically less than 10 percent. For different ratios, this error is greater. To maintain good match for unequal device sizes, linear IC designers typically use multiple emitters in the same base region. Figure 3-2 shows a 4× NPN transistor using four 1× emitters.

Emitter area mismatch can also be compensated for by using emitter resistors. If the voltage drop across the emitter resistors is significant then I_2 becomes more dependent on the resistor match (3 to 5 percent) than on the base-emitter characteristics. The use of emitter resistors is shown in Fig. 3-3.

The error due to finite beta is illustrated in Fig. 3-4 which shows the distribution of currents if the transistor betas were equal to 10. The currents shown are in milliamps but it really doesn't matter. The general equation for I_2 for finite betas is as follows:

$$I_2 = I_1\,(1 + 2/\beta)$$

The error due to beta is reduced in the circuit shown in Fig. 3-5. In this circuit Q1 supplies base current for both Q2 and Q3. The equation for I_2 is:

$$I_2 = I_1\,(1 + \frac{2}{\beta^2})$$

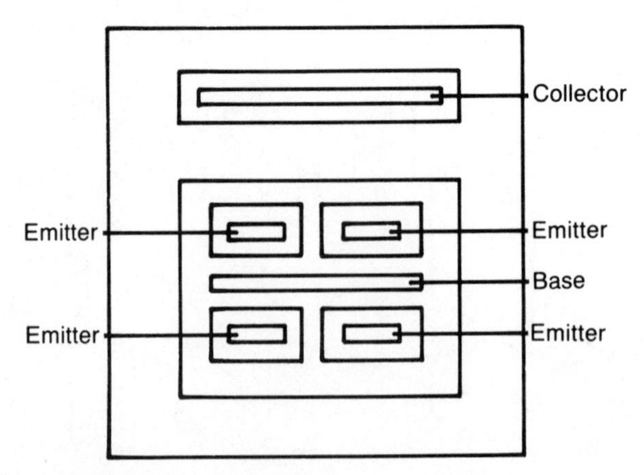

Fig. 3-2. 4× NPN transistor using four 1× emitters.

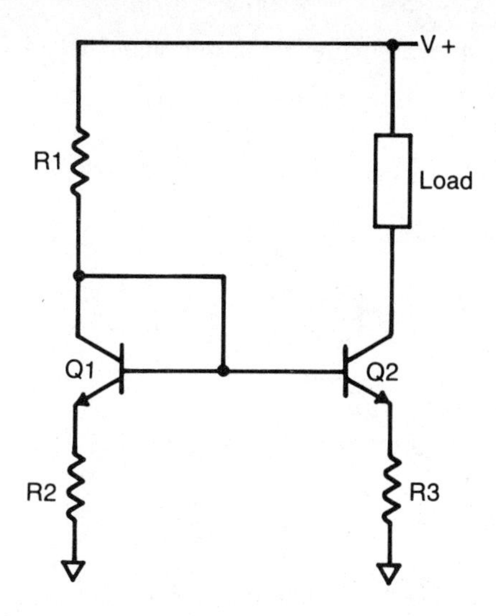

Fig. 3-3. The use of emitter resistors in a current source.

There is another error due to the fact that beta varies as a function of base-collector voltage. Figure 3-6 shows a circuit where the error due to collector voltage variation is eliminated.

Note that all of the above techniques are wasted if the current setting resistor R is a base diffusion resistor, which has a worst case tolerance of plus or minus 30 percent. When accuracy is required, the current setting resistor must be a stable low-tolerance resistor outside the IC package.

All of the above techniques apply to the PNP current mirror. The PNP current mirror, in general, is not as good as the NPN version due to lower beta and lower operating current for small geometries. The PNP, however, has the advantage over the NPN version in that emitter area mismatch is eliminated by using a sin-

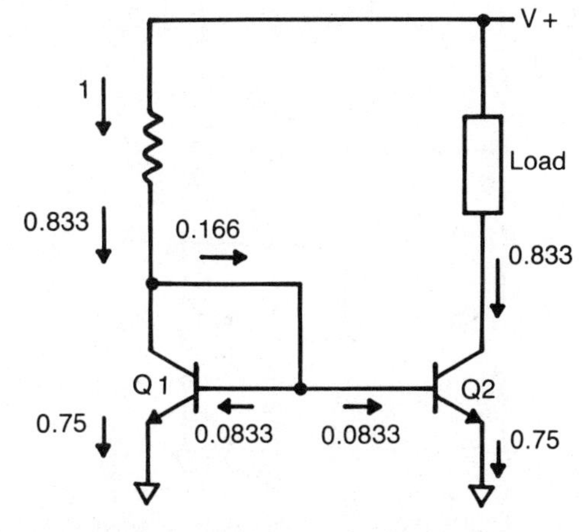

Fig. 3-4. Current source solution for betas equal to 10.

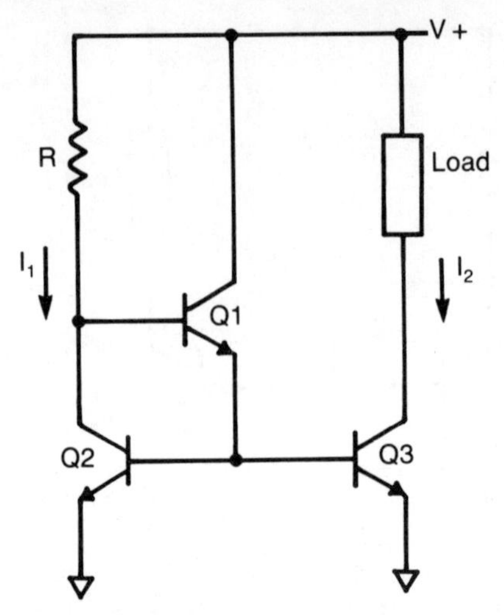

Fig. 3-5. Reduction of current mirror error due to finite betas.

gle PNP with one emitter and two collectors. A double-collector geometry is shown in Fig. 3-7. The revised PNP current mirror schematic is shown in Fig. 3-8.

Multiple current sources can be made using a single reference current. This is shown in Fig. 3-9 for NPNs and Fig. 3-10 for PNPs.

THE DIFFERENTIAL PAIR

Another fundamental linear IC building block is the differential pair. A pair of simple differential amplifiers is shown in Fig. 3-11. The current sources in this drawing are represented by a cir-

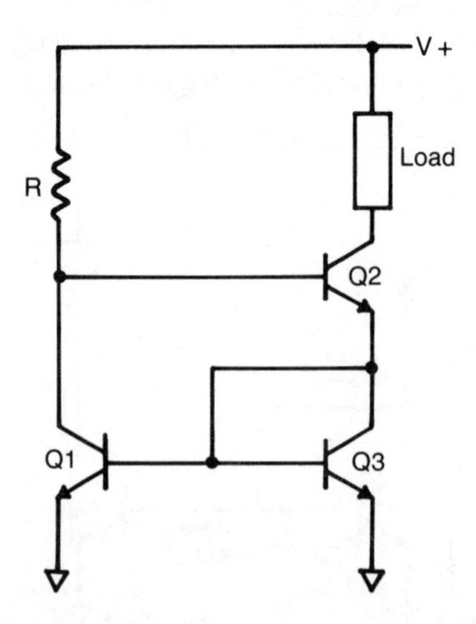

Fig. 3-6. Reduction of current mirror error due to collector voltage variation (the Wilson current source).

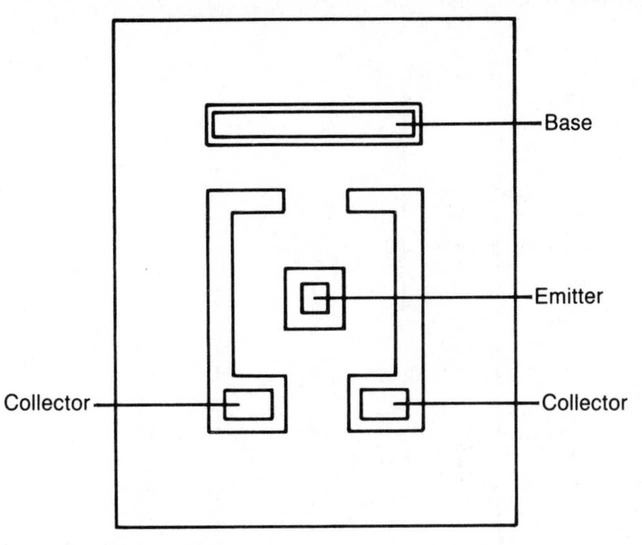

Fig. 3-7. Dual-collector PNP.

cle with an arrow in it pointing in the direction of current flow. These current sources may be as simple as those shown in Fig. 3-1 where the loads in that figure have now been replaced by differential pairs with load resistors. The NPN differential pair in Fig. 3-11 has an NPN current source, while the PNP differential pair has a PNP current source.

The gain equation for the differential pair with load resistors R_L is as follows:

$$G = V_{OUT}/V_{IN} = R_L/r_e$$

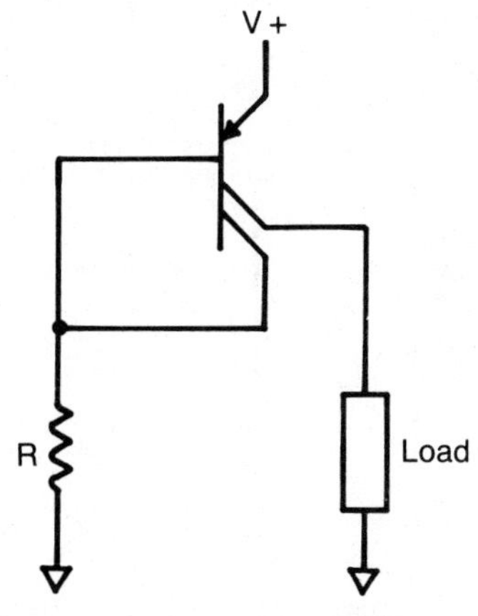

Fig. 3-8. Current source using dual-collector PNP.

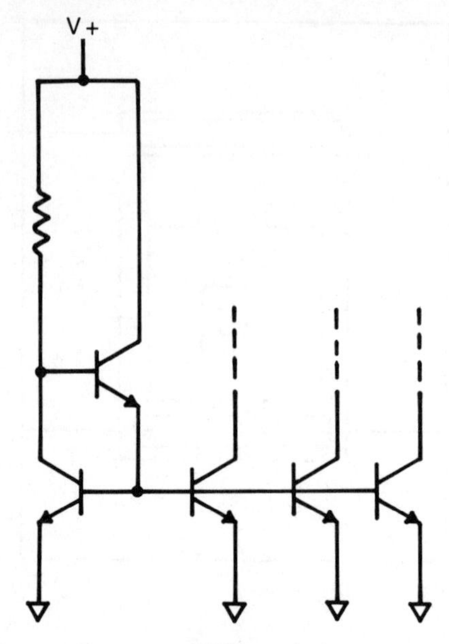

Fig. 3-9. Multiple current sources (NPN version).

where:

r_e is equal to kT/qI_C, or $0.026/I_C$ at room temperature.

The above equation is derived as follows:

1. Assuming betas of 100, the collector currents for Q1 and Q2 are approximately as follows:

$$I_{C1} = I_{ES1}e^{qV_{BE1}/kT}$$
$$I_{C2} = I_{ES2}e^{qV_{BE2}/kT}$$

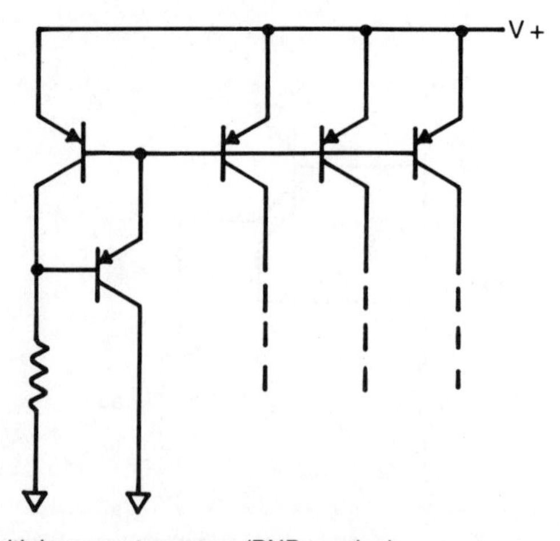

Fig. 3-10. Multiple current sources (PNP version).

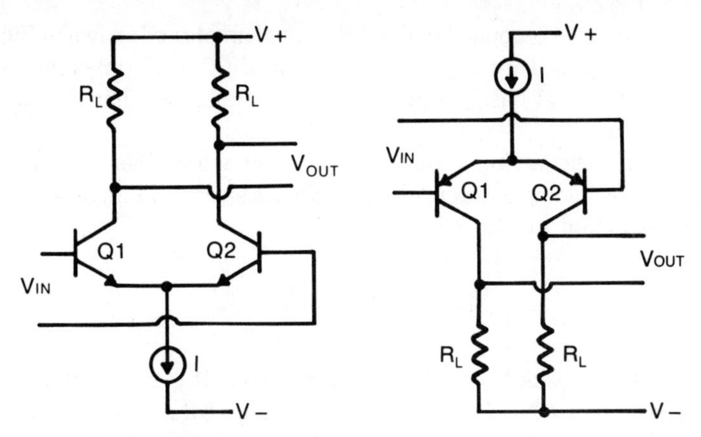

Fig. 3-11. Simple differential amplifiers.

2. Divide I_{C2} by I_{C1}. Assume that I_{ES1} equals I_{ES2}. Solve for room temperature. The result is

$$I_{C2}/I_{C1} = e^{(V_{BE2} - V_{BE1})/0.026}$$

3. Let V_{IN} equal the difference between base-emitter voltages. Solve for I_{C2}:

$$I_{C2} = I_{C1}\ e^{V_{IN}/0.026}$$

4. When the exponent of e is less than about 0.1 (or when V_{IN} is less than 2.6 mV) then the equation for I_{C2} can be simplified:

$$I_{C2} = I_{C1}\ (1 + V_{IN}/0.026)$$

5. Subtract I_{C1} from both sides:

$$I_{C2} - I_{C1} = V_{IN}\ (I_{C1}/0.026)$$

6. By inspection of Fig. 3-11,

$$V_{OUT} = R_L\ (I_{C2} - I_{C1})$$

7. Substituting (5) into (6) and dividing by V_{IN},

$$V_{OUT}/V_{IN} = R_L\ I_{C1}/0.026 = R_L/r_e$$

This equation is true only for values of V_{IN} less than 2 or 3 millivolts. Above a few millivolts, the differential amplifier is nonlinear. Recall the rule of thumb that says the current through a PN junction increases by a factor of 10 for each additional 60 millivolts.

When this rule is applied to the differential amplifiers shown in Fig. 3-11, it means that the collector current of Q2 is 10 times that of Q1 when the Q1 base voltage is 60 millivolts higher than the base voltage of Q2.

The linear region of the basic differential amplifier can be expanded by adding emitter resistors. This is shown in Fig. 3-12. The gain equation for this circuit is:

$$V_{OUT}/V_{IN} = R_L/(R_E + r_e)$$

Let's go back to Fig. 3-11 for a moment. It is generally more desirable to use NPN input transistors (higher betas and bandwidth than IC PNPs), but one of the problems associated with differential amplifiers with NPN input transistors is the reverse breakdown of the base-emitter junction when large input signals are applied, intentionally or unintentionally, to the inputs. For lateral PNP junctions and the NPN base-collector junction, it is possible to have breakdown voltages of 40 volts, but the NPN base-emitter breakdown is about 6.5 volts. Therefore, if V− was −15 volts and one of the inputs was made equal to V− while the other input was grounded, breakdown would occur and might damage the IC.

One solution to this problem is shown in Fig. 3-13. This circuit solves the breakdown problem at the expense of degraded bandwidth (due to the use of the PNPs) and degraded input offset (due to the mismatch of one additional base-emitter junction on each side). Nevertheless, the type of input stage is common in commer-

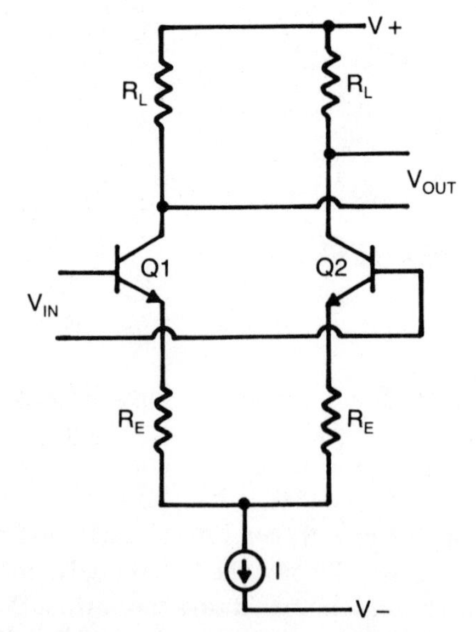

Fig. 3-12. Differential amplifier using emitter resistors.

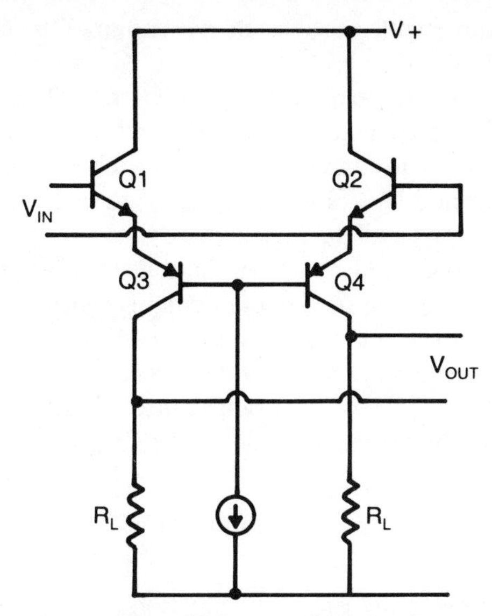

Fig. 3-13. Differential amplifier with increased input voltage range.

cial op amps in order to provide a larger maximum input voltage than can be achieved with the NPN input transistors alone.

THE ACTIVE LOAD

Figure 3-14 shows a simple op amp input stage. Basically, the load resistors of the PNP differential pair shown in Fig. 3-11 have been replaced with an NPN current mirror. The current mirror is called an active load, because transistors are active devices, i.e.,

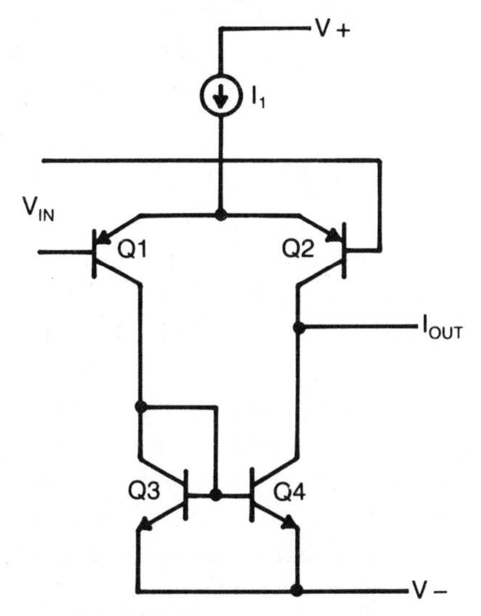

Fig. 3-14. Op amp input stage.

they have gain (while resistors do not and are thus called passive components).

Assume that the base voltages of Q1 and Q2 are both equal to zero and that Q1 is matched to Q2 and Q3 is matched to Q4. Then one-half of the current I1 flows in each side of the differential amplifier and I_{OUT} is equal to zero.

Now apply a small positive voltage to the base of Q1. The result is that the current in Q1 will be a little less and the current in Q2 will be a little more (the total remains constant). The current in Q1 flows into Q3 and is mirrored in Q4. Since the current in Q4 is less than the current in Q2, a current I_{OUT} will flow into some external load (not shown) and be equal to the difference between I_{C2} and I_{C1} according to the following equation:

$$I_{OUT} = V_{IN}\,(q/kT)\,(\tfrac{1}{2}I1)$$

This is essentially the same equation shown in step 5 in the development of the differential amplifier gain equation in the previous section. If V_{IN} is limited to 2.6 mV and I equals 20 μA (it is desirable to keep the input current, which is this current divided by beta, as low as possible) and kT/q equals 26 mV then I_{OUT} is equal to 2 μA maximum.

Because this level of current is not useful for driving significant loads, let's add another gain stage. This is shown in Fig. 3-15. Notice that Q5 has a current source load. This is another example of an active load.

The upper limit on the value of I_2 is equal to the maximum current available from the input stage (2 μA, in this example) times the minimum beta of Q5. If the minimum beta were 50, then the maximum value of I_2 is 100 μA. This is also the maximum output current. The equation for I_{OUT} (for V_{IN} less than 2.6 mV) is:

$$I_{OUT} = V_{IN}(q/kT)\,(\tfrac{1}{2}I1)\,(\beta_5)$$

Assume, for the sake of illustration, that the beta of Q5 is 200. For I_{OUT} to remain zero, all of I_2 must flow into Q5, and the base current of Q5 will be equal to 100 μA divided by 200, or 0.5 μA. (This base current represents an unwanted load on the input stage and will cause the differential pair to be slightly unbalanced. Much of the imbalance can be eliminated by using a Darlington pair in place of Q5.) Now, if the Q5 base current is decreased slightly, Q5 collector current will decrease, and a current I_{OUT} will flow into an external load (not shown). This current is equal to the difference between the current source current I_2 and Q5 collector current.

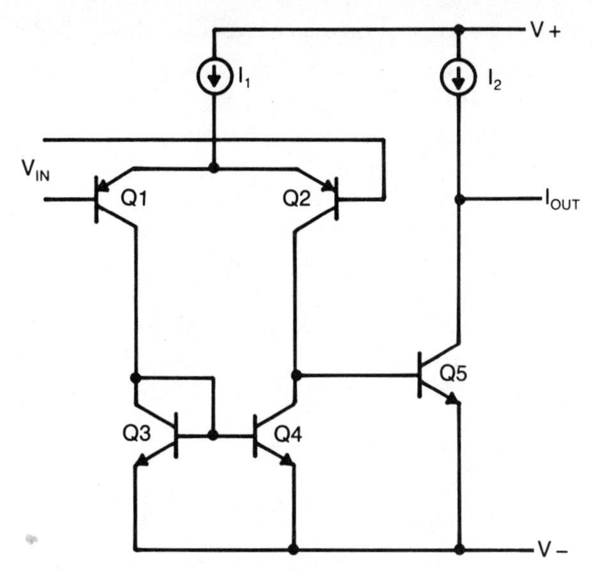

Fig. 3-15. Op amp input stage and gain stage.

THE OUTPUT STAGE

The op amp shown in Fig. 3-15 is an improvement over the one shown in Fig. 3-14, but is still not capable of driving significant loads. An op amp with an improved gain stage and an output stage is shown in Fig. 3-16.

The bias current I_2 for the gain stage (Q5,Q6) flows through Q9 and Q10 and into the collector of Q6. The purpose of Q9 and

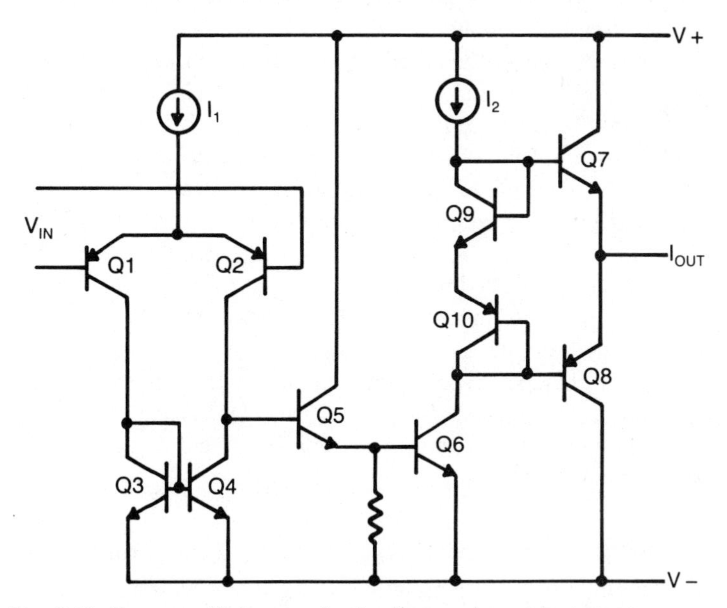

Fig. 3-16. Op amp with improved gain stage and an output stage.

Q10 (effectively diodes) is to create a voltage drop across the base-emitter junctions of Q7 and Q8, thus biasing them to be slightly on. This minimizes the cross-over distortion that can occur when Q7 is turning on and Q8 is turning off and vice versa. The equation for the output current is:

$$I_{OUT} = V_{IN}\,(q/kT)\,(\tfrac{1}{2}I1)\,(\beta_5\beta_6\beta_7)$$

If the direction (sign) of V_{IN} is such that Q8 is conducting, then replace the Q7 beta with Q8 beta in the above equation. If a load resistor is added from the emitters of Q7 and Q8, then the output current is converted to a voltage. The amplifier gain is:

$$V_{OUT}/V_{IN} = R_L\,(q/kT)\,(\tfrac{1}{2}I1)\,(\beta_5\beta_6\beta_7)$$

For example, assume the following values:

R_L = 2 kΩ
kT/q = 26 mV
$\tfrac{1}{2}$I1 = 10 μA
beta Q5, Q6 = 100
beta Q7 = 50

The resultant gain is 384,615. If the supply voltages were plus and minus 15 volts, it would take less than 40 microvolts to completely switch the output from zero to the supply voltage.

Note that the gain is dependent on the value of the load resistor. If this amplifier as shown were completely switched it is possible to burn out one or the other output transistor. Current-limiting circuitry is usually provided to protect against this possibility. A typical circuit is shown in Fig. 3-17. If either Q7 or Q8 current becomes excessive, the voltage drop across R_E will turn on a transistor which then limits the output current.

If Q9 begins to turn on, then it will begin to turn off Q7. An equilibrium point will be reached where Q7 provides only the amount of current needed to bias Q9 such that the Q9 collector current plus the Q7 base current equals the value of the current source I (Q6 is completely off for this condition).

If Q10 begins to turn on, current begins to build up in Q11 and is mirrored in Q12, thus turning off Q5 and Q6. When Q6 begins to turn off (or conduct less current), then Q8 current begins to limit as in the case of Q7.

SUMMARY

Four fundamental IC building blocks were presented in this chapter. Obviously, there are many other circuit design techniques

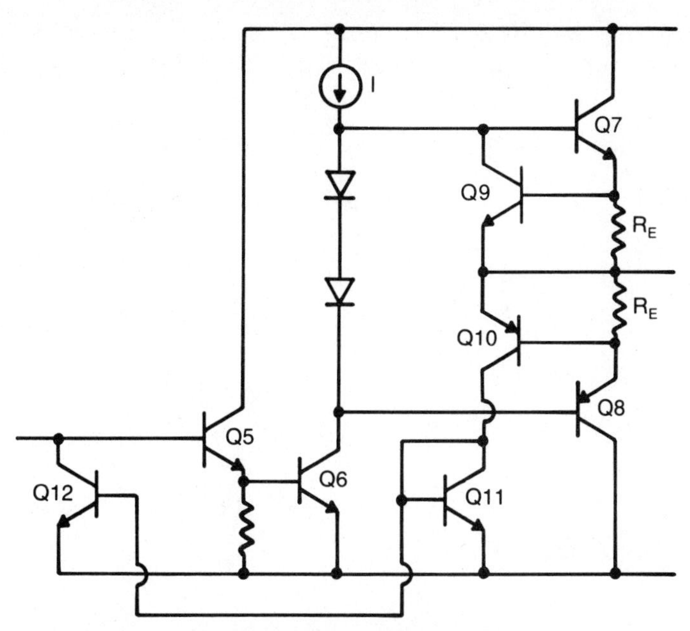

Fig. 3-17. Output stage with short-circuit protection.

used by linear IC designers to maximize the performance of their IC designs. Many of these other techniques, however, are often just variations on these basic four subcircuits.

At this point, you should be able to look at some of the schematics in the linear IC databooks and recognize these fundamental building blocks. Don't worry if you cannot immediately understand every schematic completely. For one thing, many databook schematics don't give you enough information to figure out what's going on. For another, even with extensive circuit design experience, it is often extremely difficult to second-guess another engineer's design.

Chapter 4

Operational Amplifiers

THE INTEGRATED CIRCUIT OPERATIONAL AMPLIFIER (OP AMP) is the most pervasive of all linear ICs. About a dozen manufacturers dominate the op-amp marketplace, collectively supplying hundreds of different part types ranging in price from less than 50 cents to over $100. You, the circuit designer, must choose the right op amp for your application. The "right" op amp is the lowest cost op amp that will meet the requirements. The purpose of this chapter is to explain how the op amp works internally, to explain the data sheet specifications, and to provide selection guide tables that will show the price-performance relationship for each of the major op-amp classifications.

BASIC OP AMP CIRCUITS

Before I discuss the internal operation of a typical op amp, let's look at how an op amp works in four basic circuit configurations. The open-loop configuration is shown in Fig. 4-1. This figure shows that the basic op amp is a five terminal device. There are two inputs (the noninverting input (+) and the inverting input (−)), one output, and two power supply terminals (VCC, the positive supply, and VEE, the negative supply.)

The relationship between the inputs and the output is as follows:

$$V_{OUT} = (V_{IN+} - V_{IN-}) \bullet (A_{VOL})$$

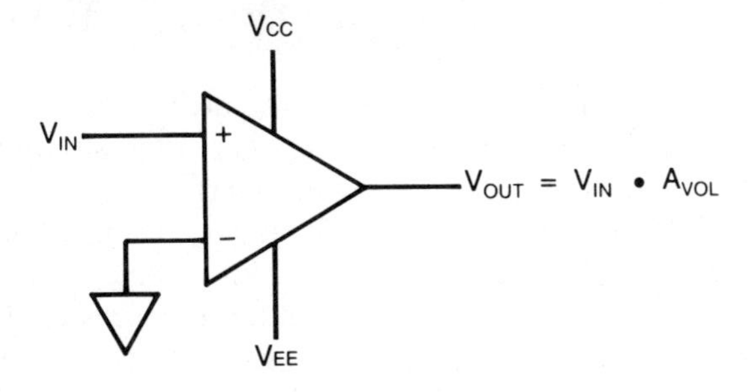

Fig. 4-1. The op amp used in the open-loop mode.

A_{VOL} is the symbol for the op amp's open loop voltage gain, typically over 100,000 volts-per-volt. Open-loop means that no external feedback networks are used to reduce the gain below the op amp's inherent gain.

The open-loop configuration is rarely, if ever, used. However, this mode is shown to point out that the basic op amp is a very high-gain amplifier that amplifies the voltage difference between the plus input and the minus input by an amount equal to its open-loop gain.

Next, Fig. 4-2 shows the op amp used as an inverting amplifier. This is a closed-loop configuration, using feedback to reduce the amplifier gain below the op amp's inherent open-loop gain. Two facts about op amps enable us to determine the amplifier's closed loop gain:

1. The op amp input resistance is very high, typically greater than 1 megohm. Ideally, no current flows into or out of the op amp's input terminals.

2. In the closed-loop configuration (a feedback path from the output back to the inverting, or minus, input), the inverting input voltage is approximately equal (within a few millivolts or less) to the noninverting input—zero volts in this case.

For illustration purposes, let's assign the following values:

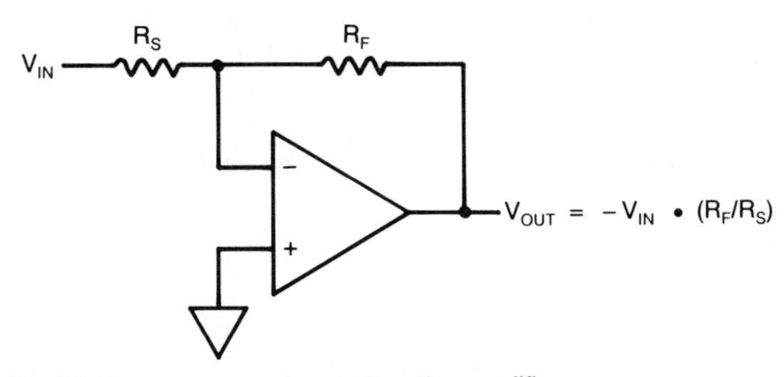

Fig. 4-2. The op amp used as an inverting amplifier.

$$V_{IN} = 0.2\text{ V} \qquad R_S = 1\text{ k}\Omega \qquad R_F = 10\text{ k}\Omega$$

The current through R_S equals V_{IN} divided by R_S, or 0.2 mA. Ideally no current flows into the op amp, then all of the 0.2 mA must flow (left to right in the figure) out of V_{IN} through R_S and through R_F and into the op amp's output terminal, thus creating a voltage drop across R_F equal to 10 kΩ times 0.2 mA, or −2 V. The sign of the voltage at the output terminal is negative because the direction of the current flow is towards the output terminal and because the inverting input terminal is approximately equal to zero volts. The output voltage is exactly 10 times the input voltage V_{IN} and the sign is negative. The amplifier gain, 10, is determined solely by the ratio of R_F to R_S.

Thus, for the inverting configuration,

$$V_{OUT} = I_F \bullet R_F$$

And since I_F is approximately equal to I_S (the current through the source resistor R_S), then

$$\begin{aligned} V_{OUT} &= I_S \bullet R_F \\ &= (V_{IN}/R_S) \bullet R_F \\ &= V_{IN} \bullet (R_F/R_S) \end{aligned}$$

The third configuration is shown in Fig. 4-3. This figure shows the op amp used as a noninverting amplifier. The circuit is analyzed as follows:

1. Since the op amp in the closed-loop configuration "wants" the voltages on the plus and minus inputs to be the same, then whatever voltage that is applied to the plus input "appears" on the minus input.
2. The voltage on the minus input generates a current equal to V_{IN} divided by R_S.
3. The current through R_S also flows through R_F, thus creating a voltage drop equal to R_F times the current. The voltage drop across the resistor is then added algebraically to the voltage on the minus input.

For example, let's assign the following values:

$$V_{IN} = 0.5\text{ V} \qquad R_S = 2\text{ k}\Omega \qquad R_F = 8\text{ k}\Omega$$

Therefore,

$$I_S = 0.5/2000 = 0.25\text{ mA}$$

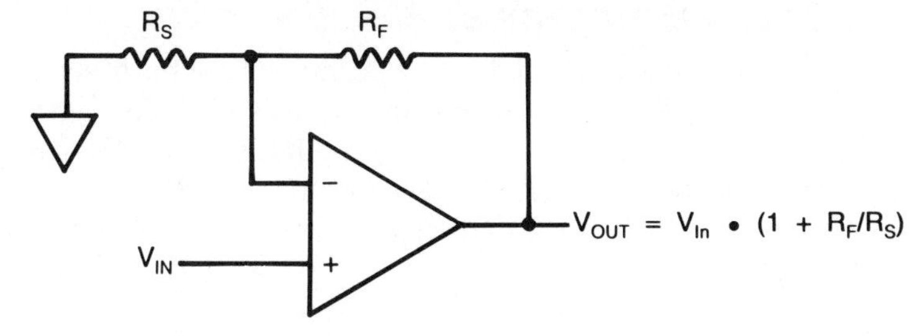

Fig. 4-3. The op amp used as a noninverting amplifier.

And because the current through R_F is approximately equal to I_S, then

$$\begin{aligned} V_{OUT} &= 0.5 + (0.25\ \text{mA}) \bullet (8000) \\ &= 0.5 + 2 = 2.5\ \text{V} \end{aligned}$$

Algebraically,

$$V_{OUT} = V_{IN} + (I_F \bullet R_F)$$

Because I_F is approximately equal to I_S, then

$$\begin{aligned} V_{OUT} &= V_{IN} + (I_S \bullet R_F) \\ &= V_{IN} + (V_{IN}/R_S) \bullet R_F \\ &= V_{IN} \bullet (1 + R_F/R_S) \end{aligned}$$

The fourth configuration is shown in Fig. 4-4. This figure shows the op amp used in the voltage follower, or buffer mode. This mode is really just a variation on the noninverting amplifier with R_S equal to infinity and R_F equal to zero. In this mode V_{OUT} is approximately equal to V_{IN}.

These four configurations illustrate the basic behavior of the op amp. For the sake of simplicity, only a few op-amp characteristics have been mentioned. A more thorough understanding of these and other op-amp characteristics is required for all but the simplest applications (a dc amplifier with a gain tolerance of ± 20 percent, for example).

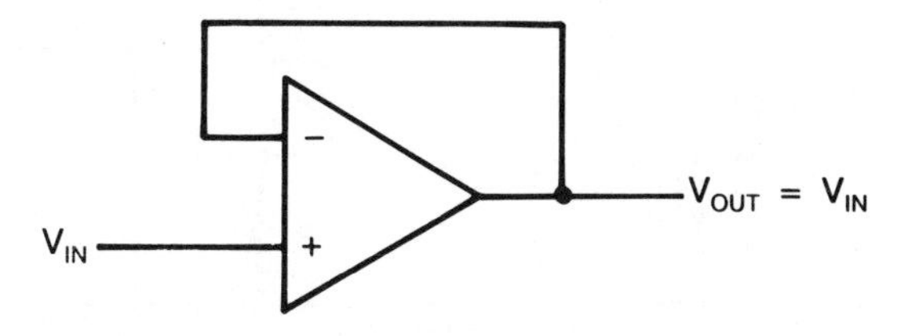

Fig. 4-4. The op amp used in the voltage-follower mode.

OP-AMP CHARACTERISTICS

A simplified schematic of the internal structure of the basic op amp is shown in Fig. 4-5. This op amp consists of three stages: an input stage (I1, Q1-Q4), a gain stage (I2, R1, C_C, Q5, Q6), and an output stage (D1, D2, Q7, Q8). The load resistor, R_L, is external. I will use this schematic to discuss several important op-amp characteristics.

Input Offset Voltage

First, to analyze the circuit shown in Fig. 4-5, let's assign the following values:

I1 = 20 μA
I2 = 200 μA
C_C = 30 pF
R1 = 200 kΩ
R_L = 2 kΩ
Beta (Q1-Q6) = 100
Beta (Q7, Q8) = 50

For completeness, let's also set VCC equal to +15 volts and VEE equal to −15 volts (typical op-amp power supply levels), although these values will not really enter into the analysis (only because it is not shown how the current sources are actually implemented). Let's also assume for the moment that Q1 and Q2

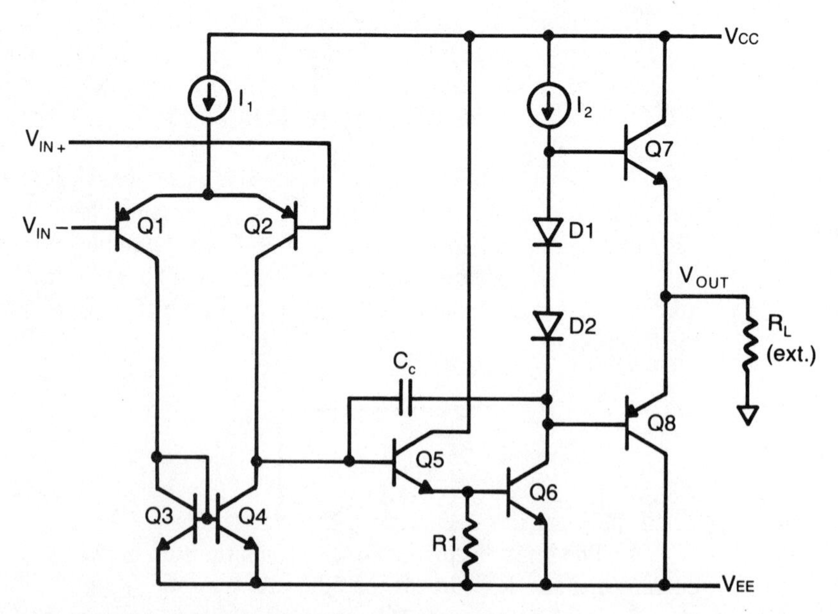

Fig. 4-5. Simplified schematic of the basic op amp.

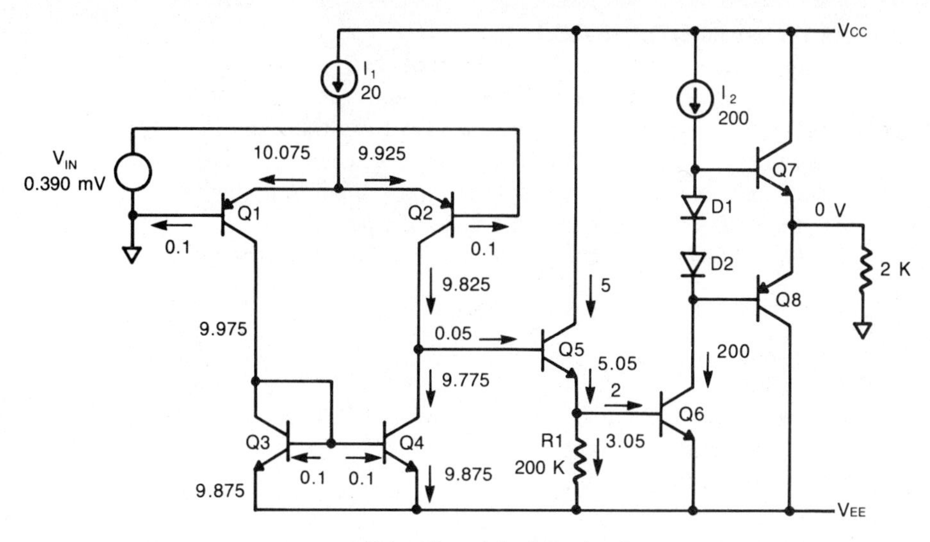

Fig. 4-6. Solution for V_{OUT} equals zero volts.

are perfectly matched to each other and that Q3 and Q4 are likewise perfectly matched (same emitter size).

Now let's find out what input voltage is required to maintain the output at exactly zero volts. This input voltage, the difference between the plus input and the minus input with the minus input grounded, is called the *input offset voltage.*

The solution is shown in Fig. 4-6. The solution is determined by working backwards from the output.

1. Since V_{OUT} equals zero, then any current in Q7 will flow into Q8. Also, any base current flowing into Q7 will be equal to the base current flowing out of Q8 (the beta of Q7 equals the beta of Q8). Thus, all 200 μA of I2 flows into the collector of Q6. The amount of Q7-Q8 current cannot be determined without knowing more about the geometries of D1, D2, Q7, and Q8. For example, if D1 were made from an NPN transistor having exactly the same emitter size as Q7, and if D2 were made from a PNP transistor having exactly the same emitter size as Q8, then the Q7-Q8 emitter currents would equal 200 μA.

2. The base current of Q6 equals 200 μA divided by 100, or 2 μA.

3. The current through R1 equals the V_{BE} of Q6 divided by 200 kΩ. For convenience sake, assume Q6 V_{BE} equals 0.61 volts. Then, 3.05 μA flows through R1.

4. The emitter current of Q5 equals the sum of the Q6 base current and the R1 current, or 5.05 μA.

5. If Q5 beta equals 100, then Q5 collector current equals 5

μA and Q5 base current equals 0.05 μA.

6. The values for the remaining currents are determined by trial and error by hand (or by computer, if you have access to a computer circuit simulation program and also know the transistor model parameters) until a balanced condition is achieved. If I1 is 20 μA then the currents in Q1-Q4 must be approximately as shown in order for 0.05 μA to flow into the base of Q5.

7. The value of V_{IN} is computed using the following formula:

$$V_{IN} = V_T e^{IE1/IE2}$$

where

$V_T = kT/q = 0.026$ at room temperature (25 °C)
IE1 = Q1 emitter current
IE2 = Q2 emitter current

For the values shown in Fig. 4-6,

$$V_{IN} = 0.026 e^{10.075/9.925} = 0.39 \text{ mV}$$

Thus, even if there were no V_{BE} mismatch between Q1 and Q2 and between Q3 and Q4, an op amp would have some input off-

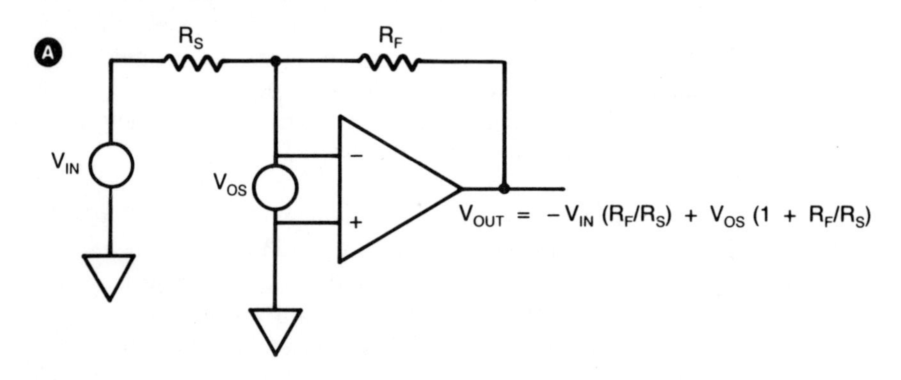

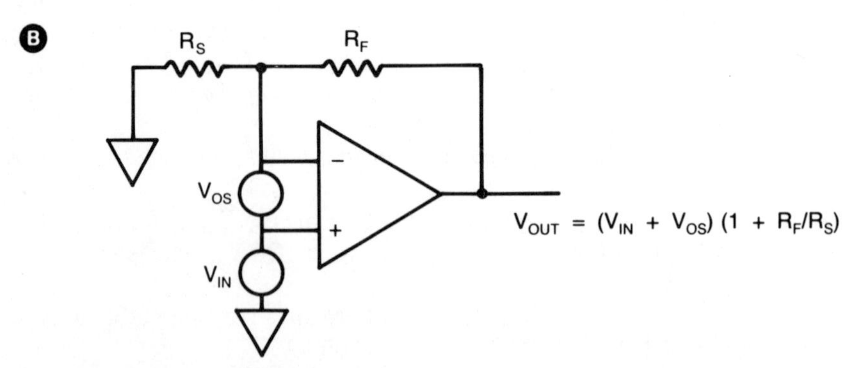

Fig. 4-7. The effect of input offset voltage.

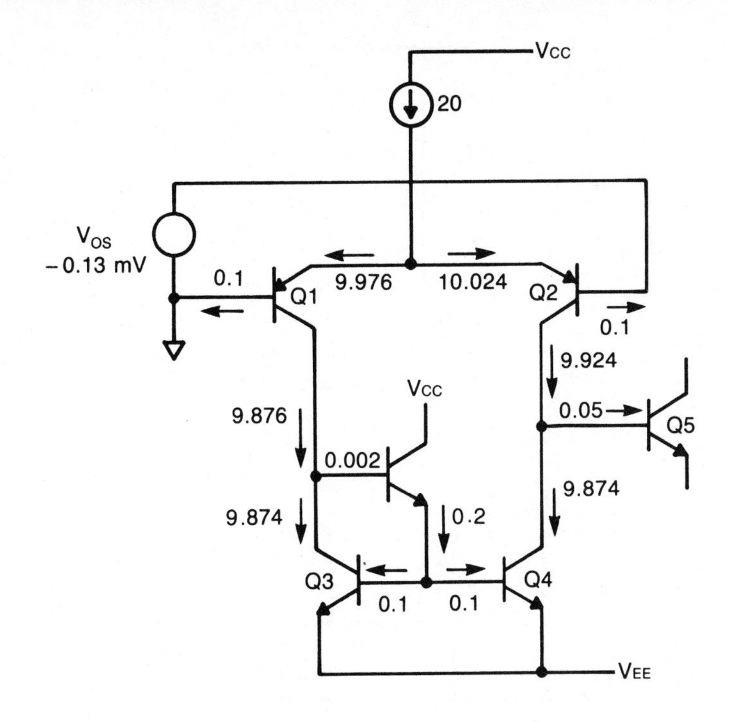

Fig. 4-8. Improved dc offset input stage.

set voltage just due to second stage bias current. Note that V_{BE} mismatch may add or subtract from the offset voltage due to biasing the gain stage because the sign of V_{BE} mismatch generally cannot be predicted.

The effect of input offset in circuit design can be seen by looking at Fig. 4-7. Any input offset voltage (positive or negative) appears as an error term in the V_{OUT} equation. This error term equals the offset voltage multiplied by one plus the ratio of R_F to R_S. Input offset voltage is reduced in actual op-amp ICs by adding a transistor that supplies base current for both Q3 and Q4. The solution is shown in Fig. 4-8.

It can be seen that input offset is important only in dc applications (this offset would not affect input and output ac capacitively-coupled amplifier circuits). If input offset is critical, then you have a choice between buying an op amp with a low-input offset (generally expensive) or buying an op amp that provides for external offset nulling (realizing that a potentiometer and its adjustment also cost money). Typical offset adjustment is shown in Fig. 4-9.

Input Bias Current

Figure 4-6 shows that some current flows out of the plus and

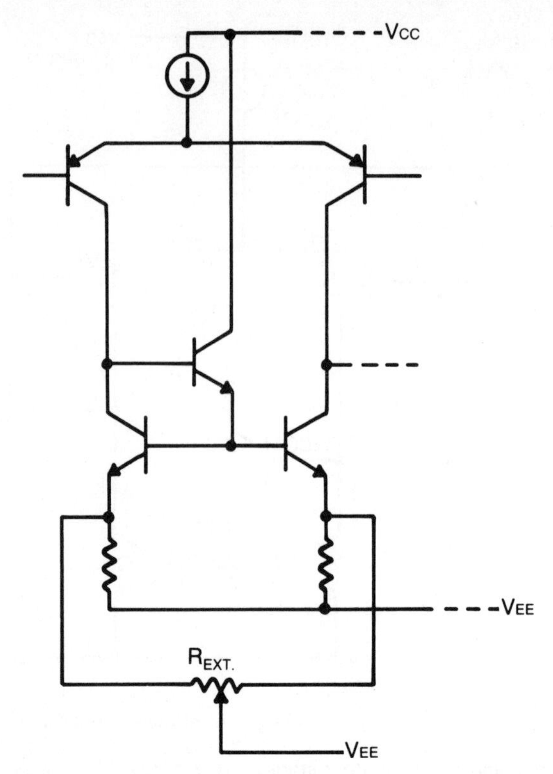

Fig. 4-9. Typical input offset nulling technique.

minus inputs (for an NPN input stage, current flows into the inputs). This current is called *input bias current*. This term used in the data sheet is defined as the average of the two input currents. This is because in real op amps both currents are not identical. The difference between the two input currents is called the *input offset current* and is almost always specified in the data sheets.

Both of these parameters are important in dc amplifier design where high value source and feedback resistors are used. An extreme case is shown in Fig. 4-10. The output should be −1 volt. For the case (not shown) of V_{IN} equal to −1 volt, the output would equal zero volts—not the desired +1 volt.

Input bias current can be reduced by replacing the PNPs in the input stage with (1) FET transistors (junction or MOS), (2) super-beta NPN transistors (LM108A, for example), (3) Darlington PNPs or NPNs, or (4) NPNs with input bias current cancellation circuitry.

Input bias current for FET-input op amps is typically below 100 picoamperes (pA). Op amps using super-beta NPN transistors have typical input bias currents below 1 nanoamp (nA). Darlington configurations can achieve super-beta performance but at the expense of degraded input offset voltage.

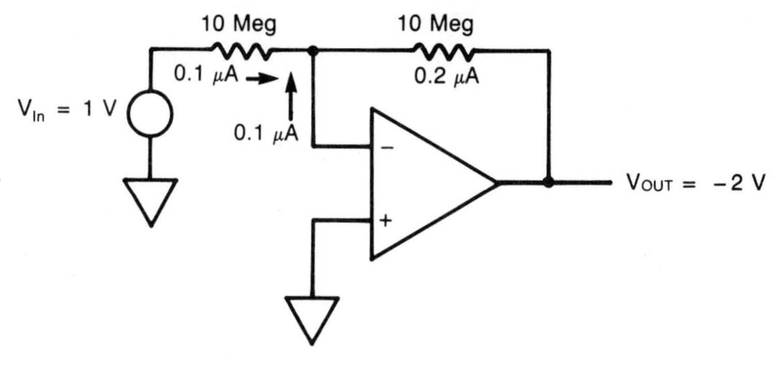

Fig. 4-10. The effect of input bias current.

Bias current cancellation techniques can achieve super-beta level bias currents without using super-beta processing (nonstandard, more expensive) and without input offset voltage degradation. One device that uses bias current cancellation is the OP-07 made by Precision Monolithics, Incorporated (PMI). A simplified schematic of the OP-07 is shown in Fig. 4-11.

Let's analyze the Q1 side of the OP-07 input stage.

1. Q3 current equals Q1 current.
2. Q3 base current flows through Q7.
3. If the Q5 emitter is identical in size to that of Q7 then the collector current of Q5 will be approximately equal to the base current of Q3.

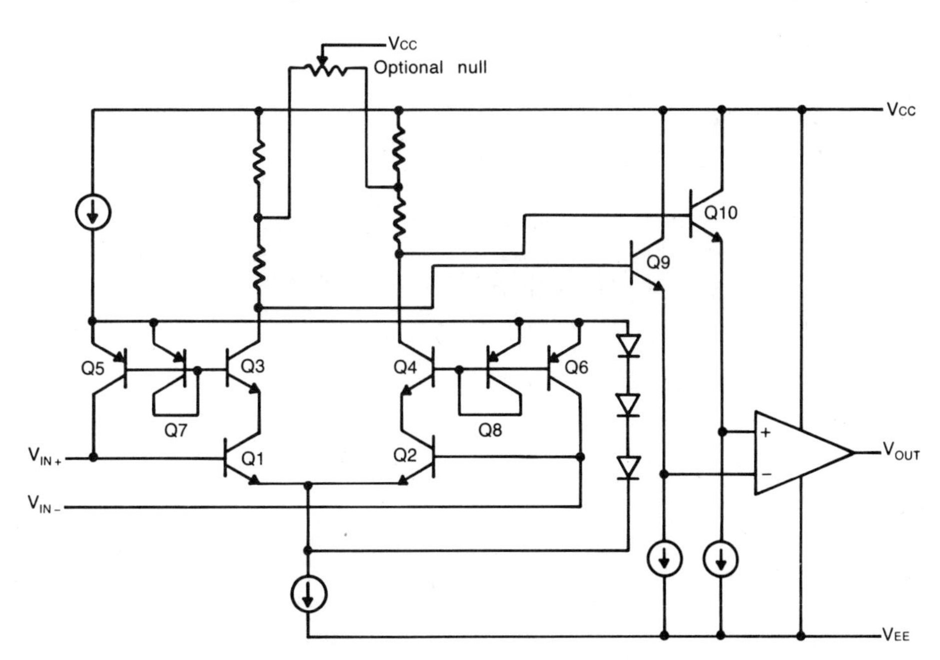

Fig. 4-11. Simplified schematic of the OP-07.

4. Q5 collector current flows into the base of Q1. If Q3 and Q1 have the same emitter size then no additional input bias current will be required.

In practice, due to transistor mismatch (both V_{BE} and beta), input bias current will still be typically around ± 1 nA.

One other technique is used to reduce the input bias current: by reducing the input stage operating current (I1 in Fig. 4-6) the input bias current is lowered. This technique is used in low-power op amps but usually at the expense of bandwidth and slew rate, which I will discuss shortly.

Open Loop Voltage Gain

Earlier I analyzed the basic op amp for the case when V_{OUT} equals zero volts (Fig. 4-6). Now let's analyze the op amp for the case when V_{OUT} equals 10 volts. The solution is shown in Fig. 4-12.

For V_{OUT} equal to zero volts, V_{IN} equals 0.39 mV (see Fig. 4-6). For V_{OUT} equal to 10 volts, V_{IN} equals 0.416 mV, a change of +0.026 mV. The change in V_{OUT} divided by the change in V_{IN} is called the *open loop voltage gain* (A_{VOL}):

$$A_{VOL} = \Delta V_{OUT}/\Delta V_{IN} = 10/0.000026 = 384{,}615$$

By comparing Fig. 4-6 with Fig. 4-12, you can see that the change in V_{OUT} is equal to the change in Q5 base current times Q5 beta times Q6 beta times Q7 beta (or times Q8 beta for negative output voltages) times R_L:

$$\Delta V_{OUT} = -\Delta I_{B5}\, \beta_5\, \beta_6\, \beta_7\, R_L \qquad \textbf{(1)}$$

The change in Q5 base current is the same as the output current from the first stage. Recall from Chapter 3 that this current is proportional to the change in input voltage:

$$-\Delta I_{B5} = \Delta V_{IN}\, (q/kT)\, (\tfrac{1}{2}I1) \qquad \textbf{(2)}$$

By substituting (2) into (1) we get the op amp gain equation developed in Chapter 3:

$$A_{VOL} = R_L\, (q/kT)\, (\tfrac{1}{2}I1)\, (\beta_5\beta_6\beta_7)$$

The expression $(q/kT)\,(\tfrac{1}{2}I1)$ is called the input stage transconductance and is designated by the symbol g_{m1}.

From this equation, you can see that the open loop voltage gain is directly dependent on (1) the load resistor R_L, (2) the operating current of the input stage (I1 current), (3) the current gain of the

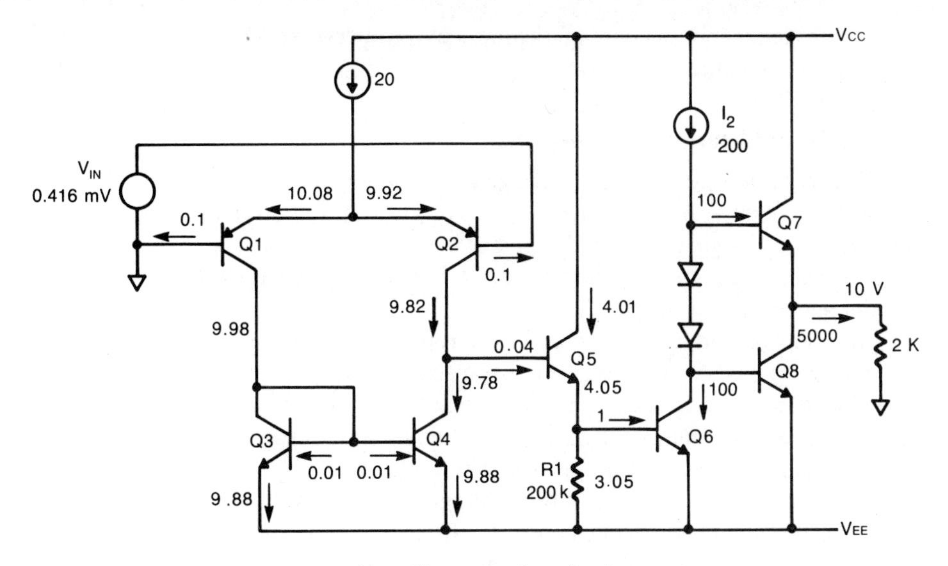

Fig. 4-12. Solution for V_{OUT} equals 10 volts.

gain stage (the betas of Q5 and Q6), and (4) the current gain of the gain stage (Q7 or Q8 beta).

Note, however, that the gain calculated here and in Chapter 3 is a factor of four times typical op amp open-loop gain measurements. The difference between the theoretical gain and actual measured gain is due to variation in temperature across the IC chip when the op amp is delivering maximum current (the case when the output voltage is 10 volts in my example). The temperature variation, or thermal gradient, across the input stage, in particular (the differential pair and the active load transistors), induces an offset voltage and, thus, reduces the gain.

Input Resistance

Input resistance is defined as the change in input voltage divided by the corresponding change in input bias current. Using the results of the solutions shown in Fig. 4-6 and Fig. 4-12, we get

$$R_{IN} = \Delta V_{IN}/\Delta I_B = \Delta V_{IN}/(\Delta I_{C2}/\beta_2)$$

$$= \frac{26\ \mu V}{0.005\ \mu A\ /\ 100} = 520{,}000 \text{ ohms}$$

For FET input op amps, input resistance in on the order of 10^{12} ohms because the gate current is a leakage current rather than a bias current dependent on the input stage operating current and transistor beta.

Small-Signal Frequency Response

The equation for open loop voltage gain developed previously is valid for dc only. At some low frequency (typically 10 hertz) the compensation capacitor C_C (see Fig. 4-5) begins to take effect. At that frequency the op amp begins to act in accordance with the model shown in Fig. 4-13.

In Fig. 4-13 the input stage has been replaced with an ac current source equal to the input stage transconductance g_{m1} times the change in input voltage V_{IN}. The output stage is not shown because it is a unity-gain voltage stage and any voltage developed on the right side of C_C also appears at the output.

Now you can see that the output voltage depends on the reactance of the compensation capacitor. Recall that reactance of a capacitor is given by the following equation:

$$X_C = 1/(2\pi fC)$$

Therefore the equation for V_{OUT} as a function of frequency is

$$\Delta V_{OUT} = \frac{g_{m1}\Delta V_{IN}}{2\pi f\ C_C}$$

and the equation for *small-signal frequency response,* or *gain-bandwidth product,* is

$$\frac{\Delta V_{OUT}}{\Delta V_{IN}} = \frac{g_{m1}}{2\pi f_1\ C_C} = 1$$

where f_1 is the frequency where the gain is equal to 1. Solving for f_1,

$$f_1 = g_{m1}/2\pi C_C$$

Based on the solutions in Fig. 4-6 and Fig. 4-12 and a compensation capacitor of 30 pF, the unity gain bandwidth is

$$f_1 = \frac{0.385 \times 10^{-3}}{(6.28)\ (30 \times 10^{-12})} = 2.04\ \text{MHz}$$

Unity gain bandwidth is directly proportional to input stage operating current and the compensation capacitor. Typical voltage gain versus frequency is shown in Fig. 4-14. Both scales are logarith-

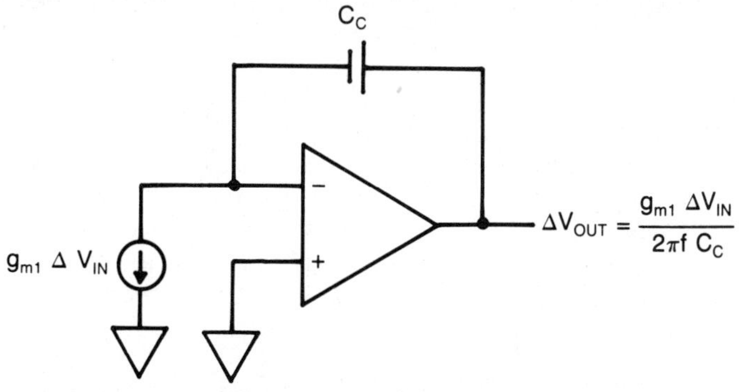

Fig. 4-13. Model of op amp used to calculate small signal frequency response.

mic. Voltage gain decreases (as expected) by a factor of 10 per decade of increasing frequency, or −20 dB per decade.

Slew Rate

Figure 4-15 shows what happens when a large voltage step (greater than 100 millivolts) is applied to the op-amp input. Assuming high betas and a perfect Q3-Q4 current mirror, Q1 turns on and all of I1 flows through Q1 into Q3 and is mirrored to Q4. Since Q2 is off, the input current to the gain stage wants to decrease by an amount equal to I1. Most of this current is not required by the gain stage and, as a result, this current will begin to charge C_C and cause the voltage at the output to rise at a rate called the *slew rate:*

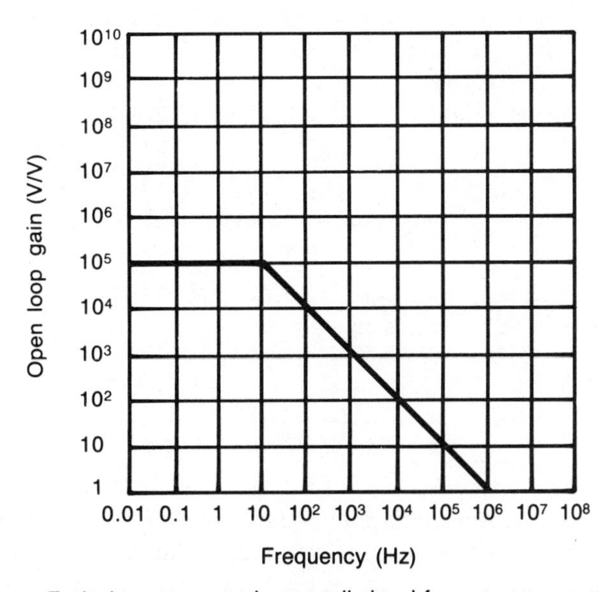

Fig. 4-14. Typical op-amp open-loop small signal frequency response.

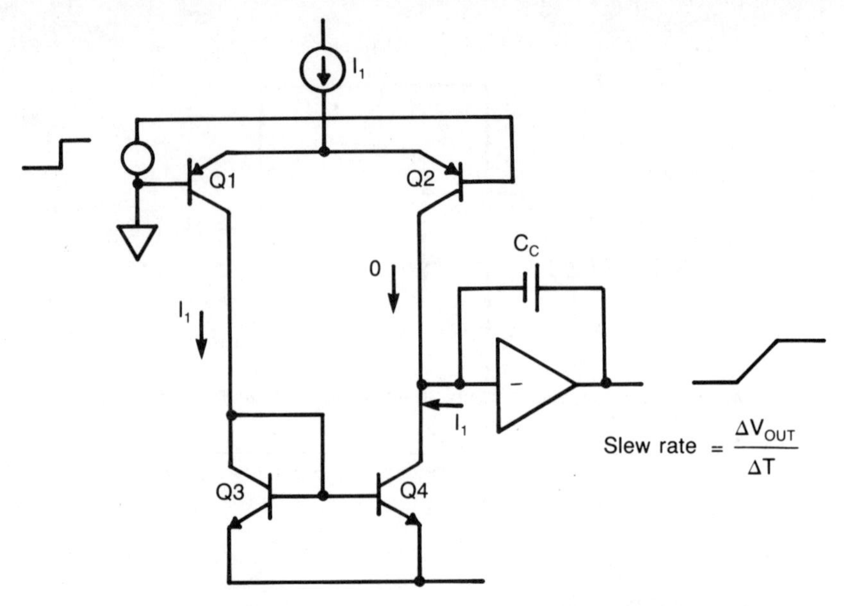

Fig. 4-15. Model of op amp used to show response to input step voltage.

$$\frac{\Delta V_{OUT}}{\Delta T} = \text{SLEW RATE} = \frac{I1}{C_C} \text{ volts/second}$$

For the values being used in this analysis,

$$\text{SLEW RATE} = 20\ \mu\text{A} / 30\ \text{pF} = 0.67\ \text{V}/\mu\text{s}$$

From this equation it can be seen that the largest sine-wave voltage that can be amplified without distortion is limited by the slew rate. This is shown graphically in Fig. 4-16. It can be shown that the maximum rate of change for a sine wave of amplitude V_P is equal to

$$\left.\frac{\Delta V}{\Delta T}\right| \text{max. for sine wave} = 2\ \pi\ f\ V_P$$

Thus, the highest frequency for a desired output voltage V_P is given by the following equation:

$$f = \left(\frac{1}{2\ \pi\ V_P}\right)\left(\frac{\Delta V_{OUT}}{\Delta T}\right) = \frac{\text{S.R.}}{2\ \pi\ V_P}$$

This frequency is called the *power bandwidth*. For example, for V_P equal to 10 volts and a slew rate equal to 0.67 volts per microsecond,

$$\text{Power Bandwidth} = (0.067 \times 10^6)/(6.28)(10) = 10.7 \text{ kHz}$$

significantly smaller than the 2.04 MHz small-signal unity gain bandwidth.

One obvious way to improve slew rate and power bandwidth is to increase the input stage operating current. Unfortunately this increases input bias current. Using JFET input transistors gets around this problem. Also, with JFET transistors, it is possible to use a low value of compensation capacitor, which is chosen to keep the bandwidth below the frequency at which phase shift though the input stage (primarily) causes the amplifier to be unstable at low gains. Because of these two factors, using JFETs can improve small signal bandwidth by a factor of 4 and the slew rate and power bandwidth by a factor of 40.

One bipolar design that rivals JFET performance without increasing the operating current of the input stage is shown in Fig. 4-17. This is a simplified schematic of the Signetics NE/SE531 input stage. Instead of a single current source for the input stage, current is provided separately to Q7 and Q8. The current provided to Q8 is expressed as

$$I_{E8} = \frac{V_{R2}}{R2} = \frac{V_{BE3} - V_{BE8} - R1(I1 - I_{E8})}{R2}$$

assuming that the voltage at the base of Q7 is equal to the voltage at the base of Q8, i.e., the balanced condition.

Without more information (values for V_{BE3}, V_{BE8}, and I1), I cannot calculate I_{E8} using this equation. However, by using some

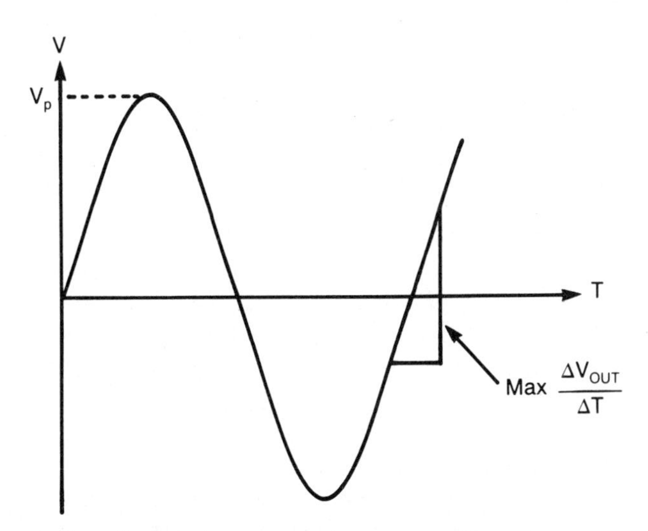

Fig. 4-16. The effect of slew rate on sine wave amplification.

data sheet information I can determine the approximate value for I_{E8}. According to the data sheet, the small-signal bandwidth is typically 1 MHz with an external C_C of 100 pF. This tells me that the transconductance g_{m1} is about 0.628×10^{-3}:

$$g_{m1} = 2 \pi f_1 C_C = (6.28)(1 \times 10^6)(100 \times 10^{12}) = 0.628 \times 10^{-3}$$

This g_{m1} is the small-signal transconductance. When emitter resistors are used the equation for g_{m1} becomes

$$g_{m1} = \frac{2}{R_E + V_T/I_E}$$

Solving for I_E,

$$I_E = \frac{V_T}{(2/g_{m1}) - R_E}$$

In Fig. 4-17 R_E equals the sum of R1 and R2. Solving for I_{E8},

$$I_{E8} = \frac{.026}{(2/0.000628) - 2400} = 33.1 \ \mu A$$

This is the current in Q8 when the plus and minus inputs are equal. This current varies only slightly for small input signals. For large signals (greater than 100 mV) the current in Q8 is approximately equal to the input voltage divided by R1 + R2 (or by R3 + R4 for voltages applied to the plus input). Thus, for a 10-volt step applied to the positive input, the charging current applied to the compensation capacitor is about 4 mA. This equates to a theoretical slew rate of 40 volts-per-microsecond. The data sheet specifies a typical slew rate of 35 volts-per-microsecond. Power bandwidth for a 10 volt (peak) sine wave is calculated to be 557 kilohertz, about one half the small-signal bandwidth.

SELECTION GUIDES

Appendix A lists the basic specifications and unit prices of the op amps of about a dozen manufacturers. Although each op amp in Appendix A is not unique, or single-sourced, there are over 1000 op amps in this appendix to choose from. (Appendix A does not list every op amp made by every manufacturer; many second-source op amps have been omitted.) The purpose of this section is to help you quickly select an op amp for your application.

There are many ways to categorize op amps. I have chosen to categorize op amps first by performance, second by operating tem-

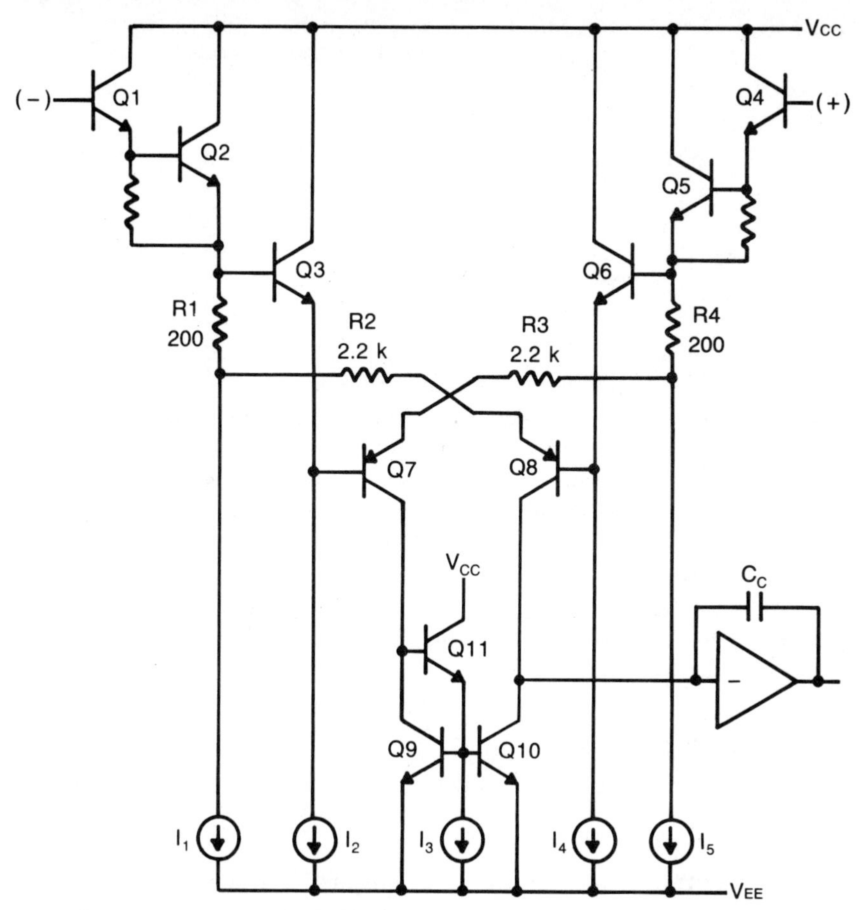

Fig. 4-17. Input stage of Signetics NE/SE531.

perature range and third by number of op amps per package.

First, op amps are classified by a performance label as follows:

- ☐ General Purpose
- ☐ Low Offset, Low Drift
- ☐ Low Input Current, High Impedance
- ☐ Wide Bandwidth
- ☐ High Slew Rate
- ☐ Low Noise
- ☐ Low Power
- ☐ High Voltage
- ☐ High Output Current (Power Op Amps)
- ☐ Programmable

Note that these performance "bins" are only an aid to selecting an op amp. Most op amps can (and are) listed under more than one heading. The LM144/LM344 is a high voltage, high slew rate op amp according to National's Linear Databook. The LM13080 is a programmable power op amp.

Second, op amps are classified by operating temperature range. Three operating temperature ranges have become standard:

- ☐ Commercial: 0 °C to 70 °C
- ☐ Industrial: −25 °C to 85 °C
- ☐ Military: −55 °C to 125 °C

There is also a fourth temperature range, −40 °C to 85 °C, which is sometimes called Industrial and sometimes Automotive. In the selection guide tables which follow, the industrial temperature range could be either the −25 °C to 85 °C range or the −40 °C to 85 °C range. Also, some op amps made by Harris are specified as having a commercial temperature range of 0 °C to 75 °C. See Appendix A for the specific minimum and maximum operating temperatures.

Third, op amps are classified by the number of op amps per package. Single, dual, and quad are standard, but Intersil, for one, offers a line of triple op amps.

General-Purpose Op Amps

Tables 4-1 through 4-9 list the lowest-cost general-purpose op amps as follows:

Table	Temperature Range	Op Amps Per Package
4-1	Commercial	Single
4-2	Commercial	Dual
4-3	Commercial	Quad
4-4	Industrial	Single
4-5	Industrial	Dual
4-6	Industrial	Quad
4-7	Military	Single
4-8	Military	Dual
4-9	Military	Quad

After studying these tables I noticed that bipolar op amps are still the lowest cost op amps, but many FET-input op amps (op amps with less than 1 nA input bias current in these tables) are also low cost. This means better performance for your money. For example, in Table 4-1, RCA's CA081 costs only 52 cents, yet has a maximum bias current of 0.05 nA, a typical bandwidth of 5 MHz, and typical slew rate of 13 V/μs—considerably better than Fairchild's 38

38 cent μA201. However, you should never buy more performance than you need.

Also, military temperature range parts are not necessarily more expensive than industrial temperature range parts (or commercial parts, as you can see in later selection guides). For example, RCA's CA741, CA748, and CA3140 single op amps are all less than the lowest cost industrial temperature range single op amps. This means that it is a good idea to see what are the prices of op amps in higher temperature ranges (higher than your application requires). You may find a lower cost part (probably due to higher volume production).

Low Offset, Low Drift Op Amps

Low offset, low drift op amps are listed in Tables 4-10 through 4-12. You can see from these tables that a low-input offset voltage is usually accompanied by low drift, or a low-offset-voltage temperature coefficient, and that as input offset increases so does drift.

Precision Monolithics Incorporated dominates this category with 58 entries over all three temperature ranges and is the only manufacturer listed with dual op amp entries (no quad op amp entries). When it comes to price, however, P M I is not the winner. The low price award goes to Intersil, whose ICL7650 and ICL7652 have an incredible offset of 0.005 millivolts. The 7650C is a factor of 8 better than its nearest competitor, the Harris HA-5130-5, and is 15 times better than the Harris HA-5135-5, the next lowest in price.

The problem (if you can call it that) with the Intersil 7650 and 7652 is that these op amps have a normal operating voltage of only ± 5 volts, while most op amps have a normal operating voltage of ± 15 volts. This fact will eliminate these op amps as candidates for some applications.

Other op amps with good price-performance ratios are (1) the μA714C (0.15 mV, $2.47), (2) the OP-21F(0.15 mV, $5.63), and (3) the CA3193B (0.075 mV, $5.44), in the commercial, industrial, and military temperature ranges, respectively. The CA3193B costs less than 37 other op amps on the commercial and industrial temperature range lists and less than 29 op amps on the military temperature range list, yet has a lower or equal input offset voltage.

Low Input Current, High Impedance Op Amps

Low bias current op amps are listed in Tables 4-13 through 4-15. These op amps are all high impedance (10^{12} ohms) FET-input op amps.

The bargains here all belong to RCA. Unless you absolutely need an input bias current of less than 1 pA, the RCA CA3450B is lower in price than all other commercial and industrial temperature range op amps (having an input bias current of 1 pA to 10 pA) except the CA3420 and CA3420A (both having an input bias current of 3 pA)—both from RCA.

Wideband Op Amps

Wideband op amps are listed in Tables 4-16 through 4-18. You can see that Harris is a leader in this category. However, one of the lowest cost op amps in this category is the Fairchild μA715 with a bandwidth of 40 MHz. The commercial temperature range version of this part is priced at a low $2.66.

Other good choices are National's LF157 (20 MHz, military, $5.05), LF257 (20 MHz, industrial, $3.60), LF357 (20 MHz, commercial, $1.10), and RCA's CA3130 (15 MHz, military, $1.10).

Analog Devices and Burr-Brown get the award for the most expensive parts so far—$105.50 (AD3554S) and $97.60 (3554S), respectively.

High Slew Rate Op Amps

High slew rate op amps are listed in Tables 4-19 through 4-21. The Analog Devices AD3554 and the Burr-Brown 3554 (Burr-Brown is the original manufacturer) have the highest slew rate—1200 volts per microsecond. This means that the output can swing 20 volts in 16.67 nanoseconds. As you can see, these parts are expensive.

As in the case of the wide bandwidth category, National's LF157, 257, and 357 are also good choices for high performance at a moderate cost. Other good choices are the Signetics SE538 (60 V/μs, military, $6.00) and the NE538 (60 V/μs, commercial, $1.26).

Low Noise Op Amps

Low noise op amps are listed in Tables 4-22 through 4-24. The noise voltage and current values shown are typical values at a frequency of 1000 Hz and a bandwidth of 1 Hz. The op amp with the best price-performance ratio is the Signetics NE5534 with an input noise voltage of 4 $nV/\sqrt{Hz}$ (nanovolts per root hertz) for a price of $1.10.

The effect of noise is shown in Fig. 4-18. This figure shows the case of a noninverting amplifier with a gain of $1 + R_F/R_{S1}$. Noise from six different sources will appear at the output. The noise sources are listed in Table 4-25 where

k = Boltzmann's constant = 1.38×10^{-23} joules/°K

T = Absolute temperature in degrees Kelvin = 298 °K at room temp

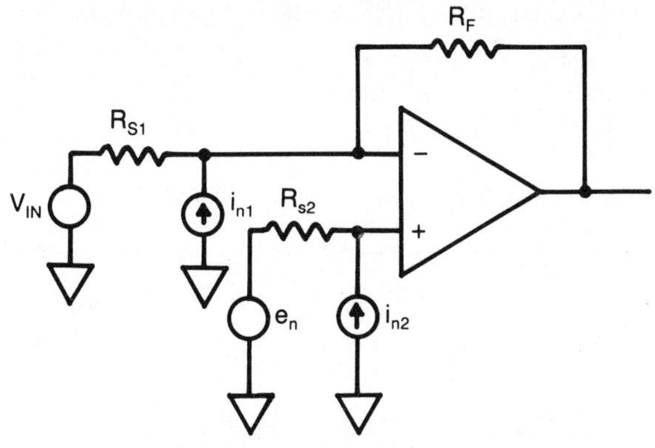

Fig. 4-18. Model used to compute output noise.

BW = Bandwidth

Total output noise is given by the equation,

$$\text{Total Output Noise} = \sqrt{e^2_{n1} + e^2_{n2} + e^2_{n3} + e^2_{n4} + e^2_{n5} + e^2_{n6}}$$

For example, let's compute the total output noise for the following component values and bandwidth:

R_{S1}	=	100 ohms	i_{n1}	=	$0.5\ pA/\sqrt{Hz}$
R_{S2}	=	100 ohms	i_{n2}	=	$0.5\ pA/\sqrt{Hz}$
R_F	=	1 kilohm	e_n	=	$10\ nV/\sqrt{Hz}$
BW	=	10 Hz to 10 kHz			

Let's also assume that the values of i_{n1}, i_{n2}, and e_n are valid for all frequencies between 10 Hz and 10 kHz. This is assumed for illustration only. This assumption is generally not true. If you look at the noise curves for op amps you will see that the noise between 0.1 Hz and 10 Hz is decreasing and flattens out somewhere between 10 Hz and 100 Hz. Thus, to accurately calculate the total noise voltage, you would have to compute the output noise at each frequency and then add by taking the root of the sum of the squares. But for this example I will assume that the noise response is flat over the range from 10 Hz to 10 kHz. Thus,

$$e_{n1} = \sqrt{(4)(1.38 \times 10^{-23})(298)(100)(9990)}\ (1000/100)$$
$$= 1.282\ \mu V\ (rms)$$
$$e_{n2} = \sqrt{(4)(1.38 \times 10^{-23})(298)(100)(9990)}\ (1 + 1000/100)$$
$$= 1.410\ \mu V\ (rms)$$
$$e_{n3} = \sqrt{(4)(1.38 \times 10^{-23})(298)(1000)(9990)}$$
$$= 0.405\ \mu V\ (rms)$$
$$e_{n4} = (0.5 \times 10^{-12})(\sqrt{9990})(1000)$$
$$= 0.050\ \mu V\ (rms)$$

$$e_{n5} = (0.5 \times 10^{-12})(\sqrt{9990})(100)(1 + 1000/100)$$
$$= 0.055\ \mu V\ (rms)$$
$$e_{n6} = (10 \times 10^{-9})(\sqrt{9990})(1 + 1000/100)$$
$$= 10.995\ \mu V\ (rms)$$

Total Output Noise (in μV rms)

$$= \sqrt{(1.282)^2 + (1.410)^2 + (0.405)^2 + (0.050)^2 + (0.055)^2 + (10.995)^2}$$

$$= 11.166\ \mu V\ (rms)$$

You can see (for these values and bandwidth) that the op amp input noise voltage is the largest component of the total output noise. As a point of comparison, general purpose op amps (or op amps where low noise is not optimized) have a typical input noise voltage of around 100 $nV/\sqrt{Hz}$.

Low-Power Op Amps

Low-power op amps are listed in Tables 4-26 through 4-34. Unless otherwise noted, the power supply current values correspond to V_S equal to ± 15 volts. Typical bandwidth values are also listed. Very-low-power op amps generally have less bandwidth than just low- or moderate-power op amps. Programmable op amps (which can be programmed for low power) are not listed in these tables.

High-Voltage Op Amps

High-voltage op amps are listed in Table 4-35. Because very-high-voltage op amps also have high price tags, I recommend that you consider designing your own high-voltage buffer (using high-voltage discrete transistors) and marrying it with a low-cost general-purpose op amp. Only at 50 volts and lower do some of these high-voltage op amps look price competitive with an op amp/buffer combination.

High-Output-Current Op Amps (Power Op Amps)

High-output-current op amps are listed in Table 4-36. Output currents are rated in three ways:

(1) Minimum output current at a specified output voltage.
(2) Minimum output voltage at a specified load resistance.
(3) Minimum output voltage at a specified output current.

The best price-performance op amp in this category is the Fairchild μA759 at \$2.09 for a 325 mA output current. For higher current capability, you may want to consider designing your own cur-

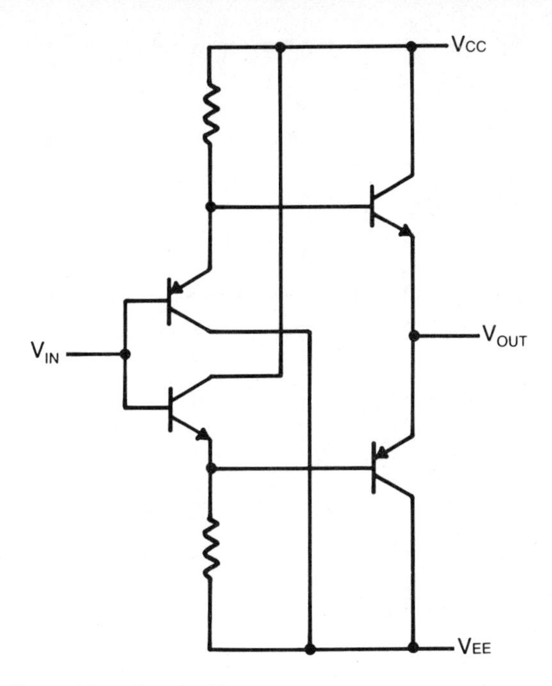

Fig. 4-19. Current booster circuit.

rent booster with discrete transistors using the circuit approach shown in Fig. 4-19. An op amp/current booster combination is shown in Fig. 4-20 (inverting amplifier).

Programmable Op Amps

Programmable op amps are listed in Table 4-37. Programmable op amps have a pin which, when connected to VCC, ground, or VEE (depends on internal design), through a resistor, sets the values of all internal current sources and thus affecting all op-amp

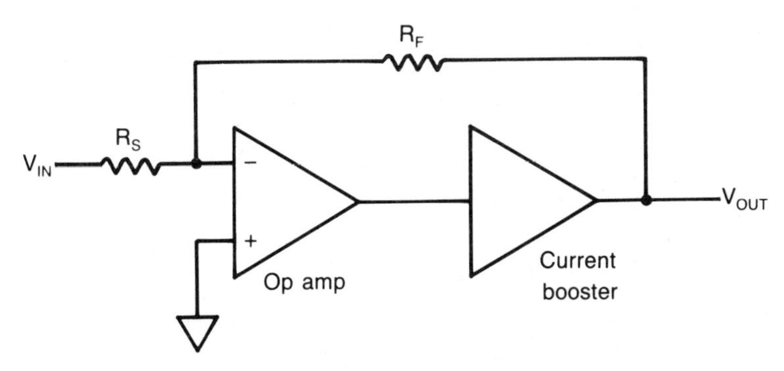

Fig. 4-20. Op amp and current booster combination.

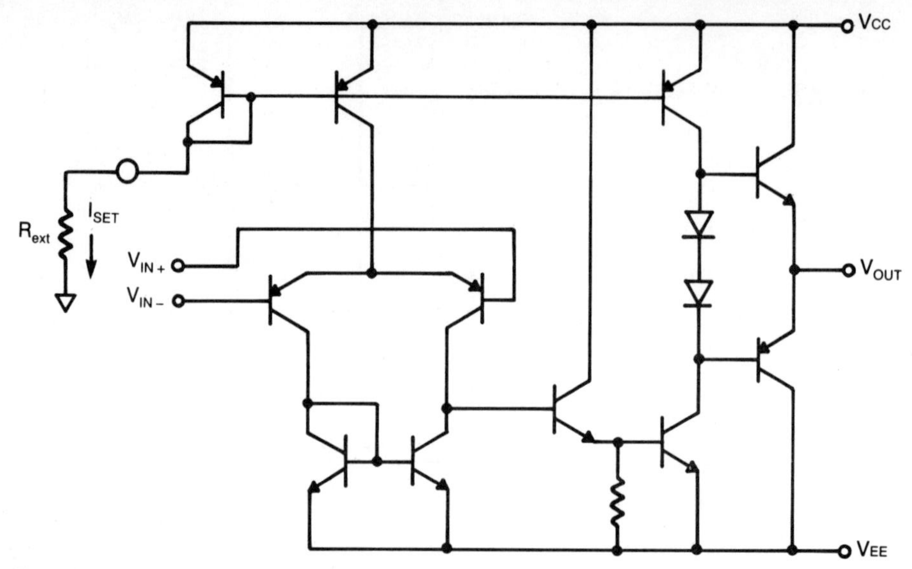

Fig. 4-21. The basic programmable op amp.

characteristics that are dependent on the values of these current sources.

The result is that you may program the small-signal bandwidth, slew rate, input bias current, input offset voltage and current, input noise, and power supply current. Manufacturers provide a set of curves and/or a set of equations for specific op-amp parameters versus set current. A common programming technique is illustrated in Fig. 4-21.

Table 4-1. Low-Cost, General-Purpose Single Op Amps, Commercial Temperature Range.

PART NUMBER	MANUFACTURER	MAXIMUM OFFSET VOLTAGE (mV)	MAXIMUM BIAS CURRENT (nA)	TYP BAND-WIDTH (MHz)	TYP SLEW RATE (V/uS)	UNIT PRICE ($)
UA201	FAIRCHILD	7.5	1500	1	0.5	.38
TL321C	T I	7	250	0.6	0.3	.41
MC1748C	MOTOROLA	6	500	1	0.8	.47
MC1741C	MOTOROLA	6	500	1	0.5	.47
LM741C	NATIONAL	6	500	1	0.5	.50
CA741C	RCA	6	500	1	0.5	.50
CA307	RCA	7.5	250	1	0.5	.50
CA301A	RCA	7.5	250	1	0.5	.50
TL081C	T I	15	0.4	3	13	.51

PART NUMBER	MANUFACTURER	MAXIMUM OFFSET VOLTAGE (mV)	MAXIMUM BIAS CURRENT (nA)	TYP BAND- WIDTH (MHz)	TYP SLEW RATE (V/uS)	UNIT PRICE ($)
TL080C	T I	15	0.4	3	13	.51
CA081E	RCA	15	0.05	5	13	.52
CA748C	RCA	6	500	1	0.5	.54
LM301A	NATIONAL	7.5	250	1	0.5	.55
LF13741	NATIONAL	15	0.2	1	0.5	.56
LM307	NATIONAL	7.5	250	1	0.5	.56
UA771L	FAIRCHILD	15	0.2	3	13	.57
UA741C	FAIRCHILD	6	500	1	0.5	.57
UA308	FAIRCHILD	7.5	7	1	0.35	.57
LM748C	NATIONAL	5	500	1	0.5	.60
LF351	NATIONAL	10	0.2	4	13	.60
MC34001	MOTOROLA	10	0.2	4	13	.60
UA301A	FAIRCHILD	7.5	250	1	0.5	.63
TL071C	T I	10	0.2	3	13	.64
TLC271C	T I	10	0.001 TYP	2.3	4.5	.64
TL070C	T I	10	0.2	3	13	.64
LM709C	NATIONAL	7.5	1500	1	0.3	.70
LM308	NATIONAL	7.5	7	1	0.3	.71
MC1709C	MOTOROLA	7.5	1500	1	0.4	.73
LF441C	NATIONAL	5	0.1	1	1	.75
UA771	FAIRCHILD	10	0.2	3	13	.76
UA709C	FAIRCHILD	7.5	1500	1	0.3	.76
CA3029	RCA	5	12000	20	3	.78
TL066C	T I	15	0.4	1	3.5	.81
TL061C	T I	15	0.2	1	3.5	.81
TL060C	T I	15	0.2	1	3.5	.81
TLC271AC	T I	5	0.001 TYP	2.3	4.5	.81
MC1741NC	MOTOROLA	6	500	1	0.5	.90
UA748C	FAIRCHILD	6	500	1	0.5	.91
CA3078	RCA	4.5	170	0.5	1.5	.92
LF411C	NATIONAL	2	0.2	4	15	.95

PART NUMBER	MANUFACTURER	MAXIMUM OFFSET VOLTAGE (mV)	MAXIMUM BIAS CURRENT (nA)	TYP BAND-WIDTH (MHz)	TYP SLEW RATE (V/uS)	UNIT PRICE ($)
CA3029A	RCA	2	4000	20	3	.96
TL081AC	T I	6	0.2	3	13	.97
TL080AC	T I	6	0.2	3	13	.97
CA081AE	RCA	6	0.04	5	13	1.00

Table 4-2. Low-Cost, General-Purpose Dual Op Amps, Commercial Temperature Range.

PART NUMBER	MANUFACTURER	MAXIMUM OFFSET VOLTAGE (mV)	MAXIMUM BIAS CURRENT (nA)	TYP BAND-WIDTH (MHz)	TYP SLEW RATE (V/uS)	UNIT PRICE ($)
NE4558	SIGNETICS	6	500	3	1	.58
MC1458C	MOTOROLA	10	700	1	0.5	.58
MC1458	MOTOROLA	6	500	1	0.5	.60
LM1458	NATIONAL	6	500	1	0.2	.60
NE532	SIGNETICS	7	250	1	0.3	.60
CA358	RCA	7	250	1	0.5	.60
CA1458	RCA	6	500	1	0.5	.62
LM358	NATIONAL	7	250	1	0.25	.65
CA747C	RCA	6	500	1	0.5	.66
TL322C	T I	10	500	1	0.6	.74
UA747C	FAIRCHILD	6	500	1	0.5	.76
UA1458C	FAIRCHILD	10	700	1.1	0.8	.76
UA1458	FAIRCHILD	6	500	1.1	0.8	.76
MC4558C	MOTOROLA	6	500	2.8	1.6	.80
MC3458	MOTOROLA	10	500	1	0.6	.82
LM747C	NATIONAL	6	500	1	0.5	.86
UA772L	FAIRCHILD	15	0.2	3	13	.89
CA082E	RCA	15	0.05	5	13	.90
TL082C	T I	15	0.4	3	13	.97
MC34002	MOTOROLA	10	0.2	4	13	.97
LF353	NATIONAL	10	0.2	4	13	.98
UA772	FAIRCHILD	10	0.2	3	13	.99
NE5512	SIGNETICS	5	20	3	1	1.00

Table 4-3. Low-Cost, General-Purpose Quad Op Amps, Commercial Temperature Range.

PART NUMBER	MANUFACTURER	MAXIMUM OFFSET VOLTAGE (mV)	MAXIMUM BIAS CURRENT (nA)	TYP BAND-WIDTH (MHz)	TYP SLEW RATE (V/uS)	UNIT PRICE ($)
UA324	FAIRCHILD	7	250	1	0.25	.68
CA324	RCA	7	250	1	0.5	.72
LM324	NATIONAL	7	250	1	0.25	.75
MC3403	MOTOROLA	10	500	1	0.6	.84
LM348	NATIONAL	6	200	1	0.5	1.10
UA3403	FAIRCHILD	8	500	1	0.6	1.10
LM349	NATIONAL	6	200	4	2	1.25
UA348	FAIRCHILD	6	200	1	0.5	1.29
NE5514	SIGNETICS	5	20	3	1	1.30
UA4136C	FAIRCHILD	6	500	3	1	1.31
MC34004	MOTOROLA	10	0.2	4	13	1.35
MC4741C	MOTOROLA	6	500	1	0.5	1.40
TL136C	T I	6	500	3	2	1.50
HA-4156-5	HARRIS	5	300	3.5	1.6	1.59
HA-4741-5	HARRIS	5	300	3.5	1.6	1.62
HA-5084-5	HARRIS	15	0.4	4	15	1.69
LF347	NATIONAL	10	0.2	4	13	1.80
TL084C	T I	15	0.4	3	13	1.84
HA-5064-5	HARRIS	15	0.4	1	4	1.85
TL074C	T I	10	0.2	3	13	1.96
TL085C	T I	15	0.4	3	13	1.96
UA774L	FAIRCHILD	15	0.2	3	13	1.96
MC34074	MOTOROLA	4.5	500	4.5	10	2.00

Table 4-4. Low-Cost, General-Purpose Single Op Amps, Industrial Temperature Range.

PART NUMBER	MANUFACTURER	MAXIMUM OFFSET VOLTAGE (mV)	MAXIMUM BIAS CURRENT (nA)	TYP BAND-WIDTH (MHz)	TYP SLEW RATE (V/uS)	UNIT PRICE ($)
SA741C	SIGNETICS	6	500	1	0.5	.76
LM201A	NATIONAL	2	75	1	0.5	.80

PART NUMBER	MANUFACTURER	MAXIMUM OFFSET VOLTAGE (mV)	MAXIMUM BIAS CURRENT (nA)	TYP BAND-WIDTH (MHz)	TYP SLEW RATE (V/uS)	UNIT PRICE ($)
TL081I	T I	6	0.2	3	13	.97
TL080I	T I	6	0.2	3	13	.97
TL071I	T I	6	0.2	3	13	1.10
TL070I	T I	6	0.2	3	13	1.10
TL066I	T I	6	0.2	1	3.5	1.27
TL061I	T I	6	0.2	1	3.5	1.27
TL060I	T I	6	0.2	1	3.5	1.27
CA3100E	RCA	5	2000	38	70	1.42
UA208	FAIRCHILD	2	2	1	0.35	1.71
UA201A	FAIRCHILD	2	75	1	0.5	1.81
LM207	NATIONAL	2	75	1	0.5	2.15
AD201A	ANALOG DEVICES	2	75	1	0.25	3.00
CA3193A	RCA	0.2	20	1.2	0.25	3.08
LF257	NATIONAL	5	0.1	20	50	3.60
LF256	NATIONAL	5	0.1	5	12	3.60
LF255	NATIONAL	5	0.1	2.5	5	3.60
UA208A	FAIRCHILD	0.5	2	1	0.35	3.80
LM208	NATIONAL	2	2	1	0.3	3.87
OP-21G	P M I	0.5	150	0.6	0.25	4.05
OPA21G	BURR-BROWN	0.5	50	0.3	0.2	4.95

Table 4-5. Low-Cost, General-Purpose Dual Op Amps, Industrial Temperature Range.

PART NUMBER	MANUFACTURER	MAXIMUM OFFSET VOLTAGE (mV)	MAXIMUM BIAS CURRENT (nA)	TYP BAND-WIDTH (MHz)	TYP SLEW RATE (V/uS)	UNIT PRICE ($)
CA2904	RCA	7	250	1	0.5	.60
SA1458	SIGNETICS	6	500	1	0.8	.88
CA258	RCA	5	150	1	0.5	.90
SA747C	SIGNETICS	6	500	1	0.5	.92
SA532	SIGNETICS	7	250	1	0.3	.94
LM2904	NATIONAL	7	250	1	0.25	1.10
TL322I	T I	8	500	1	0.6	1.15

PART NUMBER	MANUFACTURER	MAXIMUM OFFSET VOLTAGE (mV)	MAXIMUM BIAS CURRENT (nA)	TYP BAND-WIDTH (MHz)	TYP SLEW RATE (V/uS)	UNIT PRICE ($)
TL082I	T I	6	0.2	3	13	1.40
CA3240	RCA	15	0.05	4.5	9	1.48
CA258A	RCA	3	80	1	0.5	1.50
TL072I	T I	6	0.2	3	13	1.52
MC3358	MOTOROLA	8	500	1	0.6	1.55
LM258	NATIONAL	5	150	1	0.25	1.82
TL062I	T I	6	0.2	1	3.5	1.84
TL083I	T I	6	0.2	3	13	1.96
CA3240A	RCA	5	0.04	4.5	9	2.36
LM258A	NATIONAL	3	80	1	0.25	5.30
OP-221G	P M I	0.5	120	0.6	0.3	6.75
OP-220G	P M I	0.75	30	0.1	0.05	8.10
OP-227G	P M I	0.18	80	8	2.8	10.50
OP-221F	P M I	0.3	100	0.6	0.3	10.80
LH2201A	NATIONAL	2	75	1	0.5	11.40
OP-227F	P M I	0.12	55	8	2.8	13.50
OP-220F	P M I	0.3	25	0.1	0.05	14.18
OP-221E	P M I	0.15	80	0.6	0.3	14.85

Table 4-6. Low-Cost, General-Purpose Quad Op Amps, Industrial Temperature Range.

PART NUMBER	MANUFACTURER	MAXIMUM OFFSET VOLTAGE (mV)	MAXIMUM BIAS CURRENT (nA)	TYP BAND-WIDTH (MHz)	TYP SLEW RATE (V/uS)	UNIT PRICE ($)
CA224	RCA	7	250	1	0.5	.90
SA534	SIGNETICS	7	250	1	0.3	.96
LM2902	NATIONAL	7	250	1	0.25	1.10
UA2902	FAIRCHILD	7	250	1	0.25	1.24
UA3303	FAIRCHILD	8	500	1	0.6	1.33
LM224	NATIONAL	5	150	1	0.25	1.90
TL084I	T I	6	0.2	3	13	2.30
TL074I	T I	6	0.2	3	13	2.42

PART NUMBER	MANUFACTURER	MAXIMUM OFFSET VOLTAGE (mV)	MAXIMUM BIAS CURRENT (nA)	TYP BAND-WIDTH (MHz)	TYP SLEW RATE (V/uS)	UNIT PRICE ($)
MC33074	MOTOROLA	4.5	500	4.5	10	2.54
TL064I	T I	6	0.2	1	3.5	2.65
LM249	NATIONAL	6	200	4	2	3.60
LM248	NATIONAL	6	200	1	0.5	3.60
MC3303	MOTOROLA	8	500	1	0.6	4.15
LM246	NATIONAL	6	250	1.2	0.4	4.35
MC33074A	MOTOROLA	2	500	4.5	10	5.22
UA224	FAIRCHILD	5	150	1	0.25	5.32
UA248	FAIRCHILD	6	200	1	0.5	6.04
OP-421G	P M I	4	80	1	0.5	6.45
OP-421F	P M I	2.5	50	1	0.5	9.00

Table 4-7. Low-Cost, General-Purpose Single Op Amps, Military Temperature Range.

PART NUMBER	MANUFACTURER	MAXIMUM OFFSET VOLTAGE (mV)	MAXIMUM BIAS CURRENT (nA)	TYP BAND-WIDTH (MHz)	TYP SLEW RATE (V/uS)	UNIT PRICE ($)
CA741	RCA	5	500	1	0.5	.60
CA748	RCA	5	500	1	0.5	.64
CA3140	RCA	15	0.05	4.5	9	.64
CA101	RCA	5	500	1	0.5	.80
CA3160	RCA	15	0.05	4	10	1.00
CA3140A	RCA	5	0.04	4.5	9	1.00
CA3440	RCA	10	0.05	0.063	0.03	1.06
CA3130	RCA	15	0.05	15	30	1.10
UA748	FAIRCHILD	5	500	1	0.5	1.24
UA741	FAIRCHILD	5	500	1	0.5	1.33
UA709	FAIRCHILD	5	500	1	0.3	1.33
MC1741	MOTOROLA	5	500	1	0.5	1.35
MC1748	MOTOROLA	5	500	1	0.8	1.40
CA3130A	RCA	5	0.03	15	30	1.50
MC1709	MOTOROLA	5	500	1	0.4	1.50
CA3078A	RCA	3.5	12	0.04	0.5	1.56

PART NUMBER	MANUFACTURER	MAXIMUM OFFSET VOLTAGE (mV)	MAXIMUM BIAS CURRENT (nA)	TYP BAND-WIDTH (MHz)	TYP SLEW RATE (V/uS)	UNIT PRICE ($)
CA3160A	RCA	5	0.03	4	10	1.58
MC1741N	MOTOROLA	5	500	1	0.5	1.61
UA101	FAIRCHILD	5	500	1	0.5	1.62
UA709A	FAIRCHILD	2	200	1	0.3	1.71
CA3420	RCA	10	0.005	0.5	0.5	1.80
LM748	NATIONAL	5	500	1	0.5	2.00

Table 4-8. Low-Cost, General-Purpose Dual Op Amps, Military Temperature Range.

PART NUMBER	MANUFACTURER	MAXIMUM OFFSET VOLTAGE (mV)	MAXIMUM BIAS CURRENT (nA)	TYP BAND-WIDTH (MHz)	TYP SLEW RATE (V/uS)	UNIT PRICE ($)
CA747	RCA	5	500	1	0.5	.74
CA1558	RCA	5	500	1	0.5	.96
CA158	RCA	5	150	1	0.5	1.20
MC1558	MOTOROLA	5	500	1	0.5	1.56
CA158A	RCA	2	50	1	0.5	1.70
UA747	FAIRCHILD	5	500	1	0.5	2.00
CA3260	RCA	15	0.05	4	10	2.00
LM747	NATIONAL	5	500	1	0.5	2.10
SE532	SIGNETICS	5	150	1	0.3	2.20
MC1747	MOTOROLA	5	500	1	0.5	2.45
SE5512	SIGNETICS	2	10	3	1	2.70
MC4558	MOTOROLA	5	500	2.8	1.6	2.77
LM1558	NATIONAL	5	500	1	0.2	3.10
CA3260A	RCA	5	0.03	4	10	3.12
MC1537	MOTOROLA	5	500	1	0.25	3.18
UA1558	FAIRCHILD	5	500	1.1	0.8	3.33
LM158	NATIONAL	5	150	1	0.25	3.45
UA747A	FAIRCHILD	3	30	1.5	.7	3.52
MC3558	MOTOROLA	5	500	1	0.6	4.06
SE5535	SIGNETICS	4	80	1	15	4.50
MC1558S	MOTOROLA	5	500	1	12	4.60

PART NUMBER	MANUFACTURER	MAXIMUM OFFSET VOLTAGE (mV)	MAXIMUM BIAS CURRENT (nA)	TYP BAND-WIDTH (MHz)	TYP SLEW RATE (V/uS)	UNIT PRICE ($)
MC4558N	MOTOROLA	5	500	2.8	1.6	4.75
TL082M	T I	6	0.2	3	13	4.90

Table 4-9. Low-Cost, General-Purpose Quad Op Amps, Military Temperature Range.

PART NUMBER	MANUFACTURER	MAXIMUM OFFSET VOLTAGE (mV)	MAXIMUM BIAS CURRENT (nA)	TYP BAND-WIDTH (MHz)	TYP SLEW RATE (V/uS)	UNIT PRICE ($)
CA124	RCA	5	150	1	0.5	2.30
LM124	NATIONAL	5	150	1	0.25	3.19
UA124	FAIRCHILD	5	150	1	0.25	3.80
UA148	FAIRCHILD	5	100	1	0.5	3.99
LM148	NATIONAL	5	100	1	0.5	5.30
LM149	NATIONAL	5	100	4	2	5.80
MC4741	MOTOROLA	5	500	1	0.5	6.08
UA4136	FAIRCHILD	5	500	3	1	6.65
HA-4741-2	HARRIS	3	200	3.5	1.6	6.98
LM146	NATIONAL	5	100	1.2	0.4	7.20
TL084M	T I	9	0.2	3	13	7.40
SE5514	SIGNETICS	2	10	3	1	7.58
ICL7641EM	INTERSIL	20	0.05	1.4	1.6	7.60
ICL7642EM	INTERSIL	20	0.05	0.044	0.016	7.60
OP-11C	P M I	5	500	2	1	8.10
OP-421C	P M I	4	80	1	0.5	9.00
TL074M	T I	9	0.2	3	13	9.00
TLC27M4M	T I	10	0.001 TYP	0.7	0.6	9.00
TLC274M	T I	10	0.001 TYP	2.3	4.5	9.00
TLC27L4M	T I	10	0.001 TYP	0.1	0.04	9.00
MC35074	MOTOROLA	4.5	500	4.5	10	9.28
TL064M	T I	9	0.2	1	3.5	9.50
MC35004	MOTOROLA	10	0.2	4	13	9.90
TLC274AM	T I	5	0.001 TYP	2.3	4.5	10.00
TLC27L4AM	T I	5	0.001 TYP	0.1	0.04	10.00
TLC27M4AM	T I	5	0.001 TYP	0.7	0.6	10.00

Table 4-10. Low-Offset, Low-Drift Op Amps, Commercial Temperature Range.

PART NUMBER	MANUFACTURER	MAXIMUM OFFSET VOLTAGE (mV)	MAXIMUM OFFSET DRIFT (uV/°C)	UNIT PRICE ($)
ICL7650C	INTERSIL	0.005	0.05	4.15
ICL7652C	INTERSIL	0.005	0.05	6.75
AD510L	ANALOG DEVICES	0.025	0.5	24.80
HA-5130-5	HARRIS	0.025	0.6	7.52
OP-27E	P M I	0.025	0.6	15.84
OP-37E	P M I	0.025	0.6	15.84
AD510K	ANALOG DEVICES	0.050	1.0	18.00
AD517L	ANALOG DEVICES	0.050	1.3	16.90
OP-27F	P M I	0.060	1.3	9.60
OP-37F	P M I	0.060	1.3	9.60
AD517K	ANALOG DEVICES	0.075	1.8	8.30
UA714E	FAIRCHILD	0.075	1.3	5.70
HA-5135-5	HARRIS	0.075	1.3	3.99
OP-07E	P M I	0.075	1.3	5.85
OP-27G	P M I	0.100	1.8	6.60
OP-37G	P M I	0.100	1.8	6.60
* OP-207E	P M I	0.100	1.3	33.00
UA714C	FAIRCHILD	0.150	1.8	2.47
OP-06E	P M I	0.200	0.8	21.60
* OP-207F	P M I	0.200	1.8	25.50
AD547L	ANALOG DEVICES	0.250	1.0	17.00
3521K	BURR-BROWN	0.250	2.0	48.95
3521L	BURR-BROWN	0.250	1.0	72.40
OP-20F	P M I	0.250	1.5	7.80
AD504L	ANALOG DEVICES	0.500	2.0	41.80
AD504M	ANALOG DEVICES	0.500	1.0	48.80
AD547K	ANALOG DEVICES	0.500	2.0	8.70
UA725E	FAIRCHILD	0.500	2.0	5.32
OP-05E	P M I	0.500	2.0	7.80
OP-06F	P M I	0.500	2.0	18.00
* OP-10E	P M I	0.500	2.0	15.00

* DUAL OP AMPS

Table 4-11. Low-Offset, Low-Drift Op Amps, Industrial Temperature Range.

PART NUMBER	MANUFACTURER	MAXIMUM OFFSET VOLTAGE (mV)	MAXIMUM OFFSET DRIFT (uV/°C)	UNIT PRICE ($)
ICL7650I	INTERSIL	0.005	0.05	9.75
ICL7652I	INTERSIL	0.005	0.05	11.25
OP-27E	P M I	0.025	0.6	19.80

PART NUMBER	MANUFACTURER	MAXIMUM OFFSET VOLTAGE (mV)	MAXIMUM OFFSET DRIFT (uV/°C)	UNIT PRICE ($)
OP-37E	P M I	0.025	0.6	19.80
LH0044AC	NATIONAL	0.025	0.1	36.10
LH0044B	NATIONAL	0.050	0.2	20.50
3510C	BURR-BROWN	0.060	0.5	18.25
OP-27F	P M I	0.060	1.3	12.00
OP-37F	P M I	0.060	1.3	12.00
* OP-227E	P M I	0.080	1.0	22.50
LH0044C	NATIONAL	0.100	0.2	14.40
OP-21E	P M I	0.100	1.0	7.80
OP-27G	P M I	0.100	1.8	8.25
OP-37G	P M I	0.100	1.8	8.25
3510B	BURR-BROWN	0.120	1.0	11.25
* OP-227F	P M I	0.120	1.5	13.50
* OP-220E	P M I	0.150	1.5	19.88
* OP-221E	P M I	0.150	1.5	14.85
3510A	BURR-BROWN	0.150	2.0	8.90
* OP-227G	P M I	0.180	1.8	10.50
OP-21F	P M I	0.200	2.0	5.63
3527C	BURR-BROWN	0.250	2.0	33.15
OPA103D	BURR-BROWN	0.250	2.0	29.85
OP-22E	P M I	0.300	1.5	7.98
* OP-220F	P M I	0.300	2.0	14.18
* OP-221F	P M I	0.300	2.0	10.80
OP-22F	P M I	0.500	2.0	5.78
3500E	BURR-BROWN	0.500	1.0	36.25

* DUAL OP AMPS

Table 4-12. Low-Offset, Low-Drift Op Amps, Military Temperature Range.

PART NUMBER	MANUFACTURER	MAXIMUM OFFSET VOLTAGE (mV)	MAXIMUM OFFSET DRIFT (uV/°C)	UNIT PRICE
ICL7650M	INTERSIL	0.005	0.05	15.75
HA-5130-2	HARRIS	0.025	0.6	22.74
OP-07A	P M I	0.025	0.6	28.05
OP-27A	P M I	0.025	0.6	63.75
OP-37A	P M I	0.025	0.6	63.75
LH0044A	NATIONAL	0.025	0.1	68.90
LH0044	NATIONAL	0.050	0.2	24.60
AD510S	ANALOG DEVICES	0.050	1.0	45.50
OP-27B	P M I	0.060	1.3	28.73

PART NUMBER	MANUFACTURER	MAXIMUM OFFSET VOLTAGE (mV)	MAXIMUM OFFSET DRIFT (uV/°C)	UNIT PRICE ($)
OP-37B	P M I	0.060	1.3	28.73
AD517S	ANALOG DEVICES	0.075	1.8	31.10
UA714	FAIRCHILD	0.075	1.3	6.25
HA-5135-2	HARRIS	0.075	1.3	13.27
OP-07	P M I	0.075	1.3	12.75
CA3193B	RCA	0.075	2.0	5.44
* OP-227A	P M I	0.080	1.0	75.00
OP-21A	P M I	0.100	1.0	20.40
OP-27C	P M I	0.100	1.8	16.58
OP-37C	P M I	0.100	1.8	16.58
* OP-207A	P M I	0.100	1.3	63.00
3510S	BURR-BROWN	0.120	1.0	18.25
* OP-227B	P M I	0.120	1.5	37.50
OP-05A	P M I	0.150	0.9	51.00
* OP-220A	P M I	0.150	1.5	33.75
* OP-221A	P M I	0.150	1.5	27.00
* OP-227C	P M I	0.180	1.8	21.00
OP-06A	P M I	0.200	0.8	53.55
OP-21B	P M I	0.200	2.0	14.03
* OP-207B	P M I	0.200	1.8	49.50
OP-20B	P M I	0.250	1.5	20.40
OP-22A	P M I	0.300	1.5	13.58
* OP-227B	P M I	0.300	2.0	24.08
* OP-221B	P M I	0.300	2.0	18.90
AD504M	ANALOG DEVICES	0.500	2.0	39.40
UA725A	FAIRCHILD	0.500	2.0	7.41
LM725A	NATIONAL	0.500	2.0	17.30
OP-05	P M I	0.500	2.0	21.68
OP-06B	P M I	0.500	2.0	20.40
* OP-10	P M I	0.500	2.0	33.75
OP-22B	P M I	0.500	2.0	9.68

* DUAL OP AMPS

Table 4-13. Low-Bias-Current Op Amps, Commercial Temperature Range.

PART NUMBER	MANUFACTURER	MAX BIAS CURRENT (pA)	UNIT PRICE ($)
AD515L	ANALOG DEVICES	0.075	34.70
3523L	BURR-BROWN	0.10	47.60
AD515K	ANALOG DEVICES	0.15	29.00

PART NUMBER	MANUFACTURER	MAX BIAS CURRENT (pA)	UNIT PRICE ($)
3523K	BURR-BROWN	0.25	39.70
AD515J	ANALOG DEVICES	0.3	20.10
3523J	BURR-BROWN	0.5	31.75
AD545K	ANALOG DEVICES	1	14.00
AD545L	ANALOG DEVICES	1	20.00
AD545M	ANALOG DEVICES	1	32.00
3522L	BURR-BROWN	1	32.75
AD545J	ANALOG DEVICES	2	11.00
ICL8007AC	INTERSIL	4	41.65
3522K	BURR-BROWN	5	22.40
AD506L	ANALOG DEVICES	5	32.60
3522J	BURR-BROWN	10	17.00
AD503K	ANALOG DEVICES	10	20.20
AD506K	ANALOG DEVICES	10	21.00
3521L	BURR-BROWN	10	72.40
ICL7650C	INTERSIL	10	4.15

Table 4-14. Low-Bias-Current Op Amps, Industrial Temperature Range.

PART NUMBER	MANUFACTURER	MAX BIAS CURRENT (pA)	UNIT PRICE ($)
OPA104C	BURR-BROWN	0.075	29.50
3528C	BURR-BROWN	0.075	30.25
OPA104B	BURR-BROWN	0.15	23.00
3528B	BURR-BROWN	0.15	23.65
OPA104A	BURR-BROWN	0.3	17.00
3528A	BURR-BROWN	0.3	19.25
OPA103B	BURR-BROWN	1	13.80
OPA103C	BURR-BROWN	1	18.05
OPA103D	BURR-BROWN	1	29.85
OPA103A	BURR-BROWN	2	10.20
3527B	BURR-BROWN	2	20.75

PART NUMBER	MANUFACTURER	MAX BIAS CURRENT (pA)	UNIT PRICE ($)
3527A	BURR-BROWN	5	15.50
3527C	BURR-BROWN	5	33.15
LH0052C	NATIONAL	5	17.00
OPA101B	BURR-BROWN	10	43.50
OPA102B	BURR-BROWN	10	45.00
ICL7650I	INTERSIL	10	9.75

Table 4-15. Low-Bias-Current Op Amps, Military Temperature Range.

PART NUMBER	MANUFACTURER	MAX BIAS CURRENT (pA)	UNIT PRICE ($)
CA3420B	RCA	1	5.20
LH0052	NATIONAL	2.5	22.50
CA3420	RCA	3	1.80
CA3420A	RCA	3	3.30
ICL8007M	INTERSIL	4	83.25
3522S	BURR-BROWN	5	46.30
AD503S	ANALOG DEVICES	10	34.70
AD506S	ANALOG DEVICES	10	36.00
ICL7650M	INTERSIL	10	15.75
LH0062	NATIONAL	10	30.00

Table 4-16. Wideband Op Amps, Commercial Temperature Range.

PART NUMBER	MANUFACTURER	TYPICAL BANDWIDTH (MHz)	UNIT COST ($)
HA-2539-5	HARRIS	600	8.44
HA-2540-5	HARRIS	400	5.64
HA-5195	HARRIS	150	12.15
HA-2625	HARRIS	100	3.33
HA-5162-5	HARRIS	100	6.26
HA-4622-5	HARRIS	70	4.38
HA-5110-5	HARRIS	60	9.66
3551J	BURR-BROWN	50	31.80
HA-5115-5	HARRIS	50	4.93

PART NUMBER	MANUFACTURER	TYPICAL BANDWIDTH (MHz)	UNIT COST ($)
AD380J	ANALOG DEVICES	40	59.00
UA715C	FAIRCHILD	40	2.66
OP-37GP	P M I	40	6.60
AD507J	ANALOG DEVICES	35	11.90
OP-17E	P M I	30	10.13
OP-17F	P M I	28	6.00
OP-17G	P M I	26	3.60
AD509J	ANALOG DEVICES	20	14.10
3507J	BURR-BROWN	20	12.50
3550K	BURR-BROWN	20	39.75
HA-2525	HARRIS	20	3.59
LF357	NATIONAL	20	1.10
HA-5105-5	HARRIS	18	4.93
LM318	NATIONAL	15	2.50

Table 4-17. Wideband Op Amps, Industrial Temperature Range.

PART NUMBER	MANUFACTURER	TYPICAL BANDWIDTH (MHz)	UNIT COST ($)
3554A	BURR-BROWN	90	73.20
AD3554A	ANALOG DEVICES	90	79.00
LH0024C	NATIONAL	70	22.50
LH0032C	NATIONAL	70	23.80
OPA37G	BURR-BROWN	40	8.25
OP-37GZ	P M I	40	8.25
LF257	NATIONAL	20	3.60
LH0062C	NATIONAL	15	20.40
LM218	NATIONAL	15	10.60

Table 4-18. Wideband Op Amps, Military Temperature Range.

PART NUMBER	MANUFACTURER	TYPICAL BANDWIDTH (MHz)	UNIT COST ($)
HA-2539-2	HARRIS	600	33.78
HA-2540-2	HARRIS	400	18.79
HA-5190	HARRIS	150	30.01
HA-2622	HARRIS	100	10.95
HA-5160-2	HARRIS	100	27.68
AD3554S	ANALOG DEVICES	90	105.50
3554S	BURR-BROWN	90	97.60

PART NUMBER	MANUFACTURER	TYPICAL BANDWIDTH (MHz)	UNIT COST ($)
HA-4622-2	HARRIS	70	15.50
LH0024	NATIONAL	70	37.00
LH0032	NATIONAL	70	40.00
HA-5110-2	HARRIS	60	15.26
3551S	BURR-BROWN	50	56.15
AD380S	ANALOG DEVICES	40	88.00
OPA37C	BURR-BROWN	40	16.60
UA715	FAIRCHILD	40	4.28
OP-37C	P M I	40	16.58
AD507S	ANALOG DEVICES	35	28.20
OP-17A	P M I	30	32.40
OP-17B	P M I	28	16.20
OP-17C	P M I	26	8.10
AD509S	ANALOG DEVICES	20	29.15
HA-2522	HARRIS	20	15.20
LF157	NATIONAL	20	5.05
HA-5100-2	HARRIS	18	15.35
LH0062	NATIONAL	15	30.00
LM118	NATIONAL	15	11.40
CA3130	RCA	15	1.10

Table 4-19. High-Slew-Rate Op Amps, Commercial Temperature Range.

PART NUMBER	MANUFACTURER	TYPICAL SLEW RATE (V/uS)	UNIT PRICE ($)
HA-2539-5	HARRIS	600	8.44
HA-2540-5	HARRIS	400	5.64
AD380J	ANALOG DEVICES	330	59.00
3551J	BURR-BROWN	250	31.80
HA-5195	HARRIS	200	12.15
3584J	BURR-BROWN	150	94.50
HA-5160-5	HARRIS	120	11.25
AD509J	ANALOG DEVICES	120	14.10
3507J	BURR-BROWN	120	12.50
HA-2525	HARRIS	120	3.59
HA-5162-5	HARRIS	70	6.26
AD518J	ANALOG DEVICES	70	2.30
LM318	NATIONAL	70	2.50
HA-2515	HARRIS	60	2.30
OP-17E	P M I	60	10.13
NE538	SIGNETICS	60	1.26

PART NUMBER	MANUFACTURER	TYPICAL SLEW RATE (V/uS)	UNIT PRICE ($)
HA-5110-5	HARRIS	50	9.66
LF357	NATIONAL	50	1.10
OP-17F	P M I	50	6.00

Table 4-20. High-Slew-Rate Op Amps, Industrial Temperature Range.

PART NUMBER	MANUFACTURER	TYPICAL SLEW RATE (V/uS)	UNIT PRICE ($)
AD3554A	ANALOG DEVICES	1200	79.00
3554A	BURR-BROWN	1200	73.20
LH0024C	NATIONAL	500	22.50
LH0032C	NATIONAL	500	23.80
LH0062C	NATIONAL	70	20.40
LM218	NATIONAL	70	10.60
LF257	NATIONAL	50	3.60

Table 4-21. High-Slew-Rate Op Amps, Military Temperature Range.

PART NUMBER	MANUFACTURER	TYPICAL SLEW RATE (V/uS)	UNIT PRICE ($)
AD3554S	ANALOG DEVICES	1200	105.50
3554S	BURR-BROWN	1200	97.60
HA-2539-2	HARRIS	600	33.78
LH0024	NATIONAL	500	37.00
LH0032	NATIONAL	500	40.00
HA-2540-2	HARRIS	400	18.79
AD380S	ANALOG DEVICES	330	88.00
3551	BURR-BROWN	250	56.15
HA-5190	HARRIS	200	30.01
HA-5160-2	HARRIS	120	27.68
AD509S	ANALOG DEVICES	120	29.15
HA-2522	HARRIS	120	15.20
AD518S	ANALOG DEVICES	70	11.40
LH0062	NATIONAL	70	30.00
LM118	NATIONAL	70	11.40
HA-2510	HARRIS	65	22.13
HA-2512	HARRIS	60	15.59
OP-17A	P M I	60	32.40
SE538	SIGNETICS	60	6.00
HA-5110-2	HARRIS	50	15.26
LF157	NATIONAL	50	5.05
OP-17B	P M I	50	16.20

Table 4-22. Low-Noise Op Amps, Commercial Temperature Range.

PART NUMBER	MANUFACTURER	TYPICAL INPUT NOISE VOLTAGE ($nV/\sqrt{Hz}$)	TYPICAL INPUT NOISE CURRENT ($pA/\sqrt{Hz}$)	UNIT PRICE ($)
OP-27FP	P M I	3.0	0.4	9.60
OP-37FP	P M I	3.0	0.4	9.60
OP-27GP	P M I	3.2	0.4	6.60
OP-37GP	P M I	3.2	0.4	6.60
* NE5533A	SIGNETICS	3.5	0.4	3.62
NE5534A	SIGNETICS	3.5	0.4	2.15
* NE5533	SIGNETICS	4.0	0.6	2.60
NE5534	SIGNETICS	4.0	0.6	1.10
* NE5532	SIGNETICS	5.0	0.7	1.60
OP-06F	P M I	7.0	0.15	18.00
OP-06G	P M I	7.0	0.2	5.10
AD504J	ANALOG DEVICES	8.0	0.5	13.80
UA725C	FAIRCHILD	8.0	0.15	3.80
HA-4605-5	HARRIS	8.0	-	4.86
HA-4625-5	HARRIS	8.0	-	4.38
LM725C	NATIONAL	8.0	0.15	3.40
AD504M	ANALOG DEVICES	9.0	0.3	48.80
HA-4156-5	HARRIS	9.0	-	1.59
HA-4741-5	HARRIS	9.0	-	1.62
HA-5135-5	HARRIS	9.0	0.15	3.99
UA714E	FAIRCHILD	9.6	0.13	5.70
OP-05E	P M I	9.6	0.12	7.80
OP-07E	P M I	9.6	0.12	5.85
* OP-10E	P M I	9.6	0.12	26.25
* OP-207F	P M I	9.6	0.12	25.50
UA714L	FAIRCHILD	9.8	0.13	2.09
OP-05C	P M I	9.8	0.13	4.88
OP-07D	P M I	9.8	0.13	3.23
* OP-10C	P M I	9.8	0.13	15.00
AD510J	ANALOG DEVICES	10.0	0.3	10.70
UA4136C	FAIRCHILD	10.0	0.5	1.31

* DUAL OP AMPS

Table 4-23. Low-Noise Op Amps, Industrial Temperature Range.

PART NUMBER	MANUFACTURER	TYPICAL INPUT NOISE VOLTAGE ($nV/\sqrt{Hz}$)	TYPICAL INPUT NOISE CURRENT ($pA/\sqrt{Hz}$)	UNIT PRICE ($)
OP-27FZ	P M I	3.0	0.4	12.00
OP-37FZ	P M I	3.0	0.4	12.00

PART NUMBER	MANUFACTURER	TYPICAL INPUT NOISE VOLTAGE ($nV/\sqrt{Hz}$)	TYPICAL INPUT NOISE CURRENT ($pA/\sqrt{Hz}$)	UNIT PRICE ($)
* OP-227F	P M I	3.0	0.4	13.50
OP-27GZ	P M I	3.2	0.4	8.25
OP-37GZ	P M I	3.2	0.4	8.25
* OP-227G	P M I	3.2	0.4	10.50
OPA101B	BURR BROWN	8.0	0.0014	43.50
OPA102B	BURR BROWN	8.0	0.0014	45.00
OPA101A	BURR BROWN	9.0	0.002	35.00
OPA102A	BURR BROWN	9.0	0.002	37.00
LH0044C	NATIONAL	9.0	–	14.40

* DUAL OP AMPS

Table 4-24. Low-Noise Op Amps, Military Temperature Range.

PART NUMBER	MANUFACTURER	TYPICAL INPUT NOISE VOLTAGE ($nV/\sqrt{Hz}$)	TYPICAL INPUT NOISE CURRENT ($pA/\sqrt{Hz}$)	UNIT PRICE ($)
OP-27B	P M I	3.0	0.4	28.73
OP-37B	P M I	3.0	0.4	28.73
* OP-227B	P M I	3.0	0.4	37.50
OP-27C	P M I	3.2	0.4	16.58
OP-37C	P M I	3.2	0.4	16.58
* OP-227C	P M I	3.2	0.4	21.00
SE5534A	SIGNETICS	3.5	0.4	4.73
SE5534	SIGNETICS	4.0	0.6	6.00
* SE5532	SIGNETICS	5.0	0.7	8.60
OP-06B	P M I	7.0	0.15	20.40
OP-06C	P M I	7.0	0.2	15.30
AD504S	ANALOG DEVICES	8.0	0.5	39.40
UA725	FAIRCHILD	8.0	0.15	5.70
HA-4602-2	HARRIS	8.0	–	17.09
HA-4622-2	HARRIS	8.0	–	15.50
LM725	NATIONAL	8.0	0.15	17.30
HA-4741-2	HARRIS	9.0	–	6.98
HA-5135-2	HARRIS	9.0	0.14	13.27
LH0044	NATIONAL	9.0	–	24.60
UA714	FAIRCHILD	9.6	0.12	6.25
OP-05	P M I	9.6	0.2	21.68
OP-07	P M I	9.6	0.12	12.75
* OP-10	P M I	9.6	0.12	33.75
* OP-207B	P M I	9.6	0.12	49.50
AD510S	ANALOG DEVICES	10.0	0.3	45.50
UA4136	FAIRCHILD	10.0	0.5	6.65

* DUAL OP AMPS

Table 4-25. Low-Noise Components for Circuit Shown in Fig. 4-18.

NOISE COMPONENT	SOURCE	OUTPUT CONTRIBUTION
e_{n1}	R_{S1}	$\sqrt{4\ k\ T\ R_{S1}\ BW} \cdot (R_F/R_{S1})$
e_{n2}	R_{S2}	$\sqrt{4\ k\ T\ R_{S2}\ BW} \cdot (1 + R_F/R_{S1})$
e_{n3}	R_F	$\sqrt{4\ k\ T\ R_F\ BW}$
e_{n4}	i_{n1}	$i_{n1}\ R_F$
e_{n5}	i_{n2}	$(i_{n2}\ R_{S2})(1 + R_F/R_{S1})$
e_{n6}	e_n	$e_n(1 + R_F/R_{S1})$

Table 4-26. Low-Power Single Op Amps, Commercial Temperature Range.

PART NUMBER	MANUFACTURER	MAXIMUM SUPPLY CURRENT (mA)	TYPICAL BANDWIDTH (MHz)	UNIT COST ($)
ICL8021C	INTERSIL	0.05 (1)	0.27	1.90
OP-20F	P M I	0.08	0.1	7.80
OP-20G	P M I	0.085	0.1	5.63
OP-20H	P M I	0.095	0.1	4.05
LF441AC	NATIONAL	0.2	1	4.00
LF441C	NATIONAL	0.25	1	.75
ICL7614DC	INTERSIL	0.25 (2)	0.48	1.15
TL060C	T I	0.25	1	.81
TL061C	T I	0.25	1	.81
TL066C	T I	0.25	1	.81
LM316A	NATIONAL	0.6	1	12.00
OP-08E	P M I	0.6	0.8	11.70
OP-12F	P M I	0.6	0.8	7.20
UA308	FAIRCHILD	0.8	1	.57
LM11CL	NATIONAL	0.8	0.8	1.90
LM308	NATIONAL	0.8	1	.71
LM312	NATIONAL	0.8	1	3.40
LM316	NATIONAL	0.8	1	6.45
OP-08G	P M I	0.8	0.8	4.50
TL321C	T I	1	1	.41

(1) $V_S = \pm 6V$ (2) $V_S = \pm 5V$

Table 4-27. Low-Power Dual Op Amps, Commercial Temperature Range.

PART NUMBER	MANUFACTURER	MAXIMUM SUPPLY CURRENT (mA)	TYPICAL BANDWIDTH (MHz)	UNIT COST ($)
ICL8022C	INTERSIL	0.1 (1)	0.27	7.75
TL022C	T I	0.25	0.5	1.56
HA-5062A-5	HARRIS	0.4	1	2.31
LF442AC	NATIONAL	0.4	1	6.10
LF442C	NATIONAL	0.5	1	1.40
HA-5062-5	HARRIS	0.5	1	1.19
ICL7621DC	INTERSIL	0.5 (2)	0.48	2.05
TL062C	T I	0.5	1	1.38
LH2308	NATIONAL	0.8	1	11.60
LM358	NATIONAL	1.2 (3)	1	.65

(1) V_S=±6V (2) V_S=±5V (3) V_S=+5V (single supply)

Table 4-28. Low-Power Quad Op Amps, Commercial Temperature Range.

PART NUMBER	MANUFACTURER	MAXIMUM SUPPLY CURRENT (mA)	TYPICAL BANDWIDTH (MHz)	UNIT PRICE ($)
ICL7642EC	INTERSIL	0.088 (1)	0.044	3.95
TL044C	T I	0.5	0.5	2.53
OP-420H	P M I	0.6	0.15	7.80
HA-5064A-5	HARRIS	0.8	1	9.24
LF444AC	NATIONAL	0.8	1	3.80
LF444C	NATIONAL	1	1	2.25
HA-5064-5	HARRIS	1	1	3.68
TL064C	T I	1	1	2.19
UA324	FAIRCHILD	1.2 (2)	1	.68
LM324	NATIONAL	1.2 (2)	1	.75
OP-421H	P M I	3	1	3.60

(1) V_S=±5V (2) V_S=+5V (single supply)

Table 4-29. Low-Power Single Op Amps, Industrial Temperature Range.

PART NUMBER	MANUFACTURER	MAXIMUM SUPPLY CURRENT (mA)	TYPICAL BANDWIDTH (MHz)	UNIT PRICE ($)
TL060I	T I	0.25	1	1.27
TL061I	T I	0.25	1	1.27
TL066I	T I	0.25	1	1.27
OP-21E	P M I	0.3	0.6	7.80

PART NUMBER	MANUFACTURER	MAXIMUM SUPPLY CURRENT (mA)	TYPICAL BANDWIDTH (MHz)	UNIT PRICE ($)
OP-21F	P M I	0.36	0.6	5.63
OP-21G	P M I	0.42	0.6	4.05
UA208	FAIRCHILD	0.6	1	1.71
LM208	NATIONAL	0.6	1	3.87
LM212	NATIONAL	0.6	1	12.75
LM216A	NATIONAL	0.6	1	14.95
LM216	NATIONAL	0.8	1	10.45

Table 4-30. Low-Power Dual Op Amps, Industrial Temperature Range.

PART NUMBER	MANUFACTURER	MAXIMUM SUPPLY CURRENT (mA)	TYPICAL BANDWIDTH (MHz)	UNIT PRICE ($)
OP-220E	P M I	0.17	0.1	19.88
OP-220F	P M I	0.19	0.1	14.18
OP-220G	P M I	0.22	0.1	8.10
TL062I	T I	0.5	1	1.84
LH2208	NATIONAL	0.8	1	19.10
OP-221E	P M I	0.8	0.6	14.85
OP-221F	P M I	0.85	0.6	10.80
OP-221G	P M I	0.9	0.6	6.75
LM258	NATIONAL	1.2 (1)	1	1.82
LM2904	NATIONAL	1.2 (1)	1	1.10
SA532	SIGNETICS	2	1	.94

(1) V_S=+5V (single supply)

Table 4-31. Low-Power Quad Op Amps, Industrial Temperature Range.

PART NUMBER	MANUFACTURER	MAXIMUM SUPPLY CURRENT (mA)	TYPICAL BANDWIDTH (MHz)	UNIT PRICE ($)
OP-420F	P M I	0.36	0.15	18.00
OP-420G	P M I	0.46	0.15	11.70
TL064I	T I	1	1	2.65
UA224	FAIRCHILD	1.2 (1)	1	5.32
UA2902	FAIRCHILD	1.2 (1)	1	1.24
LM224	NATIONAL	1.2 (1)	1	1.90
LM2902	NATIONAL	1.2 (1)	1	1.10
OP-421F	P M I	1.8	1	9.00
OP-421G	P M I	2.3	1	6.45
SA534	SIGNETICS	3	1	.96

(1) V_S=+5V (single supply)

Table 4-32. Low-Power Single Op Amps, Military Temperature Range.

PART NUMBER	MANUFACTURER	MAXIMUM SUPPLY CURRENT (mA)	TYPICAL BANDWIDTH (MHz)	UNIT PRICE ($)
ICL8021M	INTERSIL	0.04 (1)	0.27	6.55
OP-20B	P M I	0.08	0.1	20.40
OP-20C	P M I	0.085	0.1	14.03
LF441AM	NATIONAL	0.2	1	16.15
ICL7614DM	INTERSIL	0.25 (2)	0.48	3.20
TL061M	T I	0.25	1	3.80
OP-21A	P M I	0.3	0.6	20.40
OP-21B	P M I	0.36	0.6	14.03
UA108	FAIRCHILD	0.6	1	2.28
LM11	NATIONAL	0.6	0.8	18.00
LM108	NATIONAL	0.6	1	5.75
LM112	NATIONAL	0.6	1	15.90
OP-08A	P M I	0.6	0.8	38.25
OP-12B	P M I	0.6	0.8	25.50
CA3420	RCA	0.7 (3)	0.5	1.80
OP-08C	P M I	0.8	0.8	12.75
OP-12C	P M I	0.8	0.8	12.75

(1) V_S=±6V (2) V_S=±5V (3) V_S=±10V

Table 4-33. Low-Power Dual Op Amps, Military Temperature Range.

PART NUMBER	MANUFACTURER	MAXIMUM SUPPLY CURRENT (mA)	TYPICAL BANDWIDTH (MHz)	UNIT COST ($)
ICL8022M	INTERSIL	0.08 (1)	0.27	15.55
OP-220A	P M I	0.17	0.1	33.75
OP-220B	P M I	0.19	0.1	24.08
OP-220C	P M I	0.22	0.1	17.25
LF442AM	NATIONAL	0.4	1	21.35
ICL7621DM	INTERSIL	0.5 (2)	0.48	6.20
TL062M	T I	0.5	1	6.30
LH2108	NATIONAL	0.8	1	27.90
OP-221A	P M I	0.8	0.6	27.00
OP-221B	P M I	0.85	0.6	18.90
OP-221C	P M I	0.9	0.6	13.50
LM158	NATIONAL	1.2 (3)	1	3.45
SE532	SIGNETICS	2	1	2.20

(1) V_S=±6V (2) V_S=±5V (3) V_S=+5V (single supply)

Table 4-34. Low-Power Quad Op Amps, Military Temperature Range.

PART NUMBER	MANUFACTURER	MAXIMUM SUPPLY CURRENT (mA)	TYPICAL BANDWIDTH (MHz)	UNIT PRICE ($)
ICL7642EM	INTERSIL	0.088 (1)	0.044	7.60
OP-420B	P M I	0.36	0.15	27.75
OP-420C	P M I	0.46	0.15	16.50
HA-5064-2	HARRIS	0.8	1	13.43
LF444AM	NATIONAL	0.8	1	16.60
TL064M	T I	1	1	9.50
UA124	FAIRCHILD	1.2 (2)	1	3.80
LM124	NATIONAL	1.2 (2)	1	3.19
OP-421B	P M I	1.8	1	13.50
OP-421C	P M I	2.3	1	9.00
UA148	FAIRCHILD	3.6	1	3.99
LM148	NATIONAL	3.6	1	5.30
LM149	NATIONAL	3.6	4	5.80
MC4741	MOTOROLA	4	1	6.08

(1) V_S=±5V (2) V_S=+5V (single supply)

Table 4-35. High-Voltage Op Amps.

PART NUMBER	MANUFACTURER	OPERATING TEMPERATURE	MAXIMUM SUPPLY VOLTAGE (± V)	UNIT PRICE ($)
3582J	BURR BROWN	COM	150	101.50
3583A	BURR BROWN	IND	150	100.00
3583J	BURR BROWN	COM	150	95.00
3584J	BURR BROWN	COM	150	94.50
3581J	BURR BROWN	COM	75	93.45
HA-2640	HARRIS	MIL	50	15.40
HA-2645	HARRIS	COM	50	5.67
3571A	BURR BROWN	IND	40	72.45
3572A	BURR BROWN	IND	40	83.00
MC1536	MOTOROLA	MIL	40	16.13
LM143	NATIONAL	MIL	40	23.95
LM144	NATIONAL	MIL	40	23.95
3580J	BURR BROWN	COM	35	62.00
3573A	BURR BROWN	IND	34	36.00
LM343	NATIONAL	COM	34	6.38
LM344	NATIONAL	COM	34	6.38
MC1436	MOTOROLA	COM	34	5.72
MC1436C	MOTOROLA	COM	30	3.55

Table 4-36. High-Output Current Op Amps.

PART NUMBER	MANUFACTURER	OPERATING TEMPERATURE	OUTPUT RATING	UNIT PRICE ($)
3572A	BURR BROWN	IND	V_O=30V min @ 2A	83.00
3573A	BURR BROWN	IND	I_O=2A min @ 20V	36.00
LH0021	NATIONAL	MIL	V_O=11V min @ R_L=10 Ω	52.50
LH0021C	NATIONAL	IND	V_O=10V min @ R_L=10 Ω	29.70
3571A	BURR BROWN	IND	V_O=30V min @ 1A	72.45
LH0061	NATIONAL	MIL	V_O=10V min @ R_L=20 Ω	32.40
LH0061C	NATIONAL	IND	V_O=10V min @ R_L=20 Ω	24.00
UA759	FAIRCHILD	MIL	I_O=325mA min @ 12V	3.33
UA759C	FAIRCHILD	COM	I_O=325mA min @ 12V	2.09
LM13080	NATIONAL	COM	V_O=2V min @ R_L=8 Ω	1.29
LH0041	NATIONAL	MIL	V_O=13V min @ R_L=100 Ω	27.60
LH0041C	NATIONAL	IND	V_O=13V min @ R_L=100 Ω	20.60
AD3554A	ANALOG DEVICES	COM	I_O=100mA min @ 10V	79.00
AD3554S	ANALOG DEVICES	MIL	I_O=100mA min @ 10V	105.50
3554A	BURR BROWN	IND	I_O=100mA min @ 10V	73.20
3554S	BURR BROWN	MIL	I_O=100mA min @ 10V	97.60
3583A	BURR BROWN	IND	I_O=75mA min @ 140V	100.00
3583J	BURR BROWN	COM	I_O=75mA min @ 140V	95.00
3580J	BURR BROWN	COM	I_O=60mA min @ 30V	62.00
3581J	BURR BROWN	COM	I_O=30mA min @ 70V	93.45

Table 4-37. Programmable Op Amps.

PART NUMBER	MANUFACTURER	OP AMPS PER PACKAGE	OPERATING TEMPERATURE RANGE	UNIT PRICE ($)
UA776C	FAIRCHILD	1	COM	1.24
MC3476	MOTOROLA	1	COM	1.48
MC1776C	MOTOROLA	1	COM	1.51
LM4250C	NATIONAL	1	COM	1.75
ICL8021C	INTERSIL	1	COM	1.90
LM346	NATIONAL	4	COM	2.70
OP-22H	P M I	1	COM	4.13
LM246	NATIONAL	4	IND	4.35
UA776	FAIRCHILD	1	MIL	4.75

PART NUMBER	MANUFACTURER	OP AMPS PER PACKAGE	OPERATING TEMPERATURE RANGE	UNIT PRICE ($)
MC1776	MOTOROLA	1	MIL	5.14
OP-22F	P M I	1	IND	5.78
HA-2725	HARRIS	1	COM	6.30
ICL8021M	INTERSIL	1	MIL	6.55
LM4250	NATIONAL	1	MIL	6.70
LM146	NATIONAL	4	MIL	7.20
ICL8022C	INTERSIL	2	COM	7.75
ICL8023C	INTERSIL	3	COM	8.80
OP-22B	P M I	1	MIL	9.68
HA-2740-5	HARRIS	4	COM	12.02
LH24250C	NATIONAL	2	COM	15.20
ICL8022M	INTERSIL	2	MIL	15.55
ICL8023M	INTERSIL	3	MIL	17.55
HA-2720	HARRIS	1	MIL	17.69
HA-2735	HARRIS	2	COM	18.22
LH24250	NATIONAL	2	MIL	25.50
HA-2740-2	HARRIS	4	MIL	36.04
HA-2730	HARRIS	2	MIL	42.10

Chapter 5

Comparators

IN CONTRAST TO THE IC OP AMP, THE IC VOLTAGE COMPARAtor is less pervasive. There are considerably fewer part types (dozens as compared to hundreds) and fewer manufacturers. Yet the IC voltage comparator is a standard linear IC component that has many applications. The following topics will be covered in this chapter:

- ☐ The basic comparator circuit
- ☐ Comparator response time
- ☐ Comparator selection

THE BASIC COMPARATOR

Comparators are designed to interface analog signals with standard logic families, such as TTL (Transistor-Transistor Logic), CMOS (Complementary-Metal-Oxide Semiconductor), and ECL (Emitter-Coupled Logic). The range of analog input signals will be generally limited by the comparator voltage supplies. Standard power supplies and example comparator types are listed below:

Standard Power Supplies	Examples
+5V, −0 (Single supply)	LM139, LM2901, CMP-04, CA3290

Standard Power Supplies	**Examples**
+5V, −5V	LM160, μA760
+12V, −6V	μA710, LM711, LM106
+15V, −15V	LM111, LM119, CMP-01

The output voltage levels of the comparator depend on its output structure. The two most common structures are (1) open-collector and (2) TTL. The open-collector type is the most versatile, being compatible with most logic families (not compatible with ECL which has negative voltage logic levels) when an external load (or pull-up) resistor is added between the comparator output and the logic power supply (usually 5 volts). Pure TTL-compatible types are less versatile but generally have faster response times, or shorter propagation delays.

The two most basic comparator circuits are shown in Fig. 5-1. For the sake of discussion, the comparators shown are open-collector types.

In the noninverting configuration (A), V_{IN} (plus input, or noninverting input) is compared to V_{REF} (minus input, or inverting input). If V_{IN} is less than V_{REF}, then V_{OUT} is low, or approximately equal to zero volts. If V_{IN} is greater than V_{REF}, then V_{OUT} is high, or approximately +5 volts.

In the inverting configuration (B), V_{IN} is connected to the minus, or inverting input and V_{REF} is connected to the plus input. If V_{IN} is less than V_{REF}, then V_{OUT} is high. If V_{IN} is greater than V_{REF}, then V_{OUT} is low.

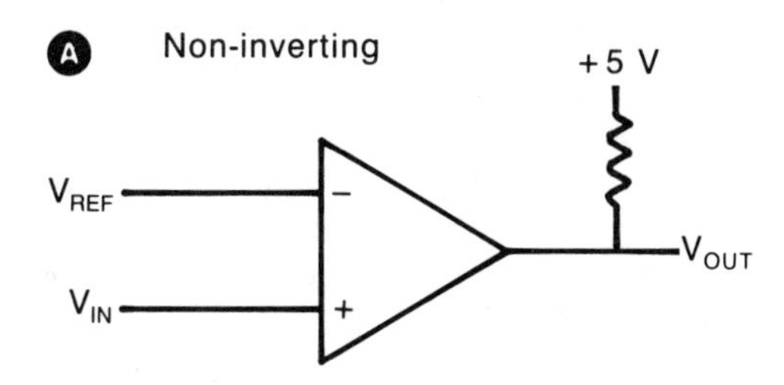

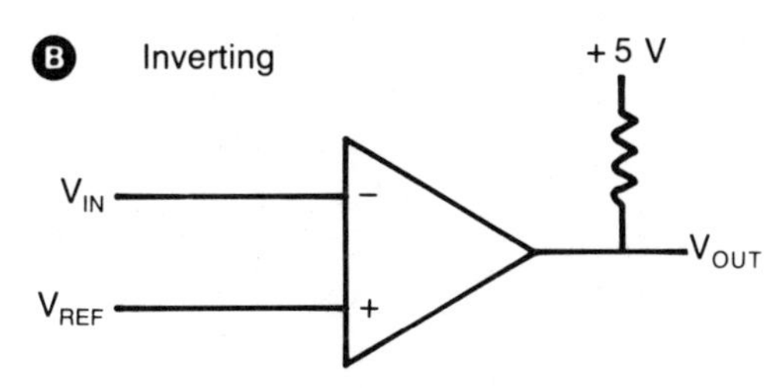

Fig. 5-1. Basic comparator circuits.

This defines the basic operation of the comparator and is true for all output structures—only the output levels are different.

RESPONSE TIME

Figure 5-2 shows a simplified schematic of an LM139 type comparator. You can see that the basic circuit design is similar to the basic op amp discussed in Chapter 4. Only the output stage and compensation capacitor are missing. Thus, the IC comparator has many of the same specifications as the op amp—input bias current, input offset current, input offset voltage, input resistance, voltage gain, and so forth.

If you look at a comparator data sheet, you will see that bandwidth and slew rate are not specified. In place of these parameters is a parameter called response time. This is the key comparator specification. This specification is tested by grounding the positive input and applying a voltage step to the negative input. Typical waveforms are shown in Fig. 5-3.

In Fig. 5-3A the initial voltage on the inverting input is +100 mV. The output is low. Then the input is switched to a few millivolts negative (the number of negative millivolts is called "overdrive"). The fall time of the input is typically 5 ns or less. Some time later the output begins to go high. The response time is defined as the time from when the input is zero to the time when the output crosses +1.4 volts.

In Fig. 5-3B, the input begins at −100 mV and is stepped to a few millivolts positive (the number of positive millivolts is also called overdrive). Some time later the output swings from the high

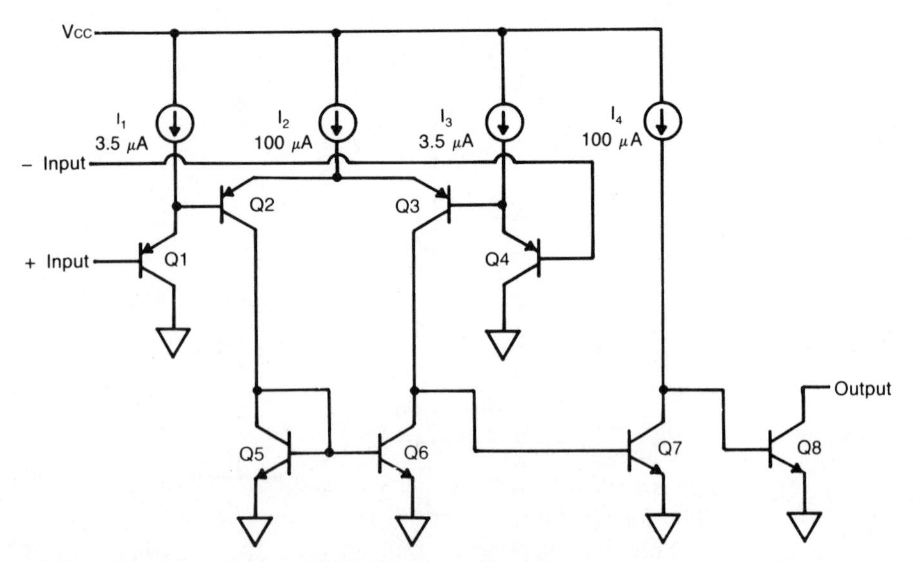

Fig. 5-2. Simplified schematic of LM139 type comparator.

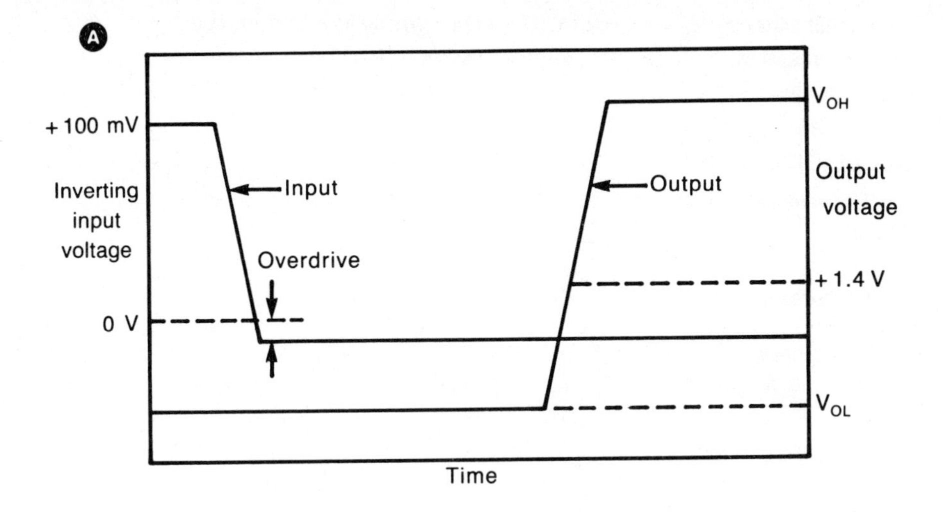

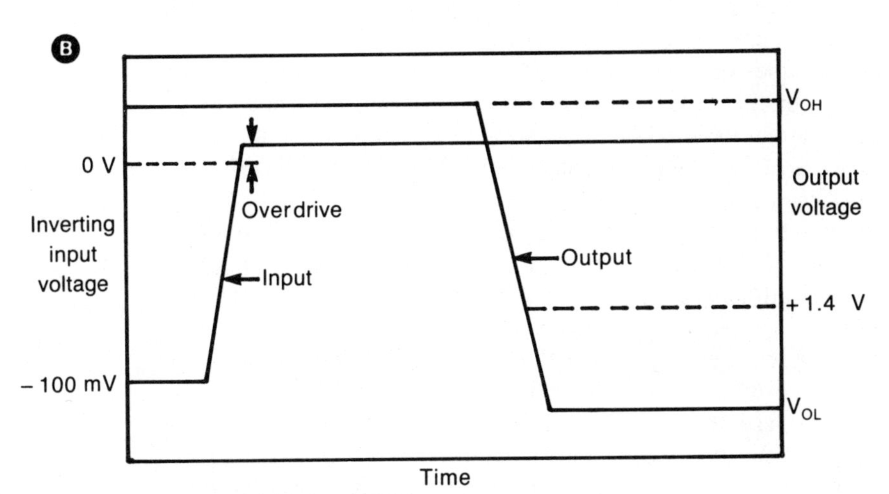

Fig. 5-3. Typical response time waveforms.

level to the low level. Again, the response time is defined as the time from when the input is zero to the time when the output crosses +1.4 volts.

The response time is inversely proportional to the amount of overdrive. Response time decreases with increasing overdrive. Most data sheets give curves showing the typical response times for different overdrives, typically 2 mV, 5 mV, 10 mV, 20 mV, and sometimes 100 mV. The most common test condition for response time is a 100 mV step with 5 mV of overdrive. This is the test condition for all response times listed in this chapter.

Note that the response time for the output to go high may not be the same as the response time for the output to go low. Some data sheets specify each response time separately, other data sheets give the response time for the slowest direction. In the latter case, the data sheet often includes curves for both directions.

SELECTION GUIDES

Comparators may be conveniently classified by their speed as follows:

Classification	Response Time
☐ Low Speed	>200 ns
☐ Medium Speed	80 - 200 ns
☐ High Speed	10 - 80 ns
☐ Very High Speed	<10 ns

Low-Speed Comparators

Low-speed comparators are listed in Tables 5-1 through 5-3. All except the RCA CA3290, CA3290A, and CA3290B have response times of 1.3 μs. The RCA parts are slightly faster at 1.2 μs. All of these devices have the same basic structure—open-collector output with PNP input stage as shown in Fig. 5-2. Again, the RCA 3290 is the exception, having MOSFET input transistors, resulting in lower input bias current at the expense of higher input offset voltage (see Appendix B for other comparator specifications).

The LM139 type comparator is open-collector and thus compatible with most logic families. The PNP input structure allows for sensing input voltages near ground potential while using a single wide-range supply voltage (2 V to 36 V). The input transistors are biased with low-value current sources (I1 and I3 are 3.5 μA, I2 and I4 are 100 μA) resulting in low power and low input bias current at the expense of slow speed, which is primarily due to the low bias currents and the use of PNPs for the differential pair (Q2, Q3). (Q1 and Q4 are substrate PNPs and are a factor of 10 faster than Q1 and Q3, which are lateral PNPs.)

Medium-Speed Comparators

Medium-speed comparators are listed in Tables 5-4 through 5-6. The industry standard in this speed range is the LM111 type comparator. (The 111, 211, and 311 are the military, industrial, and commercial temperature range versions of the same part. This numbering scheme was first established by National Semiconductor.)

A simplified schematic of the LM111 is shown in Fig. 5-4. The LM111 has PNP emitter follower input transistors, which allows for comparing voltages near the negative supply voltage, V_{EE}. The supplies, V_{CC} and V_{EE}, are typically plus and minus 15 volts, but

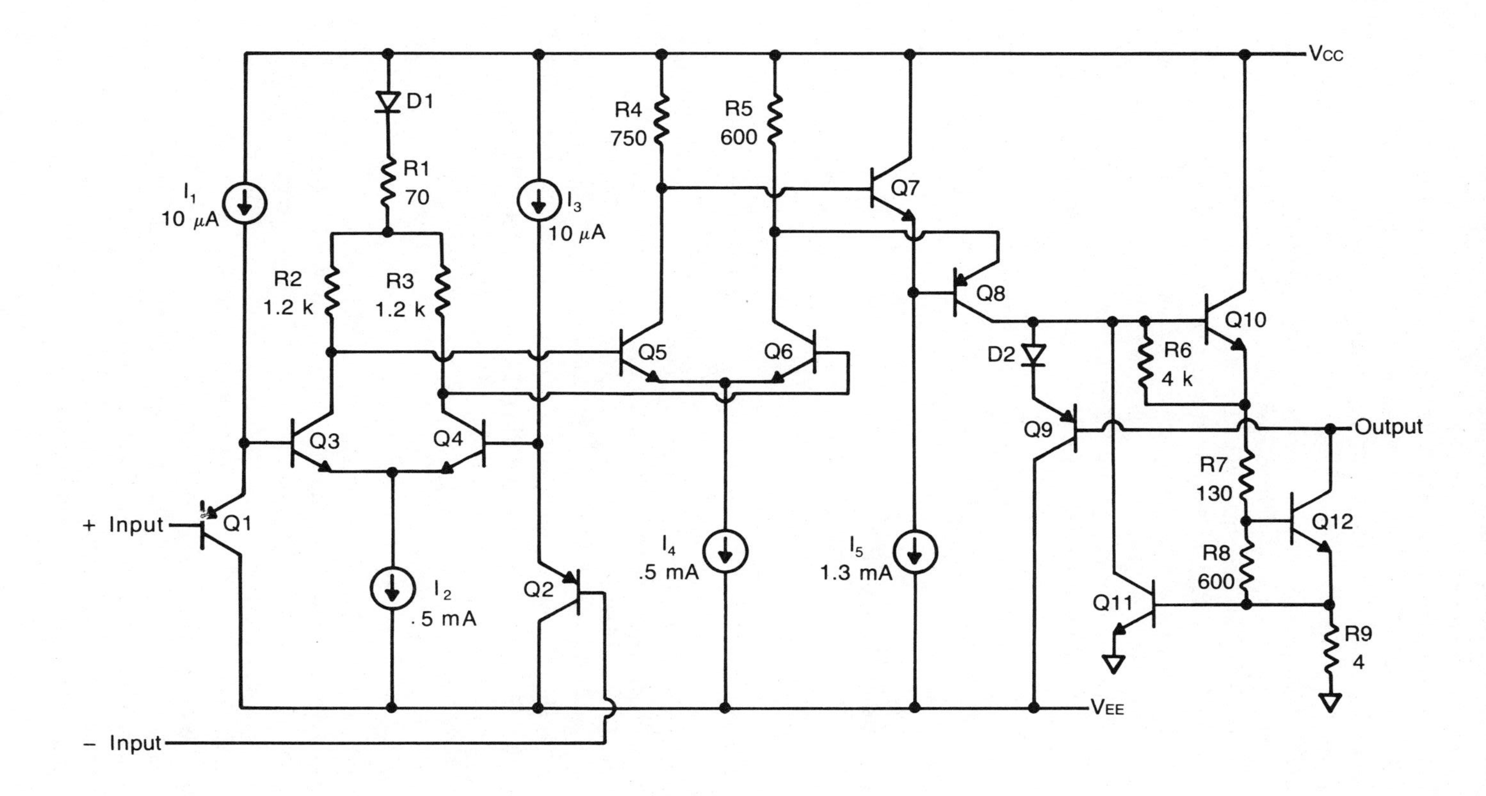

Fig. 5-4. Simplified schematic of LM111 type comparator.

the LM111 in fact may be operated with a single 5-volt supply with VEE connected to ground.

Q3-Q6 form two fast differential pairs. The collectors of Q3 and Q4 swing only 600 mV from ON to OFF. The collectors of Q5 and Q6 swing about half that amount, or approximately 300 mV.

Q7 and Q8 perform a level-shifting function. Q7 is always ON. When Q5 is OFF, the emitter of Q7 and the base of Q8 are approximately one V_{BE} below VCC. When Q5 is OFF, Q6 is ON, and the emitter of Q8 is 300 mV less than VCC. There is a net 300-400 mV across the Q8 base-emitter. This is not enough to turn Q8 ON. When Q5 is ON, the emitter of Q8 is approximately 1 volt below VCC. This will turn Q8 ON. With Q6 OFF, Q8 will pull current through R5. The amount of this current depends on the Q8 base-emitter voltage. Assuming a V_{BE} of about 0.7 volts, then the Q8 current is about 0.5 mA:

$$I_{E8} = (V_{CC} - V_{E8})/R5 \quad \textbf{(1)}$$
$$V_{E8} = V_{CC} - (I4)(R4) - V_{BE7} + V_{BE8} \quad \textbf{(2)}$$

Substituting (2) into (1),

$$I_{E8} = [(I4)(R4) - V_{BE7} + V_{BE8}]/R5$$

Q8 collector current is dumped into the base of Q10, turning Q10 and Q12 ON. The output is open-collector. Unlike the LM139, the output of the LM111 is referenced to ground whether a single supply or dual supplies are used. R9 and Q11 provide current limiting (about 150 mA). D2 and Q9 provide an excess current path when Q8 turns ON. Whatever portion of the Q8 current that Q10 doesn't need to keep Q12 saturated will flow into Q9. Q9 and D2 also establish the collector voltage of Q8 during the time between when Q12 is just turning ON, when Q12 is fully ON, and when Q12 is almost OFF. When Q8 is OFF and Q12 is OFF, the collector of Q8 is pulled to ground via R6-R9.

The speed of the LM111 is limited primarily by Q8, a PNP. As noted above, this PNP is required to level-shift from the Q5-Q6 collector voltage levels (approximately 300 mV below VCC which could be anything from 5-15 volts) to whatever voltage to which the external output load resistor is connected. Thus, the designer of the LM111 gave up some speed in order to provide a versatile output stage. Yet, the LM111 is 6 times faster than the LM139. (Actually, the LM139 quad comparator actually came later. Speed was compromised in the interest of providing four low-power, open-collector, single-supply comparators in a single package for approximately the same price as a single LM111.)

High-Speed Comparators

High-speed comparators are listed in Tables 5-7 through 5-9. All of the comparators in these tables are TTL compatible. The industry standard in this speed range is the μA710.

The schematic of the 710 is shown in Fig. 5-5. This circuit may be analyzed as follows. Transistors Q5 and Q7-Q10 are always ON. Approximately 2.4 mA flows in the Q10-Q9-Q8 path. This current is established by R8 and is mirrored and scaled down to about 2 mA in Q7 (scaling is done by the ratio of R7 to R6). Q7 will get its current from Q5 through Q1 or Q2, depending upon which one is ON.

Assume that the inverting input is positive with respect to the positive input. Therefore, Q1 is ON, Q2 is OFF. Consequently Q3 is OFF, Q4 is ON. Q4 completely saturates, causing the base of Q10 to be pulled down to about 6.3 volts, the Q4 $V_{CE(sat)}$ plus the 6.2 volt zener voltage (D1). This is dropped down to about −0.6 volts at the output by the Q10 V_{BE} and the D2 zener voltage. This volt-

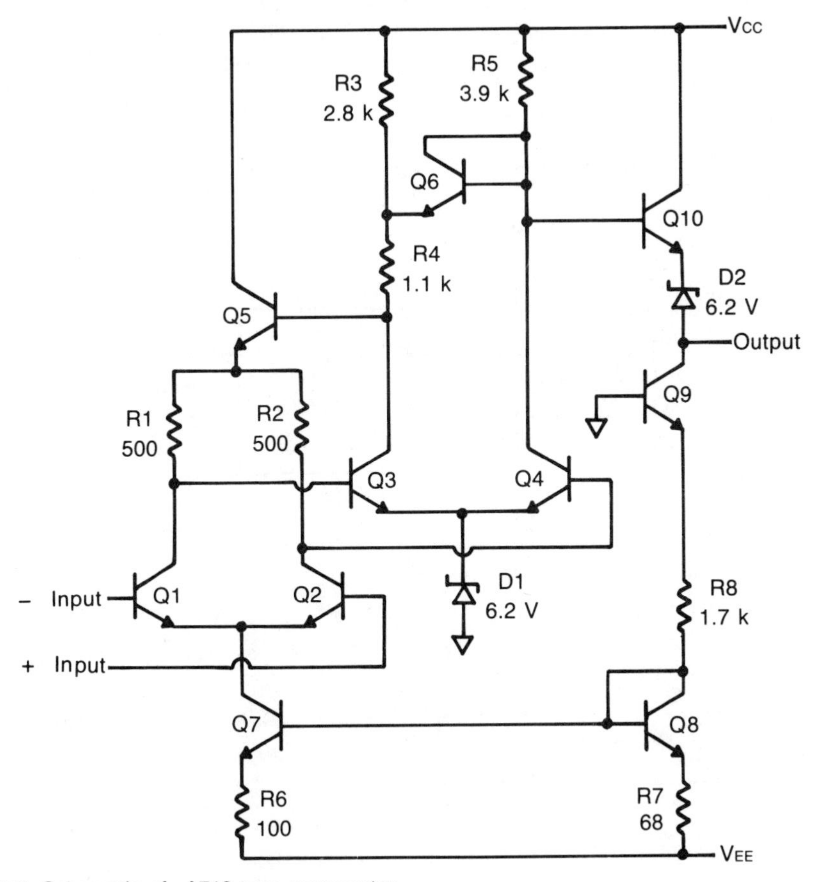

Fig. 5-5. Schematic of μA710 type comparator.

age can vary depending mainly on the tolerance of the zener. The data sheet guarantees a maximum of 0 volts and minimum of −1 volt. The output can never go more negative than the emitter voltage of Q9 (Q9 totally saturated). If the zener breakdown voltage was 8 volts, for example, Q9 would simply saturate and the voltage across the zener would equal whatever voltage was left over after the Q10 V_{BE} and the Q9 $V_{CE(sat)}$ are subtracted from the voltage at the base of Q10.

For the case when the inverting input is negative with respect to the noninverting input, Q1 is OFF, Q2 is ON, Q3 is ON, Q4 is OFF. The base of Q10 wants to go to Vcc less Q5 base current times R5 but is clamped to approximately 10 volts by the V_{BE} voltages of Q6, Q5, and Q3, the IR drops across R4 and R1, and the zener voltage of D1.

The current through R1 is Q3 base current. Therefore the voltage drop across R1 is negligible. The current through R4 is about 1.5 mA, 1 mA from the R3 path, and 0.5 mA from the Q6-R5 path. This current is set by the base of Q5 which is about 7.5 volts, the zener voltage plus the V_{BE} voltages of Q3 and Q5. The current in R4 (minus Q5 base current) flows into Q3.

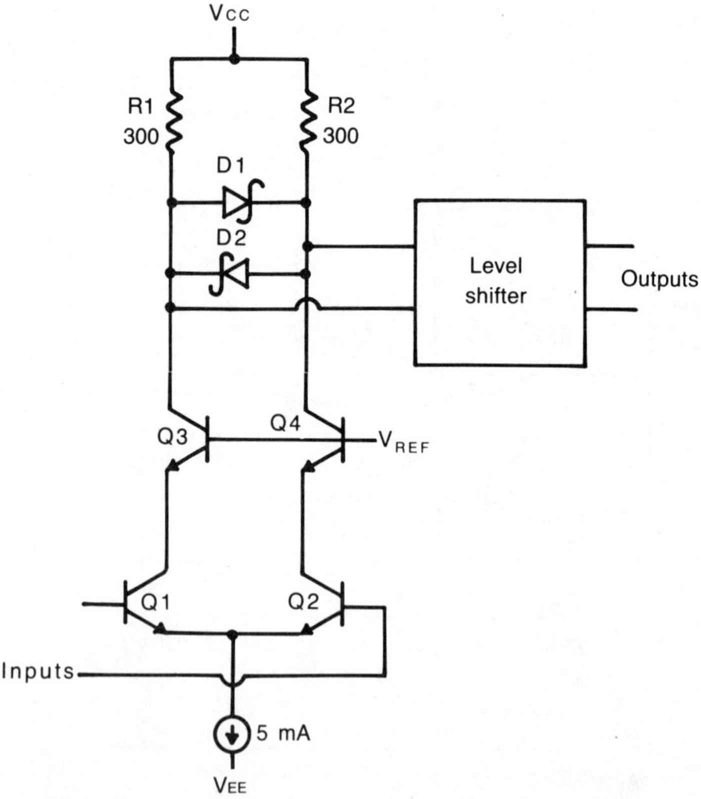

Fig. 5-6. Simplified schematic of very-high-speed differential comparator.

The output is about 3.1 volts—10 volts at the base of Q10 minus the Q10 V_{BE} minus the D2 zener voltage. The data sheet typical, minimum, and maximum values are 3.2, 2.5, and 4.0 volts.

Very-High-Speed Comparators

Table 5-10 lists very-high-speed comparators with response times less than 10 ns. These devices are all made by Advanced Micro Devices. They are less versatile in that they are designed to interface with only one logic family, either TTL (Am686) or ECL (Am685 and Am687). A simplified schematic of the input stage of these parts is shown in Fig. 5-6. High speed is achieved by using a cascode configuration and Schottky diodes.

Q3 and Q4 isolate the Q1-Q2 base-collector capacitances from the load resistors R1 and R2. Q3 and Q4 also have base-collector capacitance, but their capacitance current (which effectively slows down the comparator input stage) does not feedback to the inputs and, also, is less because the base voltage is fixed.

The Schottky diodes, D1 and D2, clamp the collector swing of Q3 and Q4 to about 900 mV. (Schottky diodes are metal-semiconductor diodes, have half the forward drop of PN junction diodes, and have about 5-10 times the speed of a typical base-emitter diode.) Assume Q2 is ON. About 3 mA will flow through R2 and about 2 mA will flow through R1 and D1. The drop across R1 is 2 mA times 300 ohms, or 600 mV. The drop across the Schottky diode is about 300 mV. This matches the drop across R2 which is 3 mA times 300 ohms.

Table 5-1. Low-Speed Comparators, Commercial Temperature Range.

PART NUMBER	MANUFACTURER	CMP PER PKG	MAXIMUM OFFSET VOLTAGE (mV)	TYPICAL RESPONSE TIME (nS)	UNIT PRICE ($)
TL331C	T I	1	5	1300	0.58
LM393	NATIONAL	2	5	1300	0.60
CA339	RCA	4	5	1300	0.66
UA339	FAIRCHILD	4	5	1300	0.67
LM339	NATIONAL	4	5	1300	0.75
LM339A	NATIONAL	4	2	1300	1.50
UA339A	FAIRCHILD	4	2	1300	1.65
CA339A	RCA	4	2	1300	1.72
LM393A	NATIONAL	2	2	1300	1.80
CMP-04FP	P M I	4	1	1300	7.80

Table 5-2. Low-Speed Comparators, Industrial Temperature Range.

PART NUMBER	MANUFACTURER	CMP PER PKG	MAXIMUM OFFSET VOLTAGE (mV)	TYPICAL RESPONSE TIME (nS)	UNIT PRICE ($)
UA3302	FAIRCHILD	4	20	1300	0.51
LM3302	NATIONAL	4	20	1300	0.67
MC3302	MOTOROLA	4	20	1300	0.73
TL331I	T I	1	5	1300	0.92
CA239	RCA	4	5	1300	0.94
LM2901	NATIONAL	4	7	1300	1.10
LM2903	NATIONAL	2	7	1300	1.15
UA2901	FAIRCHILD	4	7	1300	1.77
UA239	FAIRCHILD	4	5	1300	2.22
CA239A	RCA	4	2	1300	2.30
LM239	NATIONAL	4	5	1300	2.75
LM293	NATIONAL	2	5	1300	4.10
UA239A	FAIRCHILD	4	2	1300	5.13
LM293A	NATIONAL	2	2	1300	6.15
CMP-04FY	P M I	4	1	1300	9.75
LM239A	NATIONAL	4	2	1300	9.98

Table 5-3. Low-Speed Comparators, Military Temperature Range.

PART NUMBER	MANUFACTURER	CMP PER PKG	MAXIMUM OFFSET VOLTAGE (mV)	TYPICAL RESPONSE TIME (nS)	UNIT PRICE ($)
CA3290	RCA	2	20	1200	0.90
CA3290A	RCA	2	10	1200	1.26
CA139	RCA	4	5	1300	1.30
LM139	NATIONAL	4	5	1300	2.90
CA3290B	RCA	2	6	1200	3.20
UA139	FAIRCHILD	4	5	1300	3.23
CA139A	RCA	4	2	1300	4.42
LM193	NATIONAL	2	5	1300	5.50
UA139A	FAIRCHILD	4	2	1300	7.60
LM193A	NATIONAL	2	2	1300	9.00
LM139A	NATIONAL	4	2	1300	10.70
CMP-04B	P M I	4	1	1300	14.93

Table 5-4. Medium-Speed Comparators, Commercial Temperature Range.

PART NUMBER	MANUFACTURER	CMP PER PKG	MAXIMUM OFFSET VOLTAGE (mV)	TYPICAL RESPONSE TIME (nS)	UNIT PRICE ($)
UA311	FAIRCHILD	1	7.5	200	0.57
CA311	RCA	1	7.5	200	0.60
LM311	NATIONAL	1	7.5	200	0.65
LM319	NATIONAL	2	8	80	2.40
CMP-01C	P M I	1	2.8	110	3.75
CMP-02C	P M I	1	2.8	190	3.75
CMP-01E	P M I	1	0.8	110	5.63
CMP-02E	P M I	1	0.8	190	5.63
LF311	NATIONAL	1	10	200	6.40
LH2311	NATIONAL	2	7.5	200	10.00

Table 5-5. Medium-Speed Comparators, Industrial Temperature Range.

PART NUMBER	MANUFACTURER	CMP PER PKG	MAXIMUM OFFSET VOLTAGE (mV)	TYPICAL RESPONSE TIME (nS)	UNIT PRICE ($)
LM211	NATIONAL	1	3	200	2.70
LM219	NATIONAL	2	4	80	7.70
LH2211	NATIONAL	2	3	200	12.00
LF211	NATIONAL	1	4	200	14.25

Table 5-6. Medium-Speed Comparators, Military Temperature Range.

PART NUMBER	MANUFACTURER	CMP PER PKG	MAXIMUM OFFSET VOLTAGE (mV)	TYPICAL RESPONSE TIME (nS)	UNIT PRICE ($)
UA111	FAIRCHILD	1	3	200	2.57
LM111	NATIONAL	1	3	200	3.60
LF111	NATIONAL	1	10	200	6.40
LM119	NATIONAL	2	4	80	8.50
LH2111	NATIONAL	2	3	200	17.25
CMP-01	P M I	1	0.8	110	20.40
CMP-02	P M I	1	0.8	190	20.40

Table 5-7. High-Speed Comparators, Commercial Temperature Range.

PART NUMBER	MANUFACTURER	CMP PER PKG	MAXIMUM OFFSET VOLTAGE (mV)	TYPICAL RESPONSE TIME (nS)	UNIT PRICE ($)
TL710C	T I	1	7.5	40	0.53
LM710C	NATIONAL	1	5	40	0.70
UA710C	FAIRCHILD	1	5	40	0.74
LM711C	NATIONAL	2	5	40	0.90
TL810C	T I	1	3.5	30	1.01
TL811C	T I	2	5	33	1.31
TL514C	T I	2	3.5	30	1.33
LM1414	NATIONAL	2	5	30	1.49
UA711C	FAIRCHILD	2	5	40	1.52
TL820C	T I	2	3.5	30	1.68
MC1414	MOTOROLA	2	5	40	1.83
TL506C	T I	2	5	28	2.81
LM360	NATIONAL	1	5	14	4.50
CMP-05FP	P M I	1	0.6	37	6.00
CMP-05EP	P M I	1	0.25	37	9.15
LM306	NATIONAL	1	5	28	10.20
UA760C	FAIRCHILD	1	6	16	14.20

Table 5-8. High-Speed Comparators, Industrial Temperature Range.

PART NUMBER	MANUFACTURER	CMP PER PKG	MAXIMUM OFFSET VOLTAGE (mV)	TYPICAL RESPONSE TIME (nS)	UNIT PRICE ($)
LM260	NATIONAL	1	5	14	7.49
CMP-05FZ	P M I	1	0.6	37	7.50
CMP-05EZ	P M I	1	0.25	37	11.40
LM206	NATIONAL	1	2	28	17.20

Table 5-9. High-Speed Comparators, Military Temperature Range.

PART NUMBER	MANUFACTURER	CMP PER PKG	MAXIMUM OFFSET VOLTAGE (mV)	TYPICAL RESPONSE TIME (nS)	UNIT PRICE ($)
TL710M	T I	1	5	40	1.70
UA710	FAIRCHILD	1	2	40	1.90
UA711	FAIRCHILD	2	3.5	40	2.19
TL811M	T I	2	3.5	33	2.20
TL820M	T I	2	2	30	2.20
LM710	NATIONAL	1	2	40	2.20
LM711	NATIONAL	2	3.5	40	2.50
TL506M	T I	2	2	28	4.40
TL514M	T I	2	2	30	4.40
MC1514	MOTOROLA	2	2	40	4.92
LM1514	NATIONAL	2	2	30	4.98
CMP-05B	P M I	1	0.6	37	12.75
LM160	NATIONAL	1	5	14	14.00
UA760	FAIRCHILD	1	6	16	15.20
CMP-05A	P M I	1	0.25	37	21.68
LM106	NATIONAL	1	2	28	27.00

Table 5-10. Advanced Micro Devices (AMD) Very-High-Speed Comparators.

PART NUMBER	CMP PER PKG	OPERATING TEMP (C) MIN	MAX	OUTPUT TYPE	MAXIMUM OFFSET VOLTAGE (mV)	TYPICAL RESPONSE TIME (nS)	UNIT PRICE ($)
Am685L	1	-30	85	ECL	2	5.5	13.40
Am685M	1	-55	125	ECL	2	5.5	30.00
Am686C	1	0	70	TTL	3	9	5.25
Am686C-1	1	0	70	TTL	6	9	3.00
Am686M	1	-55	125	TTL	2	9	30.00
Am687AL	2	-30	85	ECL	3	7	28.50
Am687AM	2	-55	125	ECL	2	7	66.00
Am687L	2	-30	85	ECL	3	7	22.50
Am687M	2	-55	125	ECL	2	7	42.00

Chapter 6

Voltage References

FIGURE 6-1 SHOWS FOUR TYPICAL APPLICATIONS OF THE VOLtage reference. In all of these applications, it is possible that the well known zener diode-resistor combination shown in Fig. 6-2 would meet the requirements. The zener diode reigns in applications where a reference of about 6 volts or higher is required. For lower voltages, the basic zener voltage must either be attenuated using a resistor-divider or an op-amp circuit, which can provide both attenuation and load-driving capability. This is achieved, however, at additional expense. Also, zener diodes typically require 5 to 10 mA or more of reverse current in order to be beyond the "knee" of the breakdown curve. Hence, a 10-volt zener operating at 10 mA consumes 0.1 watt.

Although a reverse-biased base-emitter junction of an NPN transistor is often used in linear IC design, the heart of most IC voltage references and IC voltage regulators is the so-called "bandgap" voltage reference. The advantage of the bandgap reference is that it is low power, low voltage, and is easily scaled up to any desired voltage.

THE BANDGAP VOLTAGE REFERENCE

The basic bandgap reference, named after its inventor Robert Widlar, is shown in Fig. 6-3. It is called a bandgap reference be-

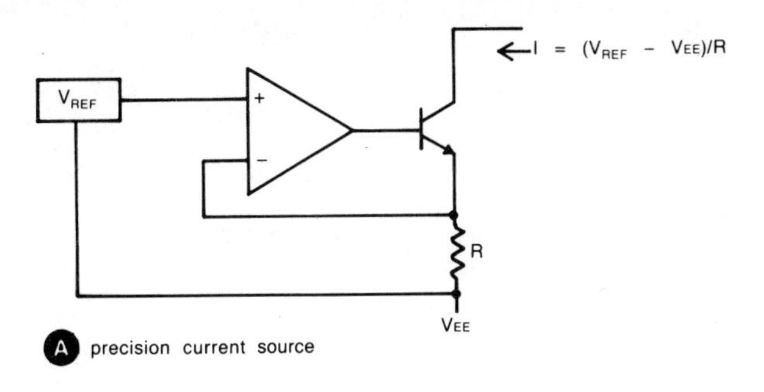

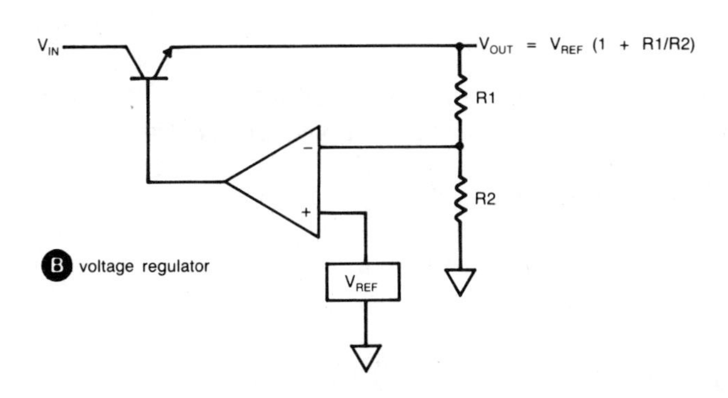

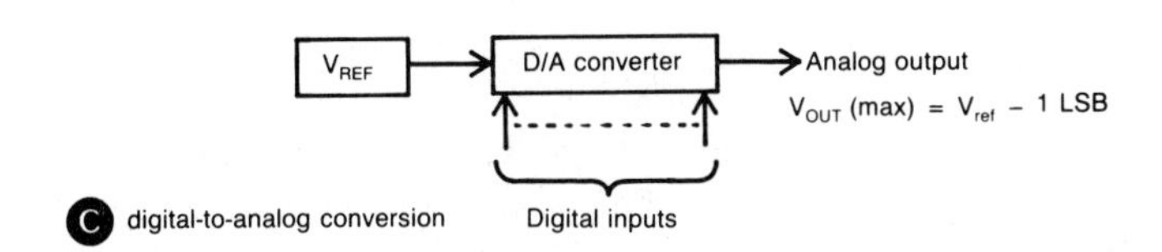

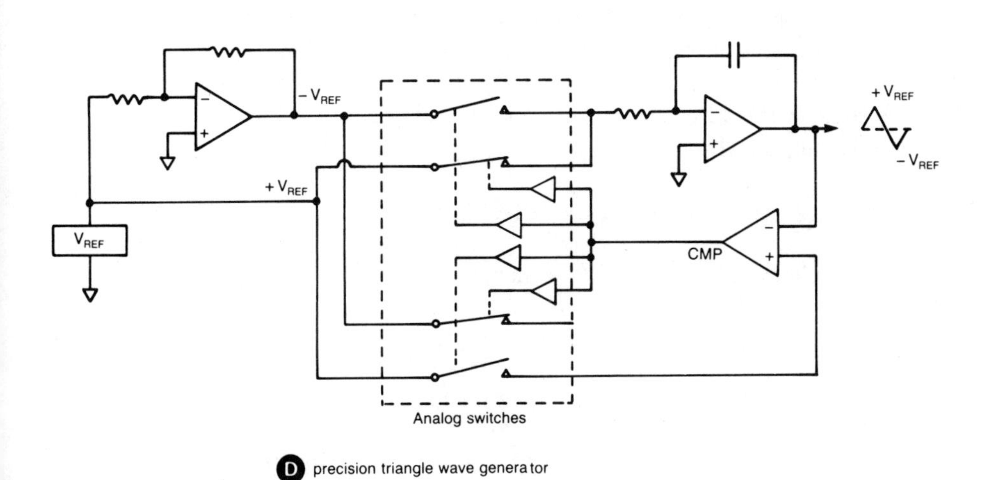

Fig. 6-1. Voltage reference applications.

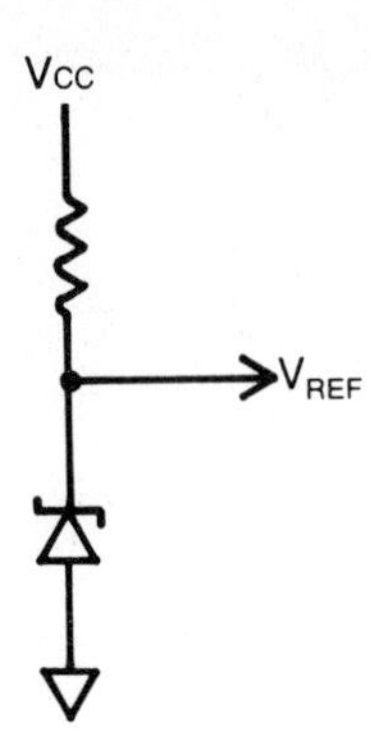

Fig. 6-2. The zener diode voltage reference.

cause the output voltage is approximately equal to 1.2 volts, which is the bandgap voltage of silicon at zero degrees Kelvin. The bandgap voltage refers to the amount of energy required to create a free electron in undoped silicon. This term appears in the temperature equation for the transistor base-emitter voltage as follows:

$$V_{BE} = V_{gO} (1 - T/T_O) + V_{BEO} (T/T_O) \qquad \textbf{(1)}$$

where:

V_{gO} is the bandgap voltage for silicon at 0 °K;

V_{BEO} is the base-emitter voltage at temperature T_O (in °K); and

V_{BE} is the base-emitter voltage at temperature T (in °K).

Now look at Fig. 6-3. The outpt voltage is the sum of the Q3 V_{BE} and the voltage drop across R2. The voltage drop across R2 is 10 times the drop across R3. Q1 and Q2 form a current mirror. Current flows into Q1 first, developing a base-emitter voltage that is applied to the base of Q2. With the values shown, and if Q1 and Q2 have the same emitter size, then the current in Q2 will be one-tenth the value of the current flowing in Q1. The voltage drop across R3 is equal to the difference in base-emitter voltages according to the following equation:

$$V_{R3} = V_{BE1} - V_{BE2} = (kT/q) \ln(I_{E1}/I_{E2}) \qquad \textbf{(2)}$$

The voltage drop across R2 is then:

$$V_{R2} = (R2/R3)\ (kT/q) \ln(I_{E1}/I_{E2}) \qquad \textbf{(3)}$$

Combining (1) and (3), the equation for the output voltage is:

$$V_{REF} = V_{gO} (1 - T/T_O) + V_{BEO} (T/T_O) + (R2/R3)\ (kT/q) \ln(I_{E1}/I_{E2}) \qquad \textbf{(4)}$$

The critical parameter for voltage references is temperature stability. Equation (4) varies with temperature as follows:

$$\Delta V_{REF}/\Delta T = -V_{gO}/T_O + V_{BEO}/T_O + (R2/R3)\,(kT_o/q)\ln(I_{E1}/I_{E2}) \quad \textbf{(5)}$$

Temperature compensation is achieved when equation (5) equals zero. Setting equation (5) equal to zero and solving for V_{gO} yields:

$$V_{gO} = V_{BEO} + (R2/R3)\,(kT_O/q)\ln(I_{E1}/I_{E2}) \quad \textbf{(6)}$$

Thus, when the sum of the Q3 base-emitter voltage and the drop across R2 is equal to the bandgap voltage, then the output voltage is temperature compensated, or has a zero temperature coefficient. This is the theory and basis (and name) for the bandgap reference. In practice, of course, a zero temperature coefficient is never achieved, as will be seen in the tables in this chapter.

Put more simply, the purpose of the bandgap reference shown in Fig. 6-3 is to compensate the negative 2 mV/°C temperature coefficient of the V_{BE} of Q3 with a positive 2 mV/°C. It is known that the difference of base-emitter voltages (same emitter size, but different currents; or different emitter sizes, but the same emitter current) has a positive temperature coefficient. Because it is impractical to generate approximately 600 mV by ratioing current (a current ratio of 10^{10} would be required), a voltage drop proportional to the V_{BE} difference is generated across R2.

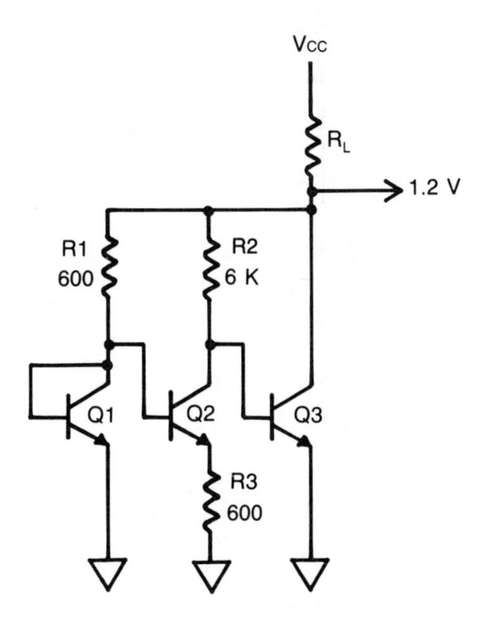

Fig. 6-3. Widlar bandgap voltage reference.

Another bandgap voltage reference circuit is shown in Fig. 6-4. In this circuit, identical currents are made to flow in Q1 and Q2, but the emitter area of Q1 is some factor (let's say 8) times Q2. As a result, a voltage is developed across R1 as follows:

$$V_{R1} = (kT/q) \ln (A_{E1}/A_{E2}) = 0.026 \ln 8 = 54 \text{ mV}$$

If R1 is set equal to 540 ohms, then 0.1 mA will flow in Q1. With R3 = R4 the op amp will cause 0.1 mA to flow in Q2. It only remains to choose R2 so that the voltage drop across R2 plus Q2 V_{BE} equals 1.2 volts. For a given process and geometry size, the Q2 V_{BE} could be calculated. For the sake of illustration, I will assume (as I did in the discussion of the Widlar circuit) that the V_{BE} is about 600 mV. Then R2 must be 600 mV/0.2 mA, or 3000 ohms.

In practice, the function of the op amp and R3 and R4 is performed by a few transistors as shown in Fig. 6-5. In this circuit, Q3 and Q4 provide equal currents to Q1 and Q2. The 1.2 volts at the bases of Q1 and Q2 is furthermore scaled up to any desired valued by Q5, R3 and R4.

PARAMETER DEFINITIONS

Note that the differential pair type bandgap reference is a three-terminal device. The Widlar reference is a two-terminal device, with

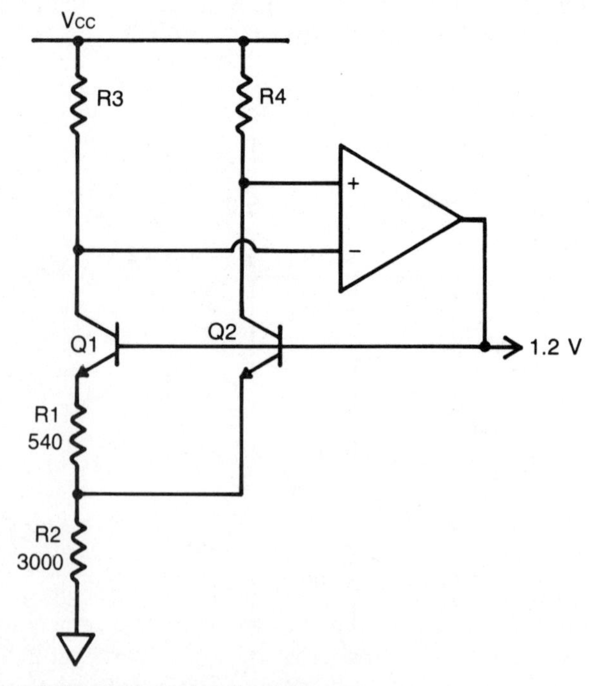

Fig. 6-4. Differential pair bandgap reference.

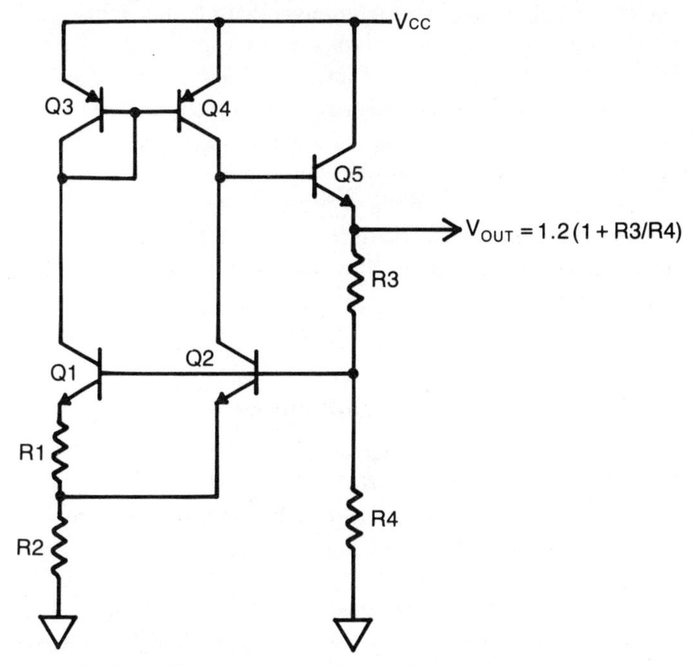

Fig. 6-5. The basic three-terminal voltage reference.

an external current-setting resistor. The Widlar type reference operates like a zener with a low-voltage breakdown. In actuality, of course, no junction breakdown occurs.

Two-terminal type references (some of which are not Widlar type references) are sometimes called "reference diodes" by the manufacturers. For example, the National LM103, LM113/LM313, LM136 series, and LM185 series voltage references are called reference diodes.

Two-terminal Parameter Definitions

The basic performance parameters for two-terminal references are defined as follows:

Breakdown Voltage. The breakdown voltage is the voltage across the reference at a specified reverse current, usually 1 mA. Hence, the reference is tested in the same way that a zener is tested. The "anode" of the reference is connected to ground, the "cathode" connected to a current source. The "cathode" voltage is the breakdown voltage. Minimum, typical, and maximum values are specified.

Temperature Stability. The temperature stability is the variation in breakdown voltage over the operating temperature range; expressed in either millivolts (mV) or in parts-per-million per-degree Centigrade (PPM/ °C). When expressed in mV, this parameter is called breakdown voltage drift; when expressed in PPM/°C, this parameter is the temperature coefficient of the breakdown voltage. Typical and maximum values are specified.

Breakdown Voltage Change with Current. The breakdown voltage change with current is the change in breakdown voltage over a specified operating current range; expressed in mV. Typical and maximum values are usually specified.

Reverse Dynamic Resistance. The reverse dynamic resistance is the change in breakdown voltage divided by the change in current (that caused the change in voltage). Typical and maximum values are specified.

Three-terminal Parameter Definitions

The basic performance parameters for three-terminal voltage references are defined as follows:

Output Voltage Tolerance. The output voltage tolerance is the maximum deviation from nominal output voltage; expressed in either ±mV or ±%.

Temperature Stability. The temperature stability is the output voltage change over the operating temperature range; expressed in mV or PPM/°C. Maximum value is specified.

Line Regulation. The line regulation is the output voltage change over a specified input voltage range; expressed in mV, or as a percent of nominal output voltage divided by the change in input voltage (%/V). Typical and maximum values are specified. Line regulation is sometimes specified over more than one range of input voltages. (In Appendix C, line regulation is specified in %/V over the range that resulted in the best regulation.)

Load Regulation. The load regulation is the output voltage change over a specified load current range; expressed in either mV, or as a percent change in the nominal output voltage divided by the change in load current (%/mA). Usually the output voltage is measured at zero output current (no load) and maximum recommended (not necessarily absolute maximum) output current (full load). Most references can source up to 5 or 10 mA.

SELECTION GUIDES

Tables 6-1 through 6-8 list voltage references by standard values: 1.2 V, 2.5 V, 5 V, and 10 V. Commercial and military temperature range parts are listed. See Appendix C for industrial range parts, which were too few to warrant separate tables.

In most applications, output voltage tolerance (or breakdown voltage range) and temperature stability are the critical parameters for voltage references. For this reason, only these two parameters are listed in the tables. See Appendix C for additional data.

Table 6-1. 1.2-Volt Reference Diode ICs, Commercial Temperature Range.

PART NUMBER	MANUFACTURER	BREAKDOWN VOLTAGE (V) MIN	TYP	MAX	MAXIMUM TEMPCO (PPM/°C)	UNIT PRICE ($)
ICL8069DC	INTERSIL	1.2	1.23	1.25	100	1.15
ICL8069CC	INTERSIL	1.2	1.23	1.25	50	1.60
LM385-1.2	NATIONAL	1.205	1.235	1.26	20 TYP	1.75
AD589J	ANALOG DEVICES	1.2	1.235	1.25	100	2.20
LM385B-1.2	NATIONAL	1.223	1.235	1.247	20 TYP	2.45
AD589K	ANALOG DEVICES	1.2	1.235	1.25	50	2.90
LM313	NATIONAL	1.16	1.22	1.28	100 TYP	6.75
ICL8069B	INTERSIL	1.2	1.23	1.25	25	7.15
AD589L	ANALOG DEVICES	1.2	1.235	1.25	25	8.10
ICL8069A	INTERSIL	1.2	1.23	1.25	10	15.85
AD589M	ANALOG DEVICES	1.2	1.235	1.25	10	18.10

Table 6-2. 1.2-Volt Reference Diode ICs, Military Temperature Range.

PART NUMBER	MANUFACTURER	BREAKDOWN VOLTAGE (V) MIN	TYP	MAX	MAXIMUM TEMPCO (PPM/°C)	UNIT PRICE ($)
AD589S	ANALOG DEVICES	1.2	1.235	1.25	100	4.40
ICL8069DM	INTERSIL	1.2	1.23	1.25	100	4.65
ICL8069CM	INTERSIL	1.2	1.23	1.25	50	5.55
AD589T	ANALOG DEVICES	1.2	1.235	1.25	50	6.50
LM113	NATIONAL	1.16	1.22	1.28	100 TYP	11.25
LM113-2	NATIONAL	1.195	1.22	1.245	100 TYP	15.00
LM185-1.2	NATIONAL	1.223	1.235	1.247	20 TYP	17.70
AD589U	ANALOG DEVICES	1.2	1.235	1.25	25	18.40
LM113-1	NATIONAL	1.21	1.22	1.232	100 TYP	22.50

Table 6-3. 2.5-Volt References, Commercial Temperature Range.

PART NUMBER	MANUFACTURER	TOLERANCE (±%)	MAXIMUM TEMPCO (PPM/°C)	UNIT PRICE ($)
*LM336-2.5	NATIONAL	4	24	1.10
*LM385-2.5	NATIONAL	3	20 TYP	1.75
*LM336B-2.5	NATIONAL	2	24	2.00

PART NUMBER	MANUFACTURER	TOLERANCE (±%)	MAXIMUM TEMPCO (PPM/°C)	UNIT PRICE ($)
*LM385B-2.5	NATIONAL	1.5	20 TYP	2.25
AD1403	ANALOG DEVICES	1	40	2.50
MC1403	MOTOROLA	1	40	2.71
AD580J	ANALOG DEVICES	3	85	3.90
AD584J	ANALOG DEVICES	0.3	30	5.70
AD1403A	ANALOG DEVICES	0.4	25	6.00
MC1403A	MOTOROLA	1	25	6.24
MC1400G2	MOTOROLA	0.2	25	6.38
AD580K	ANALOG DEVICES	1	40	8.00
MC1400AG2	MOTOROLA	0.2	10	10.43
AD580L	ANALOG DEVICES	0.4	25	11.00
AD584K	ANALOG DEVICES	0.14	15	12.20
AD580M	ANALOG DEVICES	0.4	10	17.00
AD584L	ANALOG DEVICES	0.1	10	21.80

* DENOTES REFERENCE DIODE TYPE

Table 6-4. 2.5-Volt References, Military Temperature Range.

PART NUMBER	MANUFACTURER	TOLERANCE (±%)	MAXIMUM TEMPCO (PPM/°C)	UNIT PRICE ($)
MC1503	MOTOROLA	1	55	7.61
*LM136-2.5	NATIONAL	2	72	8.64
*LM136A-2.5	NATIONAL	1	72	12.50
MC1503A	MOTOROLA	1	25	14.25
AD580S	ANALOG DEVICES	1	55	16.00
AD584S	ANALOG DEVICES	0.3	30	17.10
MC1500G2	MOTOROLA	0.2	40	17.63
*LM185-2.5	NATIONAL	1.5	20 TYP	18.45
AD580T	ANALOG DEVICES	0.4	25	25.10
AD584T	ANALOG DEVICES	0.14	20	25.30
MC1500AG2	MOTOROLA	0.2	10	36.75
AD580U	ANALOG DEVICES	0.4	10	53.80

* DENOTES REFERENCE DIODE TYPE

Table 6-5. 5-Volt References, Commercial Temperature Range.

PART NUMBER	MANUFACTURER	TOLERANCE (±%)	MAXIMUM TEMPCO (PPM/°C)	UNIT PRICE ($)
*LM336-5.0	NATIONAL	4	48	1.10
*LM336B-5.0	NATIONAL	2	48	2.00
REF-02D	P M I	2	250	2.10
MC1404U5	MOTOROLA	1	40	2.73
REF-02C	P M I	1	65	3.75
MC1404AU5	MOTOROLA	1	25	3.85
REF-02H	P M I	0.5	25	4.88
AD584J	ANALOG DEVICES	0.3	30	5.70
MC1400G5	MOTOROLA	0.2	25	6.38
REF-02E	P M I	0.3	8.5	9.60
MC1400AG5.	MOTOROLA	0.2	10	10.43
AD584K	ANALOG DEVICES	0.12	15	12.20
AD584L	ANALOG DEVICES	0.06	5	21.80

* DENOTES REFERENCE DIODE TYPE

Table 6-6. 5-Volt References, Military Temperature Range.

PART NUMBER	MANUFACTURER	TOLERANCE (±%)	MAXIMUM TEMPCO (PPM/°C)	UNIT PRICE ($)
*LM136-5.0	NATIONAL	2	144	8.64
MC1504U5	MOTOROLA	1	55	9.12
*LM136A-5.0	NATIONAL	1	144	12.50
MC1504AU5	MOTOROLA	1	25	14.38
AD584S	ANALOG DEVICES	0.3	30	17.10
MC1500G5	MOTOROLA	0.2	40	17.63
REF-02	P M I	0.5	25	19.13
AD584T	ANALOG DEVICES	0.12	15	25.30
MC1500AG5	MOTOROLA	0.2	10	36.75
REF-05B	P M I	0.5	25	37.50
REF-02A	P M I	0.3	8.5	38.25
REF-05A	P M I	0.3	8.5	75.00

* DENOTES REFERENCE DIODE TYPE

Table 6-7. 10-Volt References, Commercial Temperature Range.

PART NUMBER	MANUFACTURER	TOLERANCE (±%)	MAXIMUM TEMPCO (PPM/°C)	UNIT PRICE ($)
MC1404U10	MOTOROLA	1	40	2.73
REF-01C	P M I	1	65	3.75
MC1404AU10	MOTOROLA	1	25	3.85
REF-01H	P M I	0.5	25	4.88
AD581J	ANALOG DEVICES	0.3	30	5.60
AD584J	ANALOG DEVICES	0.3	30	5.70
MC1400G10	MOTOROLA	0.2	25	6.38
AD581K	ANALOG DEVICES	0.1	15	9.20
REF-01E	P M I	0.3	8.5	9.60
MC1400AG10	MOTOROLA	0.2	10	10.43
AD584K	ANALOG DEVICES	0.1	15	12.20
AD581L	ANALOG DEVICES	0.05	5	21.80
AD584L	ANALOG DEVICES	0.05	5	21.80
AD2710K	ANALOG DEVICES	0.01	5	35.85
AD2710L	ANALOG DEVICES	0.01	5	45.25

Table 6-8. 10-Volt References, Military Temperature Range.

PART NUMBER	MANUFACTURER	TOLERANCE (±%)	MAXIMUM TEMPCO (PPM/°C)	UNIT PRICE ($)
LH0070-0	NATIONAL	0.1	40	6.60
MC15004U10	MOTOROLA	1	55	9.12
LH0070-1	NATIONAL	0.1	20	10.20
MC15004AU10	MOTOROLA	1	25	14.38
AD584S	ANALOG DEVICES	0.3	30	17.10
AD581S	ANALOG DEVICES	0.3	30	17.20
MC1500G10	MOTOROLA	0.2	40	17.63
LH0070-2	NATIONAL	0.05	8	18.50
REF-01	P M I	0.5	25	19.13
AD584T	ANALOG DEVICES	0.1	15	25.30
AD581T	ANALOG DEVICES	0.1	15	26.00
MC1500AG10	MOTOROLA	0.2	10	36.75

PART NUMBER	MANUFACTURER	TOLERANCE (±%)	MAXIMUM TEMPCO (PPM/°C)	UNIT PRICE ($)
REF-10B	P M I	0.5	25	37.50
REF-01A	P M I	0.3	8.5	38.25
AD581U	ANALOG DEVICES	0.05	10	44.80
AD2700S	ANALOG DEVICES	0.05	3	53.00
REF-10A	P M I	0.3	8.5	75.00
AD2700T	ANALOG DEVICES	0.025	3	83.00

Chapter 7

Voltage Regulators

THERE ARE TWO BASIC TYPES OF VOLTAGE REGULATOR ICs: *linear*, and *switchmode*. The basic linear regulator is shown in Fig. 7-1. This is a conventional series regulator comprised of the pass transistor Q1, sensing resistors R1 and R2, a voltage reference, and an operational amplifier. The equation for the regulated output voltage is as follows:

$$V_{OUT} = V_{REF}\,(1 + R1/R2)$$

All of the load current passes through Q1. As a result Q1 continuously dissipates power equal to the load current times the Q1 collector to emitter voltage drop. Thus, although the linear regulator is simple, it wastes power.

The switchmode regulator, on the other hand, is highly efficient (power efficiencies around 80 percent are typical), but is more complicated to design. Circuit designers who do not regularly design switching regulators are apt to be depressed at the idea of computing the values for L1 and C1 and then probably having to buy toroids and wind their own inductors! Fortunately, the manufacturers of switchmode power supply controller ICs provide design equations and one or more examples including step-by-step design procedures.

Figures 7-2 through 7-5 show four basic switchmode voltage

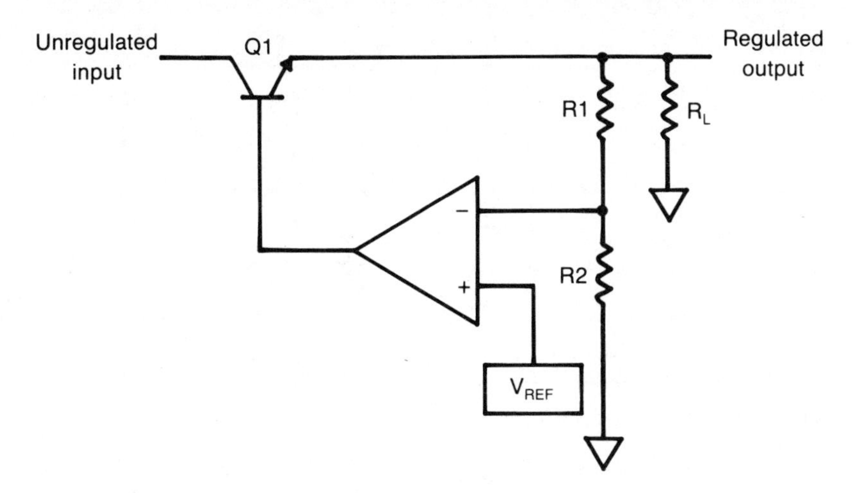

Fig. 7-1. Basic linear regulator.

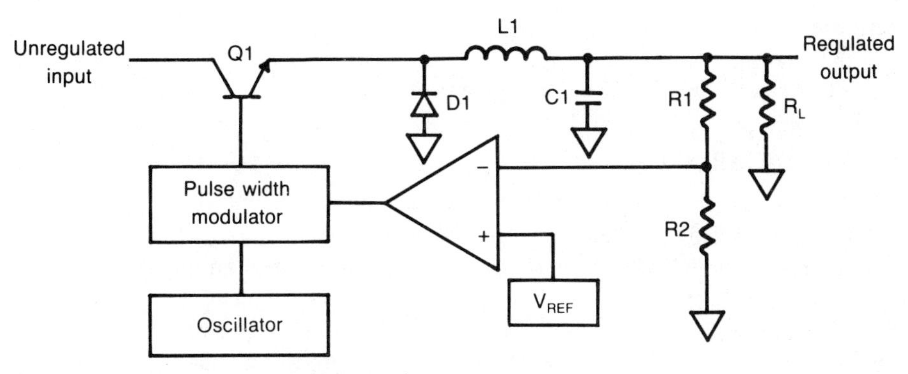

Fig. 7-2. Basic step-down switching regulator.

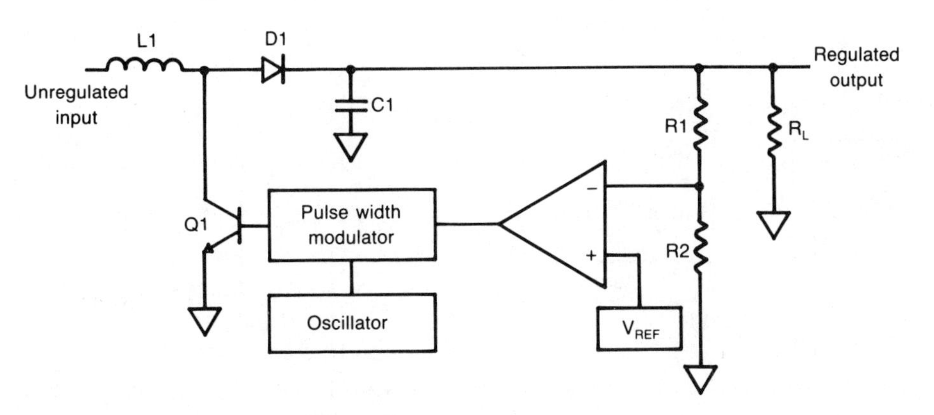

Fig. 7-3. Basic step-up switching regulator.

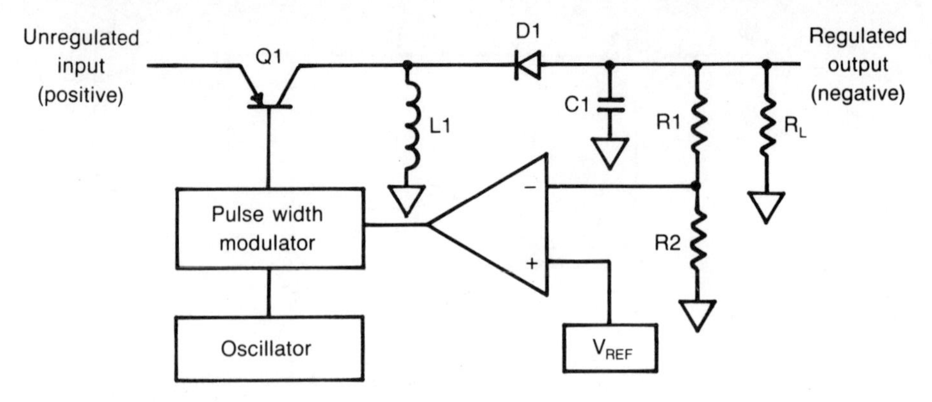

Fig. 7-4. Basic switchmode dc-dc inverter.

regulators: step-down, step-up, inverter, and push-pull converter. In each of these applications, Q1 is switched on and off usually at a constant frequency (20 kHz, typically) with a variable duty cycle (ratio of ON time to total period). Q1 dissipates little power because of the low collector-emitter voltage drop when the transistor is saturated.

In this chapter we will look at the internal operation and characteristics of both linear and switchmode regulator ICs.

LINEAR REGULATORS

As was stated in Chapter 6, the heart of most linear voltage regulators is the bandgap reference. A popular exception is the 723 voltage regulator IC, which contains a zener reference, op amp, and Darlington pass transistor. These circuits are not connected to each other on the chip. Instead, the inputs and outputs are available to the designer who then has the freedom to wire these basic blocks to produce both positive and negative voltage regulators.

In contrast to the 723 is the 7800 series three-terminal voltage regulators that require only one input capacitor and one output ca-

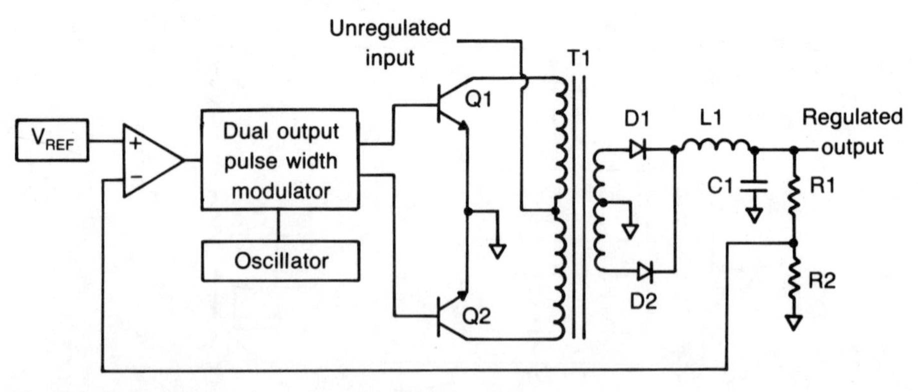

Fig. 7-5. Basic switchmode push-pull dc-dc converter.

pacitor to be a complete linear regulator.

The 7800 series and other standard regulators like the LM109/209/309 are based on the Widlar bandgap reference which was explained in Chapter 6. A stripped down 7805 is shown in Fig. 7-6. At first glance, this circuit may seem complicated (it is generally difficult to figure out someone else's design anyway), and it may be difficult to recognize the Widlar reference. Recall that the Widlar circuit presented in Chapter 6 had only three transistors and that the output voltage was the sum of one V_{BE} and the voltage drop across a resistor. In Fig. 7-6, the 5-volt output is the sum of four V_{BE}'s (Q3, Q4, Q5, and Q6) and the voltage drop across R2.

The voltage drop across R1 sets the current in Q1. Q1, in turn, establishes a voltage across R3 and the Q2 base-emitter junction. The voltage drop across R3 is the difference between the V_{BE}'s of Q1 and Q2 and has a positive temperature coefficient. The voltage drop across R3 is R2/R3 times the drop across R3. The values of R1, R2, and R3 are selected such that the drop across R3 is about 2.5 volts. To this is added the V_{BE}'s of Q3 through Q6, for a total of 5 volts. The temperature coefficient across the junctions is −8

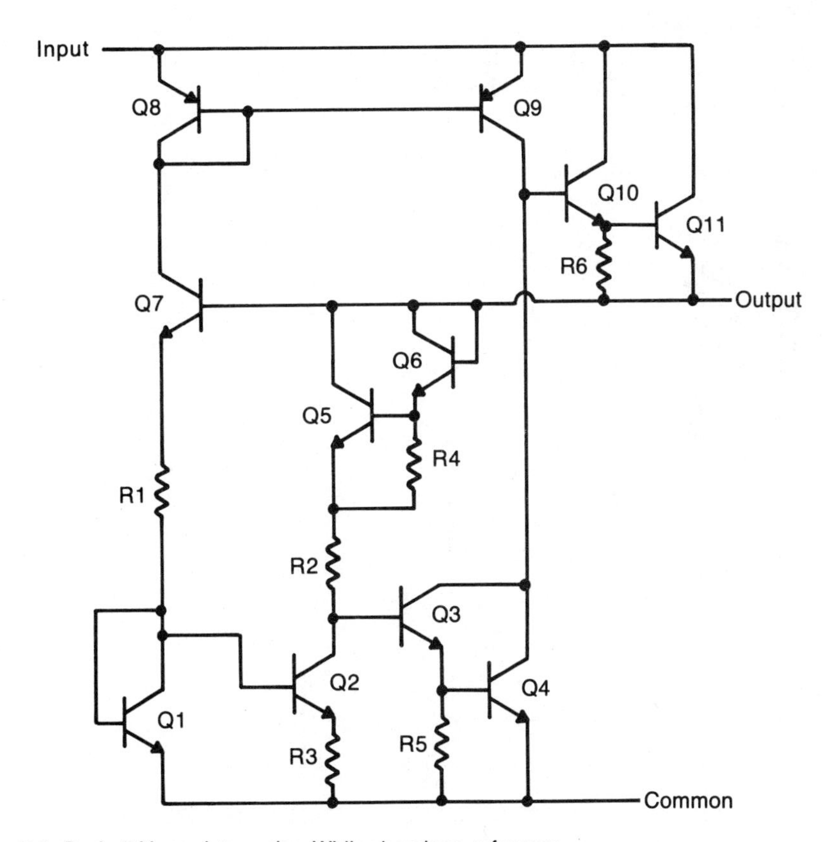

Fig. 7-6. Basic 5-V regulator using Widlar bandgap reference.

mV/°C (2 mV for each junction). The temperature coefficient across R2 is about +8 mV/°C.

The current in Q1 is mirrored via Q8 over to Q9, which provides base current for the Q10-Q11 pass transistor combination and bias current for Q3 and Q4. Q11, the output transistor, provides both load current to the external world, base current for Q7, and bias current for the Q6-Q5-R2-Q2-R3 chain. Q7 maintains a constant voltage on the top side of R1. R1 could have been connected directly to Q8, but this would make the current in the regulator vary with input voltage. This would unnecessarily increase line regulation, something we don't want to do.

The circuit as shown in Fig. 7-6 will not necessarily get into the desired state when powered up. Figure 7-7 shows the additional circuitry required to ensure proper start-up. The zener reference circuit composed of D1, Q13, R7, R8, and R9 will always have the same state when powered-up. When power is applied, about 2.5 volts is applied to the base of Q14. Current will flow through Q8, Q14, R1, and Q1. Eventually (microseconds), the output rises towards 5 volts. When the emitter of Q5 gets to about 2 volts (on its way to about 3.7 volts) Q14 begins to turn off. In the steady-state, Q14 is always off.

Notice that I have redrawn the connections between R1, Q5, and Q7, and added Q12 to conform to the standard 7800 series

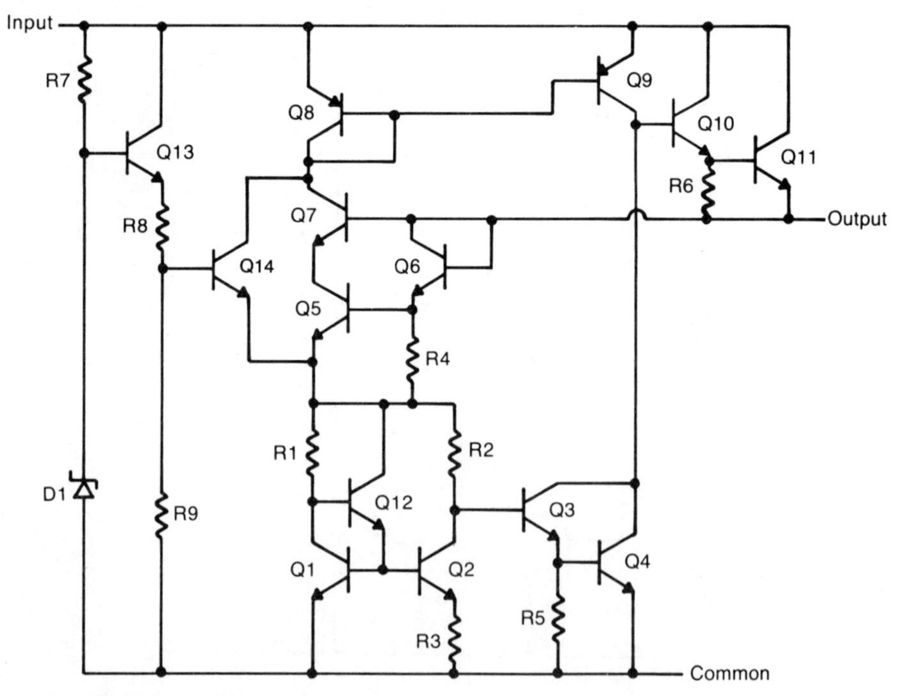

Fig. 7-7. Regulator with start-up circuitry.

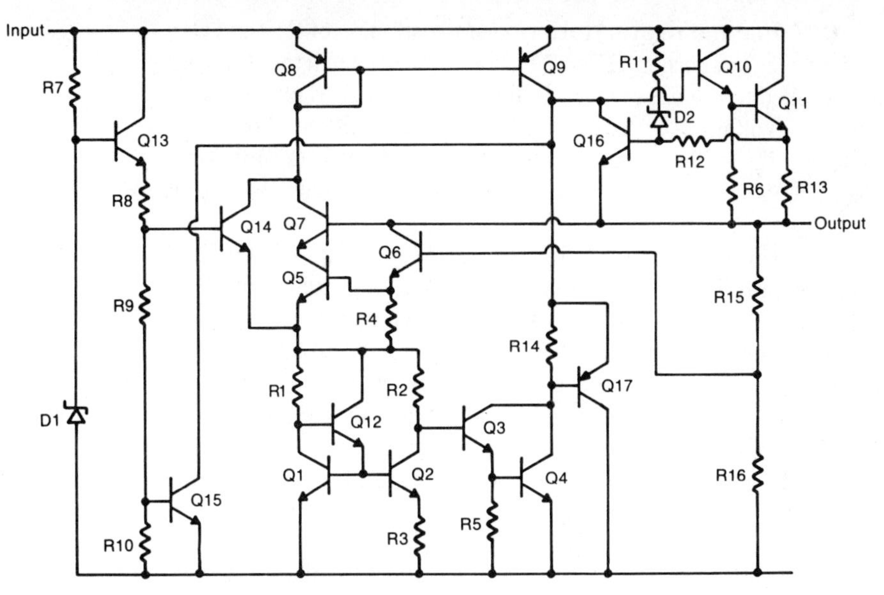

Fig. 7-8. Regulator with protection circuitry.

schematic. This implementation reduces the size of R1 for the same current and makes the Q1-Q2 current mirror more accurate and less sensitive to beta.

Figure 7-8 shows the regulator with protection circuitry added. This circuit now has three kinds of protection: protection against excessive output current, protection against excessive combined input voltage and output current (output transistor safe operating area, i.e., power), and protection against excessive chip temperature.

(1) Current protection. R13 senses the load current being delivered to the outside world. The value of R13 (let's say 0.3 ohm) is selected such that when maximum current is being approached (let's say 1.5 A) then Q16 will start turning on, which will in turn begin to limit and even shut down Q10-Q11.

(2) Power protection. R11, R12, and D2 work with Q16 and R13 to provide power protection. First, consider the case when the voltage drop across R13 is very low, corresponding to light-load condition. As the input voltage is increased beyond the breakdown voltage of D1, current flows through R11, D1 and R12. When the voltage drop across R12 is about 400 mV, Q16 begins to turn on. Note that the voltage across the Q16 base-emitter is the sum of the drop across R13 *and* R12. Thus, if the voltage drop across R13 increases (higher output current), then it will take less input voltage to make the voltage drop across R12 sufficient to shut down the regulator. This is equivalent to limiting the power dissipation of Q11 to some desired value which in turn determines the values of R11 and R12.

(3) Thermal protection. Notice the addition of R10 and Q15 in the lower left corner of Fig. 7-8. R10 is chosen such that the voltage applied to the base of Q15 is about 200 mV. At room temperature, this voltage is insufficient to turn on Q15. But at about 175° C, Q15's V_{BE} will be reduced about 300 mV below its room temperature value and, hence, will turn on and shut down the output stage.

A few other components have been added in Fig. 7-8. Q17 and R14 regulate the current to Q3 and Q4. With no load, most of the Q9 current will flow into Q17. As the load current increases, Q10 requires more base current, reducing the current through Q17, but not significantly reducing the Q3-Q4 current.

Finally, the base of Q6 is tied to the output voltage sensing resistors R15 and R16, which enable the output to be scaled up from 5 volts. In fact, the entire 7800 series is based on one basic die. With metal mask or bonding pad options, values of 5, 8, 12, 15, 18, and 24 volts may be generated from one die.

The basic 7900 series type negative voltage regulator is shown in Fig. 7-9. I have deliberately left off the protection circuitry and other so-called "bells and whistles" so that you can clearly see the design approach of this circuit, which is not the same as the 7800 series positive voltage regulators.

This circuit uses a zener diode reference, which is divided down and applied to the base of Q5. If you go back to Chapter 3 and look

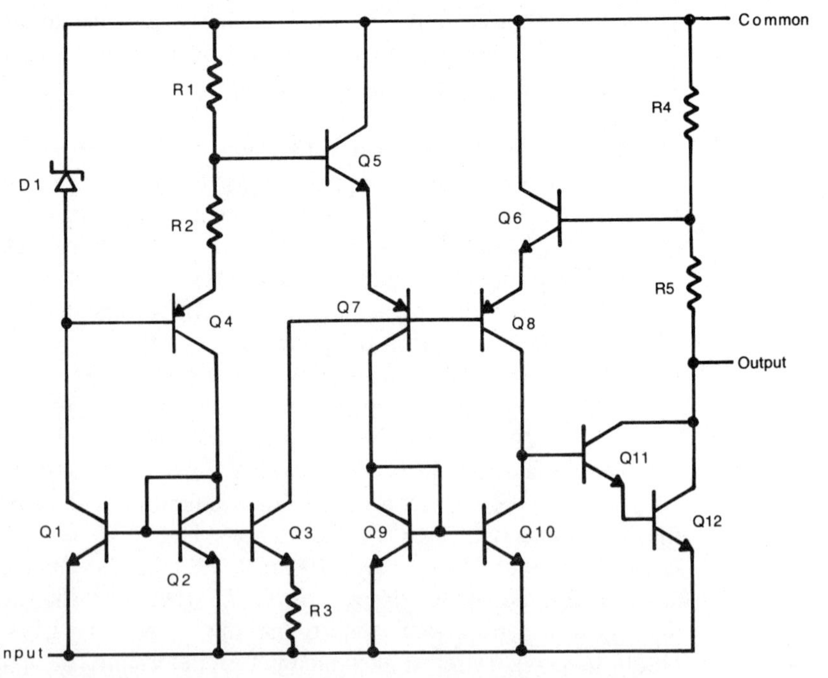

Fig. 7-9. Basic negative voltage regulator using zener diode reference.

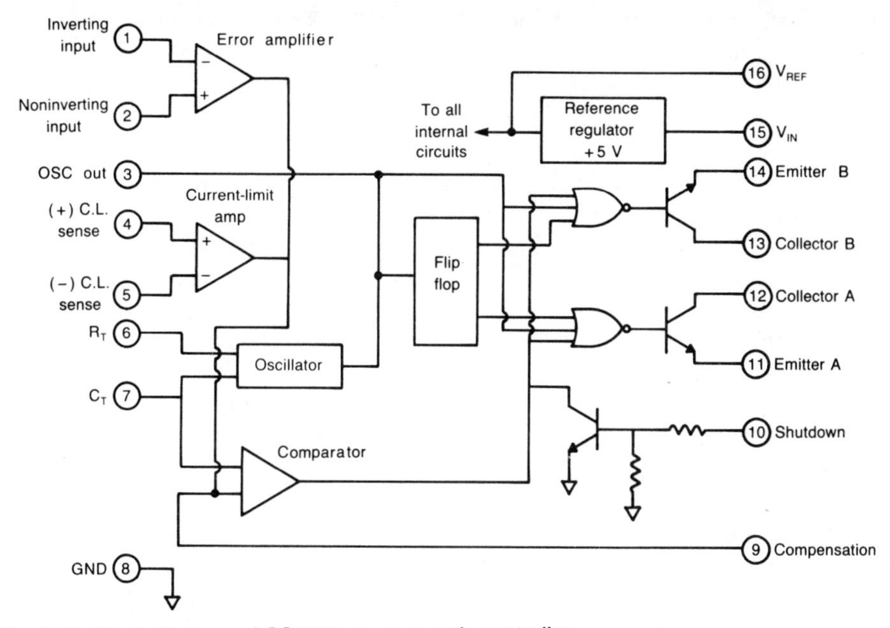

Fig. 7-10. Block diagram of SG1524 power-supply controller.

at Fig. 3-13, you will see that Q5 through Q10 form a differential amplifier with a high-input voltage range. Q3 and R3 form a low-value current source that biases Q7 and Q8. The operating current of Q5 is Q7 beta times approximately one-half of the Q3 current. The pass transistor Darlington Q11-Q12 maintains the output voltage such that the bases of Q5 and Q6 are the same (less offsets due to mismatched transistors). Essentially, the output voltage is equal to (1 + R4/R5) times the voltage drop across R1.

As in the case of the 7800 series, 7900 series negative regulators are variations of the same basic die. R5 scales up the reference voltage between ground and the base of Q5. For the user adjustable version of this chip, R4 and R5 are off-chip. The reference voltage is −5 volts and is the minimum negative voltage. The maximum is −30 volts.

Selection guides for linear voltage regulator ICs are given in Tables 7-1 through 7-19. Additional data is available in Appendix D.

SWITCHMODE REGULATORS

Figure 7-10 shows the block diagram and pin connections for the Silicon General SG1524 pulse-width modulator-type power-supply-controller IC, which can be used to make a variety of switchmode regulators. The SG1524 is second sourced by several manufacturers. Compare this block diagram to Figs. 7-2 through 7-5 which showed typical switchmode regulator configurations.

The SG1524 has four main components: reference, oscillator,

error op amp, and pulse-width modulator (PWM). The PWM is composed of the comparator, flip-flop, NOR gates and output transistors. The current-limit amplifier and shutdown transistor are auxiliary circuits that may or may not be used.

The +5-volt reference circuit is shown in Fig. 7-11. This is a zener reference regulator. Q9 supplies constant current to the zener D2 and temperature compensation diode D3. The voltage across D2 and D3 is divided down to 2.5 volts by Q2, R2, R3, and Q3. Also, these four components along with R6 and Q6 make up a bias chain for the current source transistors Q9, Q10, and Q4. The Q4 current is twice the Q10 current. Q1, R1, and D1 is the start-up circuit for the bias chain and associated current sources.

Q4, 5, 6, and 10 form a high-gain op amp. The voltage at the collector of Q6 (the output of the op amp) will be whatever it takes to maintain 2.5 volts at the base of Q6 (the negative input of the op amp) to balance the 2.5 volts at the base of Q5 (the positive input of the op amp). The 2.5 volts at the base of Q6 is scaled up to 5 volts by Q13, R4, R5, Q14 and the pass-transistor combination Q11-Q12. The voltages across Q13 and Q14 are used to bias other current sources in other sections of the chip. R10 and Q15 provide current limiting for the reference circuit.

The SG1524 oscillator circuit is shown in Fig. 7-12. The oscillator consists of two subcircuits: a ramp generator, and a resistor circuit. The ramp generator is formed by Q1-Q2, an external resistor R_T, and an external capacitor C_T. R_T sets the current in Q1 which is mirrored over to Q2 and flows through Q3 to charge the capacitor C_T at a rate equal to

$$dV/dT = I/C_T = (5 - V_{BE2} - V_{BE3})/(R_T C_T)$$

where "dV/dT" means the change in voltage across the capacitor divided by the change in time.

The reset circuit is composed of a comparator (Q4 – Q13), a pair of transistors (Q15, Q16) to discharge the capacitor C_T, and a hysteresis transistor (Q14). (Note that bias for the Q10 current source comes from the reference circuit section shown in Fig. 7-11.)

Assume that Q14 – Q16 are off and that the capacitor is charging. The capacitor charges at a linear rate according to the above equation until it reaches the reference voltage of about 3.5 volts established by the resistor divider R1-R2 at the base of Q9. Then Q8 begins to turn on. The current in Q8 is mirrored into Q5, which is greater than the current in Q12, which is approximately equal to the current in Q9 (which is turning off). The Q5 current minus the Q12 current begins to flow into Q13, Q14, and Q15.

When Q14 turns on, the reference at the base of Q9 drops to about 2 volts. This action causes the comparator to switch com-

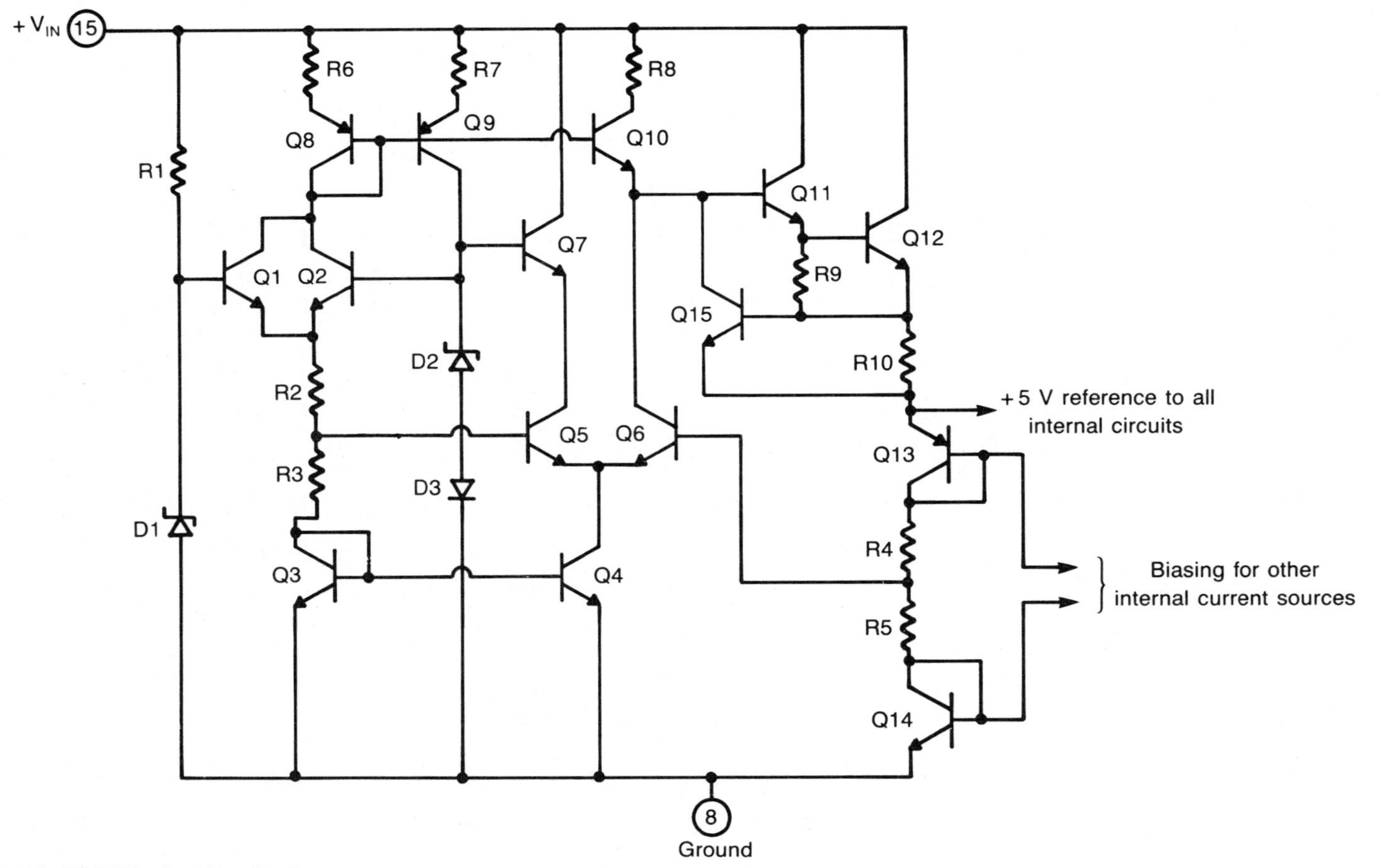

Fig. 7-11. SG1524 reference circuit.

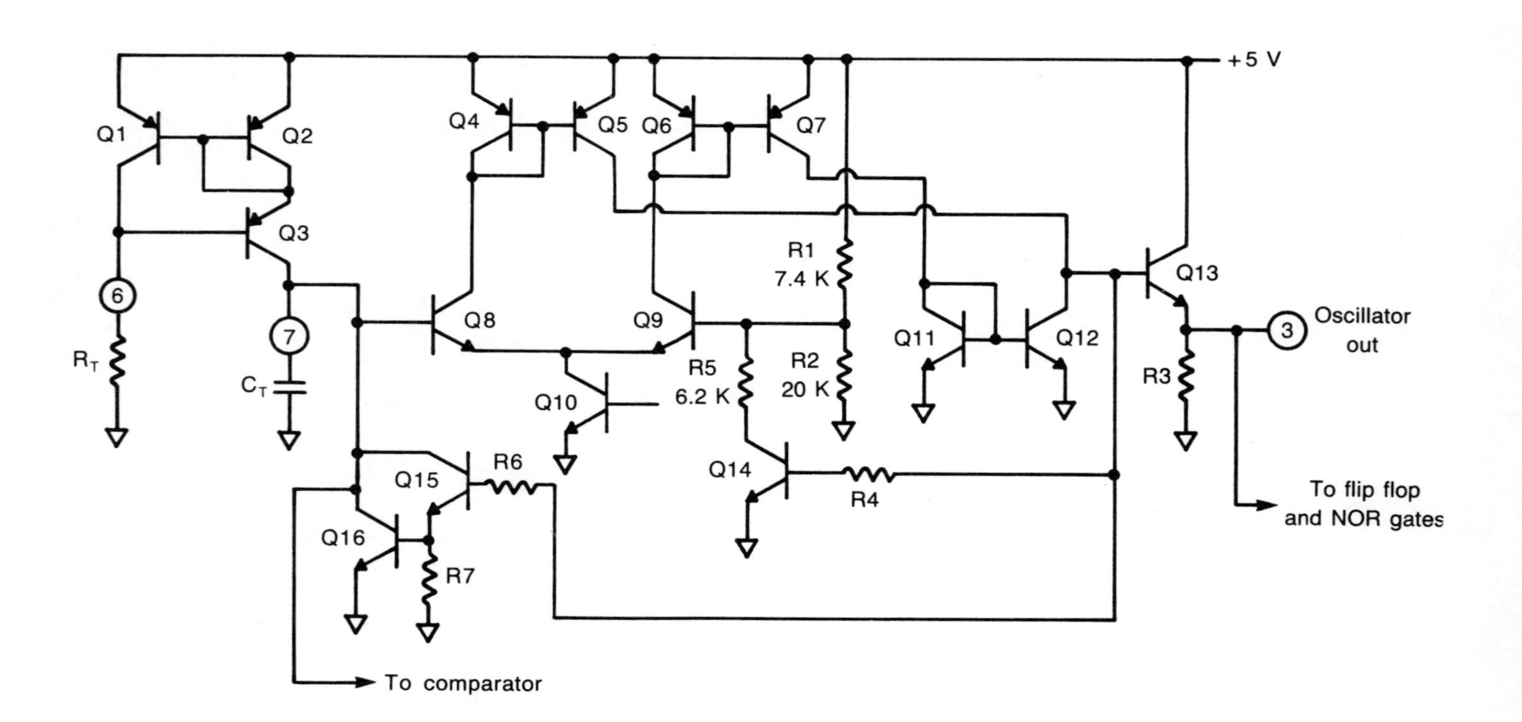

Fig. 7-12. SG1524 oscillator circuit.

pletely (if it hasn't already). By this time the Darlington Q15-Q16 is discharging the capacitor down to the saturation voltage of Q15 plus the V_{BE} of Q16, or about 1 volt. Of course, as soon as the base of Q8 is below 2 volts, the comparator has switched back, Q8 is off, Q9 is on, Q14-Q15 are off. All this happens in about 200 nanoseconds. Then the capacitor starts charging all over again. The result is a sawtooth waveform across the capacitor and pulse waveform at the collector of Q12. The pulse is due to the finite time it takes to discharge the capacitor and the time it takes for the comparator to recover. The pulse waveform is buffered by Q13 and sent to the flip-flop and NOR gates as well as to pin 3 to the outside world. These waveforms are shown in Fig. 7-13.

The SG1524 error amplifier is shown in Fig. 7-14. This is a simple op amp where the Q3 current is twice the current in Q5-Q7 (and mirrored into Q4). The output of this amplifier is connected to pin 9 (called the "compensation" pin, because open-loop gain and frequency compensation components may be connected to the error amplifier at that node) and one side of the comparator shown in Fig. 7-15. The error amp output voltage may be anywhere between 1 volt and 3.5 volts, the upper and lower values of the sawtooth oscillator voltage, which is applied to the other side of the comparator. The output of the comparator is a constant-frequency pulse waveform with a duty cycle determined by the error amplifier voltage, which in turn will be determined by the external circuit configuration of the overall switchmode regulator. The error amplifier senses changes in the overall regulator output voltage and adjusts its (the error amp's) output voltage in order to decrease or increase the pulse-width output of the comparator to bring the regulator output back to the desired value.

The SG1524 NOR gate and output stage is shown in Fig. 7-16.

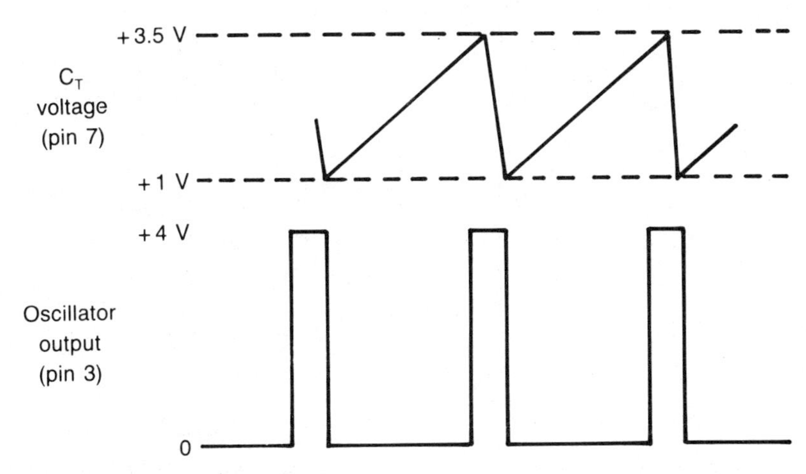

Fig. 7-13. SG1524 oscillator waveforms.

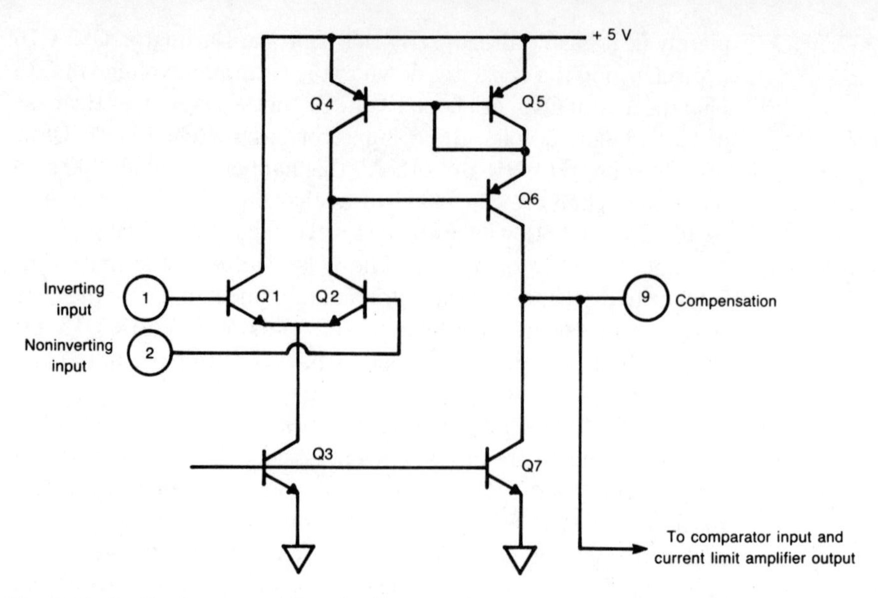

Fig. 7-14. SG1524 error amplifier circuit.

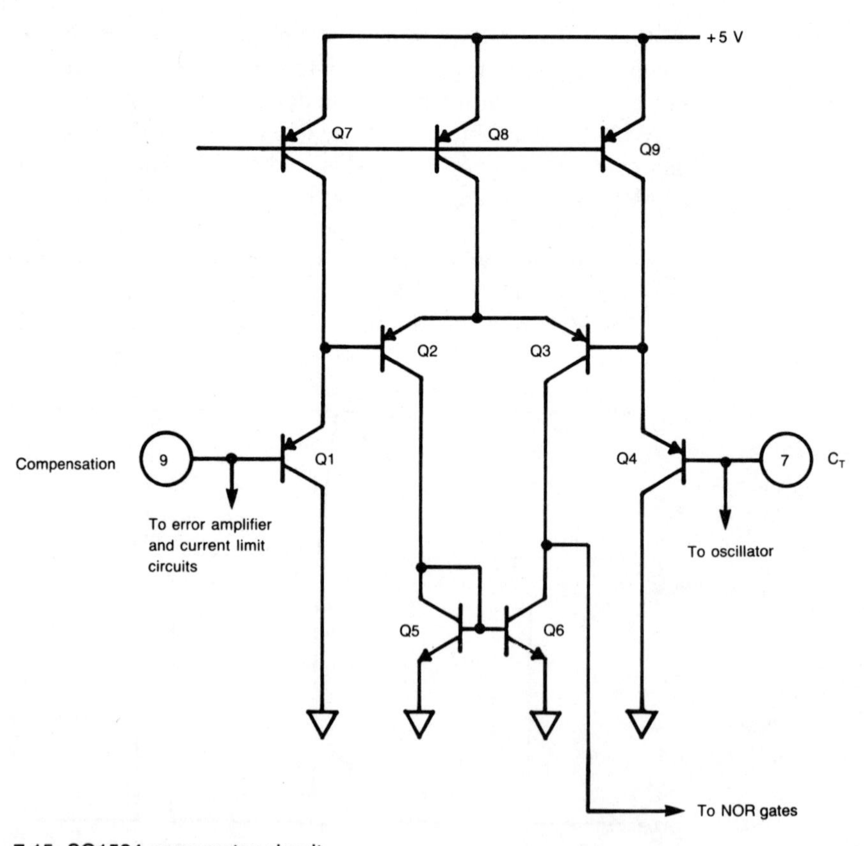

Fig. 7-15. SG1524 comparator circuit.

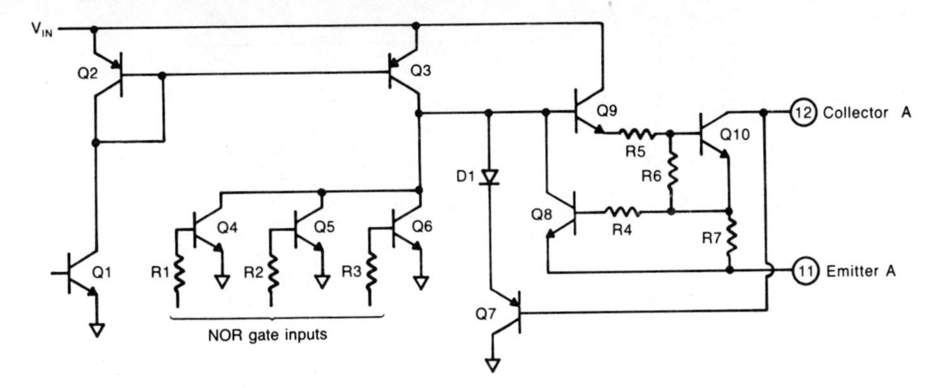

Fig. 7-16. SG1524 NOR gate and output stage.

The NOR gate is a resistor-transistor logic (RTL) gate with a constant-current source load (Q3). The output of this gate is connected to the base of an output transistor, which, in fact, is a circuit in itself. The output "transistor" is composed of transistors Q7-Q10, resistors R4-R7, and diode D1.

Q9 and Q10 are the basic output Darlington transistor. Q8 provides current limiting. D1 and Q7 are used to keep Q10 from saturating. When the voltage drop from the base of Q9 to the collector of Q10 approaches about 1.4 volts then Q7 begins to turn on and robs base current from Q9, thus keeping Q10 from going into saturation. This improves the speed of the output stage.

The SG1524 current-limit circuit is shown in Fig. 7-17. As-

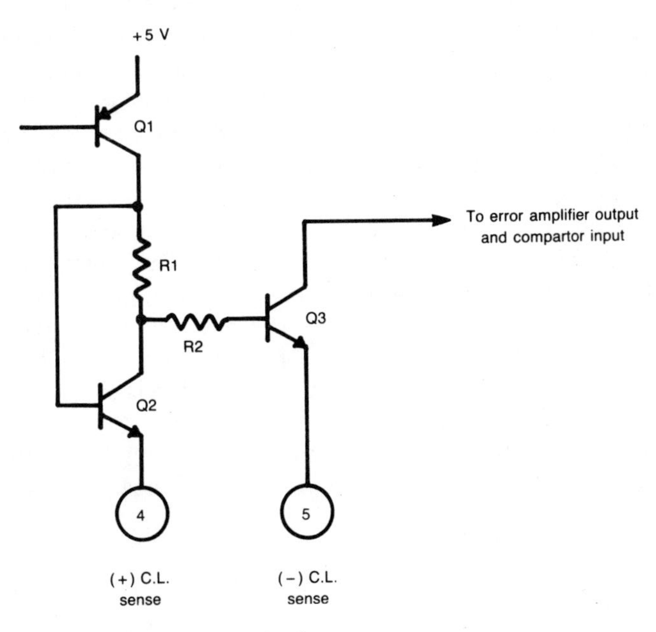

Fig. 7-17. SG1524 current limit circuit.

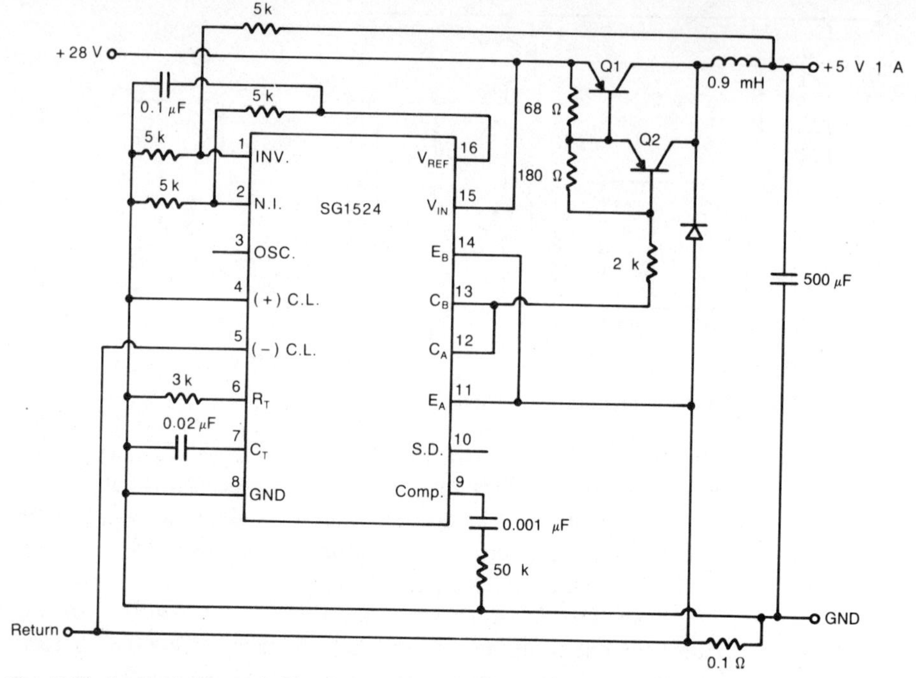

Fig. 7-18. 5-volt, 1-amp switching regulator using the SG1524.

sume that pin 4 is grounded. Q2 operates at constant current. If pin 5 is at ground potential, no current will flow in Q3 due to the voltage drop (about 200 mV according to the manufacturer's literature) across R1. Thus, current will flow in Q3 when the emitter of Q3 is about 200 mV below ground. When this happens, Q3 begins to divert current from the error amplifier and will shutdown the pulse-width modulator if the Q3 emitter voltage continues to go negative. This is an optional circuit that may be used by connecting pins 4 and 5 across a load current-sensing resistor.

Figure 7-18 shows a typical application of the SG1524. In this application, both outputs are connected in parallel and drive an external high-current Darlington transistor pair. The noninverting side of the error amplifier is biased at 2.5 V, V_{REF} divided down by two 5 kΩ resistors. The inverting side is connected to two other 5 kΩ resistors that sense changes in the output voltage. A 0.1-ohm resistor between the 5 V ground and the 28 V return is used to monitor the load current.

Selection guides for switchmode power-supply controller ICs are given in Tables 7-20 through 7-22. Additional information is supplied in Appendix D.

Table 7-1. +5-Volt Regulators, Commercial Temperature Range.

PART NUMBER	MANUFACTURER	OUTPUT VOLTAGE (V) MIN	MAX	MAXIMUM OUTPUT CURRENT (A)	UNIT PRICE ($)
UA78L05C	FAIRCHILD	4.8	5.2	0.1	0.40
LM78L05AC	NATIONAL	4.8	5.2	0.1	0.43
LM78L05C	NATIONAL	4.6	5.4	0.1	0.43
MC78L05C	MOTOROLA	4.6	5.4	0.1	0.52
MC78L05AC	MOTOROLA	4.8	5.2	0.1	0.56
LM342-5	NATIONAL	4.8	5.2	0.25	0.68
LM340L-5	NATIONAL	4.9	5.1	0.1	0.75
LM341-5	NATIONAL	4.8	5.2	0.5	0.75
LM78M05C	NATIONAL	4.8	5.2	0.5	0.75
UA78M05C	FAIRCHILD	4.8	5.2	0.5	0.76
UA7805C	FAIRCHILD	4.8	5.2	1.5	0.86
MC78M05C	MOTOROLA	4.8	5.2	1.5	0.86
MC7805C	MOTOROLA	4.8	5.2	1.5	0.95
LM340T-5	NATIONAL	4.8	5.2	1.5	0.95
LM7805C	NATIONAL	4.8	5.2	1.5	0.95
LM330	NATIONAL	4.8	5.2	0.15	1.00
TL780-05C	T I	4.95	5.05	1.5	1.03
MC7805AC	MOTOROLA	4.9	5.1	1.5	1.14
LM340AT-5	NATIONAL	4.9	5.1	1.5	1.50
UA309	FAIRCHILD	4.8	5.2	1.5	2.09
LM340-5	NATIONAL	4.8	5.2	1.5	2.10
LM309K	NATIONAL	4.8	5.2	1.5	2.20
LM309H	NATIONAL	4.8	5.2	0.5	3.10
LM340A-5	NATIONAL	4.9	5.1	1.5	3.65
MC78T05C	MOTOROLA	4.8	5.2	3	5.15
MC78T05AC	MOTOROLA	4.9	5.1	3	5.70
LM323	NATIONAL	4.7	5.3	3	6.95
UA78H05C	FAIRCHILD	4.85	5.25	5	7.81
UA78H05AC	FAIRCHILD	4.85	5.25	5	8.23
UA78P05C	FAIRCHILD	4.85	5.25	10	14.38

Table 7-2. +5-Volt Regulators, Industrial Temperature Range.

PART NUMBER	MANUFACTURER	OUTPUT VOLTAGE (V) MIN	MAX	MAXIMUM OUTPUT CURRENT (A)	UNIT PRICE ($)
LM2930-5	NATIONAL	4.5	5.5	0.15	1.15
LM2931A-5	NATIONAL	4.75	5.25	0.15	1.15
MC78M05B	MOTOROLA	4.8	5.2	0.5	1.23
MC7805B	MOTOROLA	4.8	5.2	1.5	1.27
LM209K	NATIONAL	4.7	5.3	1.5	5.75
LM209H	NATIONAL	4.7	5.3	0.5	11.25
LM223	NATIONAL	4.7	5.3	3	15.35

Table 7-3. +5-Volt Regulators, Military Temperature Range.

PART NUMBER	MANUFACTURER	OUTPUT VOLTAGE (V) MIN	MAX	MAXIMUM OUTPUT CURRENT (A)	UNIT PRICE ($)
LM140L-5	NATIONAL	4.9	5.1	0.1	3.00
UA78M05A	FAIRCHILD	4.8	5.2	0.5	3.80
UA7805	FAIRCHILD	4.8	5.2	1.5	5.70
UA109	FAIRCHILD	4.7	5.3	1.5	6.65
LM140-5	NATIONAL	4.8	5.2	1.5	8.45
MC7805	MOTOROLA	4.8	5.2	1.5	11.69
LM109H	NATIONAL	4.7	5.3	0.5	13.00
LM140A-5	NATIONAL	4.9	5.1	1.5	13.45
MC7805A	MOTOROLA	4.9	5.1	1.5	13.58
LM109K	NATIONAL	4.7	5.3	1.5	15.40
MC78T05	MOTOROLA	4.8	5.2	3	26.50
LM123	NATIONAL	4.7	5.3	3	30.75
MC78T05A	MOTOROLA	4.9	5.1	3	31.25
UA78H05	FAIRCHILD	4.85	5.25	5	43.23
UA78H05A	FAIRCHILD	4.85	5.25	5	47.92
UA78P05	FAIRCHILD	4.85	5.25	10	58.33

Table 7-4. +6-Volt Regulators.

PART NUMBER	MANUFACTURER	TEMP RANGE	OUTPUT VOLTAGE (V) MIN	MAX	MAXIMUM OUTPUT CURRENT (A)	UNIT PRICE ($)
UA78M06C	FAIRCHILD	COM	5.75	6.25	0.5	0.76
MC78M06C	MOTOROLA	COM	5.75	6.25	0.5	0.86
UA7806C	FAIRCHILD	COM	5.75	6.25	1.5	0.86
MC7806C	MOTOROLA	COM	5.75	6.25	1.5	0.95
MC7806AC	MOTOROLA	COM	5.88	6.12	1.5	1.14
MC78M06B	MOTOROLA	IND	5.75	6.25	0.5	1.23
MC7806B	MOTOROLA	IND	5.75	6.25	1.5	1.27
UA78M06	FAIRCHILD	MIL	5.75	6.25	0.5	3.80
MC78T06C	MOTOROLA	COM	5.75	6.25	3	6.85
MC7806	MOTOROLA	MIL	5.75	6.25	1.5	11.69
MC78T06	MOTOROLA	MIL	5.75	6.25	3	26.50

Table 7-5. +8-Volt Regulators.

PART NUMBER	MANUFACTURER	TEMP RANGE	OUTPUT VOLTAGE (V) MIN	MAX	MAXIMUM OUTPUT CURRENT (A)	UNIT PRICE ($)
MC78L08C	MOTOROLA	COM	7.36	8.84	0.1	0.52
MC78L08AC	MOTOROLA	COM	7.7	8.8	0.1	0.56
UA78M08C	FAIRCHILD	COM	7.7	8.3	0.5	0.76
UA7808C	FAIRCHILD	COM	7.7	8.3	1.5	0.86
MC78M08C	MOTOROLA	COM	7.7	8.8	0.5	0.86
MC7808C	MOTOROLA	COM	7.7	8.3	1.5	0.95
MC7808AC	MOTOROLA	COM	7.84	8.16	1.5	1.14
LM2930-8	NATIONAL	IND	7.2	8.8	0.15	1.15
MC78M08B	MOTOROLA	IND	7.7	8.8	0.5	1.23
MC7808B	MOTOROLA	IND	7.7	8.3	1.5	1.27
UA78M08	FAIRCHILD	MIL	7.7	8.3	0.5	3.80
MC78T08C	MOTOROLA	COM	7.7	8.8	3	5.15
UA7808	FAIRCHILD	MIL	7.7	8.3	1.5	5.70
MC7808	MOTOROLA	MIL	7.7	8.8	1.5	11.69
MC7808A	MOTOROLA	MIL	7.84	8.16	1.5	13.58
MC78T08	MOTOROLA	MIL	7.7	8.8	3	26.50

Table 7-6. +12-Volt Regulators.

PART NUMBER	MANUFACTURER	TEMP RANGE	OUTPUT VOLTAGE (V) MIN	MAX	MAXIMUM OUTPUT CURRENT (A)	UNIT PRICE ($)
UA78L12C	FAIRCHILD	COM	11.5	12.5	0.1	0.40
LM78L12AC	NATIONAL	COM	11.5	12.5	0.1	0.43
LM78L12C	NATIONAL	COM	11.1	12.9	0.1	0.43
MC78L12C	MOTOROLA	COM	11.1	12.9	0.1	0.52
MC78L12AC	MOTOROLA	COM	11.5	12.5	0.1	0.56
LM342-12	NATIONAL	COM	11.5	12.5	0.25	0.68
LM341-12	NATIONAL	COM	11.5	12.5	0.5	0.75
LM340L-12	NATIONAL	COM	11.75	12.25	0.1	0.75
LM78M12C	NATIONAL	COM	11.5	12.5	0.5	0.75
UA78M12C	FAIRCHILD	COM	11.5	12.5	0.5	0.76
UA7812C	FAIRCHILD	COM	11.5	12.5	1.5	0.86
MC78M12C	MOTOROLA	COM	11.5	12.5	0.5	0.86
MC7812C	MOTOROLA	COM	11.5	12.5	1.5	0.95
LM7812C	NATIONAL	COM	11.5	12.5	1.5	0.95
LM340T-12	NATIONAL	COM	11.5	12.5	1.5	0.95
TL780-12C	T I	COM	11.88	12.12	1.5	1.03
MC7812AC	MOTOROLA	COM	11.75	12.25	1.5	1.14
MC78M12B	MOTOROLA	IND	11.5	12.5	0.5	1.23
MC7812B	MOTOROLA	IND	11.5	12.5	1.5	1.27
LM340AT-12	NATIONAL	COM	11.75	12.25	1.5	1.50
LM340-12	NATIONAL	COM	11.5	12.5	1.5	2.10
LM140L-12	NATIONAL	MIL	11.75	12.25	0.1	3.00
LM340A-12	NATIONAL	COM	11.75	12.25	1.5	3.65
UA78M12	FAIRCHILD	MIL	11.5	12.5	0.5	3.80
MC78T12C	MOTOROLA	COM	11.5	12.5	3	5.15
UA7812	FAIRCHILD	MIL	11.5	12.5	1.5	5.70
MC78T12AC	MOTOROLA	COM	11.75	12.25	3	5.70
LM140-12	NATIONAL	MIL	11.5	12.5	1.5	8.45
UA78H12AC	FAIRCHILD	COM	11.5	12.5	5	8.75
MC7812	MOTOROLA	MIL	11.5	12.5	1.5	11.69

PART NUMBER	MANUFACTURER	TEMP RANGE	OUTPUT VOLTAGE (V) MIN	MAX	MAXIMUM OUTPUT CURRENT (A)	UNIT PRICE ($)
LM140A-12	NATIONAL	MIL	11.75	12.25	1.5	13.45
MC7812A	MOTOROLA	MIL	11.75	12.25	1.5	13.58
MC78T12	MOTOROLA	MIL	11.5	12.5	3	26.50
MC78T12A	MOTOROLA	MIL	11.75	12.25	3	31.25
UA78H12A	FAIRCHILD	MIL	11.5	12.5	5	47.92

Table 7-7. +15-Volt Regulators.

PART NUMBER	MANUFACTURER	TEMP RANGE	OUTPUT VOLTAGE (V) MIN	MAX	MAXIMUM OUTPUT CURRENT (A)	UNIT PRICE ($)
UA78L15C	FAIRCHILD	COM	14.4	15.6	0.1	0.40
LM78L15AC	NATIONAL	COM	14.4	15.6	0.1	0.43
LM78L15C	NATIONAL	COM	13.8	16.2	0.1	0.43
MC78L15C	MOTOROLA	COM	13.8	16.2	0.1	0.52
MC78L15AC	MOTOROLA	COM	14.4	15.6	0.1	0.56
LM342-15	NATIONAL	COM	14.4	15.6	0.25	0.68
LM341-15	NATIONAL	COM	14.4	15.6	0.5	0.75
LM340L-15	NATIONAL	COM	14.7	15.3	0.1	0.75
UA78M15C	FAIRCHILD	COM	14.4	15.6	0.5	0.76
UA7815C	FAIRCHILD	COM	14.4	15.6	1.5	0.86
MC78M15C	MOTOROLA	COM	14.4	15.6	0.5	0.86
MC7815C	MOTOROLA	COM	14.4	15.6	1.5	0.95
LM340T-15	NATIONAL	COM	14.4	15.6	1.5	0.95
LM7815C	NATIONAL	COM	14.4	15.6	1.5	0.95
TL780-15C	T I	COM	14.85	15.15	1.5	1.03
MC7815AC	MOTOROLA	COM	14.7	15.3	1.5	1.14
MC78M15B	MOTOROLA	IND	14.4	15.6	0.5	1.23
MC7815B	MOTOROLA	IND	14.4	15.6	1.5	1.27
LM340AT-15	NATIONAL	COM	14.7	15.3	1.5	1.50
LM340-15	NATIONAL	COM	14.4	15.6	1.5	2.10
LM140L-15	NATIONAL	MIL	14.7	15.3	0.1	3.00
LM340A-15	NATIONAL	COM	14.7	15.3	1.5	3.65

PART NUMBER	MANUFACTURER	TEMP RANGE	OUTPUT VOLTAGE (V) MIN	MAX	MAXIMUM OUTPUT CURRENT (A)	UNIT PRICE ($)
UA78M15	FAIRCHILD	MIL	14.4	15.6	0.5	3.80
MC78T15C	MOTOROLA	COM	14.4	15.6	3	5.15
MC78T15AC	MOTOROLA	COM	14.7	15.3	3	5.70
UA7815	FAIRCHILD	MIL	14.4	15.6	1.5	5.70
LM140-15	NATIONAL	MIL	14.4	15.6	1.5	8.45
MC7815	MOTOROLA	MIL	14.4	15.6	1.5	11.69
LM140A-15	NATIONAL	MIL	14.4	15.6	1.5	13.45
MC7815A	MOTOROLA	MIL	14.7	15.3	1.5	13.58
MC78T15	MOTOROLA	MIL	14.4	15.6	3	26.50
MC78T15A	MOTOROLA	MIL	14.7	15.3	3	31.25

Table 7-8. +18-Volt Regulators.

PART NUMBER	MANUFACTURER	TEMP RANGE	OUTPUT VOLTAGE (V) MIN	MAX	MAXIMUM OUTPUT CURRENT (A)	UNIT PRICE ($)
MC78L18C	MOTOROLA	COM	16.6	19.4	0.1	0.52
MC78L18AC	MOTOROLA	COM	17.3	18.7	0.1	0.56
UA7818C	FAIRCHILD	COM	17.3	18.7	1.5	0.86
MC78M18C	MOTOROLA	COM	17.3	18.7	0.5	0.86
MC7818C	MOTOROLA	COM	17.3	18.7	1.5	0.95
MC7818AC	MOTOROLA	COM	17.64	18.36	1.5	1.14
MC78M18B	MOTOROLA	IND	17.3	18.7	0.5	1.23
MC7818B	MOTOROLA	IND	17.3	18.7	1.5	1.27
MC78T18C	MOTOROLA	COM	17.3	18.7	3	5.15
UA7818	FAIRCHILD	MIL	17.3	18.7	1.5	5.70
MC7818	MOTOROLA	MIL	17.3	18.7	1.5	11.69
MC7818A	MOTOROLA	MIL	17.64	18.36	1.5	13.58
MC78T18	MOTOROLA	MIL	17.3	18.7	3	26.50

Table 7-9. +24-Volt Regulators.

PART NUMBER	MANUFACTURER	TEMP RANGE	OUTPUT VOLTAGE (V) MIN	MAX	MAXIMUM OUTPUT CURRENT (A)	UNIT PRICE ($)
MC78L24C	MOTOROLA	COM	22.1	25.9	0.1	0.52
MC78L24AC	MOTOROLA	COM	23	25	0.1	0.56
UA7824C	FAIRCHILD	COM	23	25	1.5	0.86
MC78M24C	MOTOROLA	COM	23	25	0.5	0.86
MC7824C	MOTOROLA	COM	23	25	1.5	0.95
MC7824AC	MOTOROLA	COM	23.5	24.5	1.5	1.14
MC78M24B	MOTOROLA	IND	23	25	0.5	1.23
MC7824B	MOTOROLA	IND	23	25	1.5	1.27
UA7824	FAIRCHILD	MIL	23	25	1.5	5.70
MC78T24C	MOTOROLA	COM	23	25	3	5.70
MC7824	MOTOROLA	MIL	23	25	1.5	11.69
MC7824A	MOTOROLA	MIL	23.5	24.5	1.5	13.58
MC78T24	MOTOROLA	MIL	23	25	3	26.50

Table 7-10. Adjustable Positive Voltage Regulators, Commercial Temperature Range.

PART NUMBER	MANUFACTURER	OUTPUT VOLTAGE (V) MIN	MAX	MAXIMUM OUTPUT CURRENT (A)	UNIT PRICE ($)
UA723C	FAIRCHILD	2	37	0.05	0.49
MC1723C	MOTOROLA	2	37	0.05	0.54
LM376	NATIONAL	5	37	0.025	0.56
UA376	FAIRCHILD	5	37	0.025	0.57
CA723C	RCA	2	37	0.05	0.60
LM723C	NATIONAL	2	37	0.05	0.64
LM317LZ	NATIONAL	1.2	37	0.1	0.75
UA305	FAIRCHILD	4.5	40	0.012	0.76
TL317C	T I	2	37	0.1	0.83
UA78MGC	FAIRCHILD	5	30	0.5	0.95
LM305	NATIONAL	4.5	30	0.012	1.00
LM317M	NATIONAL	1.2	37	0.5	1.05

PART NUMBER	MANUFACTURER	OUTPUT VOLTAGE (V) MIN	MAX	MAXIMUM OUTPUT CURRENT (A)	UNIT PRICE ($)
UA78GC	FAIRCHILD	5	30	1.5	1.14
LM317T	NATIONAL	1.2	37	1.5	1.50
UA317	FAIRCHILD	1.2	37	1.5	1.52
UA305A	FAIRCHILD	4.5	40	0.045	1.79
LM317K	NATIONAL	1.2	37	1.5	2.88
TL783C	T I	1.25	125	0.7	2.94
LM305A	NATIONAL	4.5	30	0.045	3.40
LM317H	NATIONAL	1.2	37	0.5	4.18
MC1469G	MOTOROLA	2.5	32	0.2	4.30
LM350T	NATIONAL	1.2	33	3	4.75
LM350K	NATIONAL	1.2	33	3	5.50
MC1469R	MOTOROLA	2.5	32	0.5	5.65
LM317HVH	NATIONAL	1.2	37	0.5	6.20
LM317HVK	NATIONAL	1.2	37	1.5	6.20
UA78HGAC	FAIRCHILD	5	24	5	9.90
LM396	NATIONAL	1.25	15	10	18.50

Table 7-11. Adjustable Positive Voltage Regulators, Industrial Temperature Range.

PART NUMBER	MANUFACTURER	OUTPUT VOLTAGE (V) MIN	MAX	MAXIMUM OUTPUT CURRENT (A)	UNIT PRICE ($)
SA723C	SIGNETICS	2	37	0.05	0.76
LM2931T	NATIONAL	3	24	0.15	1.75
LM205	NATIONAL	4.5	40	0.012	4.70
LM217H	NATIONAL	1.2	37	0.5	11.00
LM217K	NATIONAL	1.2	37	1.5	11.50
LM250	NATIONAL	1.2	33	3	17.28
LM217HVK	NATIONAL	1.2	57	1.5	20.49
LM238	NATIONAL	1.2	32	5	22.75

Table 7-12. Adjustable Positive Voltage Regulators, Military Temperature Range.

PART NUMBER	MANUFACTURER	OUTPUT VOLTAGE (V) MIN	MAX	MAXIMUM OUTPUT CURRENT (A)	UNIT PRICE ($)
CA723	RCA	2	37	0.05	0.74
CA3085	RCA	1.8	26	0.012	0.94
CA3085A	RCA	1.7	36	0.1	1.20
UA723	FAIRCHILD	2	37	0.05	1.71
MC1723	MOTOROLA	2	37	0.05	2.00
UA105	FAIRCHILD	4.5	40	0.012	2.00
LM723	NATIONAL	2	37	0.05	3.20
LM105	NATIONAL	4.5	40	0.012	3.85
UA78G	FAIRCHILD	5	30	1.5	5.70
CA3085B	RCA	1.7	46	0.1	6.72
MC1569G	MOTOROLA	2.5	37	0.2	6.88
UA117	FAIRCHILD	1.2	37	1.5	7.98
MC1569R	MOTOROLA	2.5	37	0.5	10.49
LM117H	NATIONAL	1.2	27	0.5	15.00
LM117K	NATIONAL	1.2	37	1.5	16.00
LM117HVH	NATIONAL	1.2	57	0.5	30.00
LM117HVK	NATIONAL	1.2	57	1.5	30.00
LM150	NATIONAL	1.2	33	3	34.60
UA78HGA	FAIRCHILD	5	24	5	47.92
LM138	NATIONAL	1.2	32	5	48.75
LM196	NATIONAL	1.25	15	10	90.00

Table 7-13. −5-Volt Regulators.

PART NUMBER	MANUFACTURER	TEMP RANGE	OUTPUT VOLTAGE (V) MIN	MAX	MAXIMUM OUTPUT CURRENT (A)	UNIT PRICE ($)
MC79L05C	MOTOROLA	COM	-4.6	-5.4	0.1	0.73
UA79M05C	FAIRCHILD	COM	-4.8	-5.2	0.5	0.76
MC79L05AC	MOTOROLA	COM	-4.8	-5.2	0.1	0.82
UA7905C	FAIRCHILD	COM	-4.8	-5.2	1.5	0.86
MC7905C	MOTOROLA	COM	-4.8	-5.2	1.5	0.95

PART NUMBER	MANUFACTURER		OUTPUT VOLTAGE (V) MIN	MAX	MAXIMUM OUTPUT CURRENT (A)	UNIT PRICE ($)
LM320L-5	NATIONAL	COM	-4.8	-5.2	0.1	1.00
LM79L05AC	NATIONAL	COM	-4.8	-5.2	0.1	1.00
LM7905C	NATIONAL	COM	-4.8	-5.2	1.5	1.08
LM79M05C	NATIONAL	COM	-4.8	-5.2	0.5	1.19
LM320ML-5	NATIONAL	COM	-4.8	-5.2	0.25	1.20
LM320MP-5	NATIONAL	COM	-4.8	-5.2	0.5	1.85
LM320T-5	NATIONAL	COM	-4.8	-5.2	1.5	2.49
LM320K-5	NATIONAL	COM	-4.8	-5.2	1.5	3.55
UA79M05	FAIRCHILD	MIL	-4.8	-5.2	0.5	3.80
LM320H-5	NATIONAL	COM	-4.8	-5.2	0.5	5.50
UA7905	FAIRCHILD	MIL	-4.8	-5.2	1.5	5.70
LM345-5	NATIONAL	COM	-4.8	-5.2	3	9.35
LM120H-5	NATIONAL	MIL	-4.9	-5.1	0.5	14.20
LM120K-5	NATIONAL	MIL	-4.9	-5.1	1.5	14.95
LM145-5	NATIONAL	MIL	-4.9	-5.1	3	52.50

Table 7-14. −8-Volt Regulators.

PART NUMBER	MANUFACTURER	TEMP RANGE	OUTPUT VOLTAGE (V) MIN	MAX	MAXIMUM OUTPUT CURRENT (A)	UNIT PRICE ($)
UA79M08C	FAIRCHILD	COM	-7.7	-8.3	0.5	0.76
UA7908C	FAIRCHILD	COM	-7.7	-8.3	1.5	0.86
MC7908C	MOTOROLA	COM	-7.7	-8.3	1.5	0.95
UA79M08	FAIRCHILD	MIL	-7.7	-8.3	0.5	3.80
UA7908	FAIRCHILD	MIL	-7.7	-8.3	1.5	5.70

Table 7-15. −12-Volt Regulators.

PART NUMBER	MANUFACTURER	TEMP RANGE	OUTPUT VOLTAGE (V) MIN	MAX	MAXIMUM OUTPUT CURRENT (A)	UNIT PRICE ($)
MC79L12C	MOTOROLA	COM	-11.1	-12.9	0.1	0.73
UA79M12C	FAIRCHILD	COM	-11.5	-12.5	0.5	0.76
MC79L12AC	MOTOROLA	COM	-11.5	-12.5	0.1	0.82

PART NUMBER	MANUFACTURER	TEMP RANGE	OUTPUT VOLTAGE (V) MIN	MAX	MAXIMUM OUTPUT CURRENT (A)	UNIT PRICE ($)
UA7912C	FAIRCHILD	COM	-11.5	-12.5	1.5	0.86
MC7912C	MOTOROLA	COM	-11.5	-12.5	1.5	0.95
LM320L-12	NATIONAL	COM	-11.5	-12.5	0.1	1.00
LM79L12AC	NATIONAL	COM	-11.5	-12.5	0.1	1.00
LM7912C	NATIONAL	COM	-11.5	-12.5	1.5	1.08
LM79M12C	NATIONAL	COM	-11.5	-12.5	0.5	1.19
LM320ML-12	NATIONAL	COM	-11.5	-12.5	0.25	1.20
LM320MP-12	NATIONAL	COM	-11.5	-12.5	0.5	1.85
LM320T-12	NATIONAL	COM	-11.6	-12.4	1.5	2.49
LM320K-12	NATIONAL	COM	-11.6	-12.4	1.5	3.55
UA79M12	FAIRCHILD	MIL	-11.5	-12.5	0.5	3.80
LM320H-12	NATIONAL	COM	-11.6	-12.4	0.5	5.50
UA7912	FAIRCHILD	MIL	-11.5	-12.5	1.5	5.70
LM120H-12	NATIONAL	MIL	-11.7	-12.3	0.5	14.20
LM120K-12	NATIONAL	MIL	-11.7	-12.3	1.5	14.95

Table 7-16. – 15-Volt Regulators.

PART NUMBER	MANUFACTURER	TEMP RANGE	OUTPUT VOLTAGE (V) MIN	MAX	MAXIMUM OUTPUT CURRENT (A)	UNIT PRICE ($)
MC79L15C	MOTOROLA	COM	-13.8	-16.2	0.1	0.73
UA79M15C	FAIRCHILD	COM	-14.4	-15.6	0.5	0.76
MC79L15AC	MOTOROLA	COM	-14.4	-15.6	0.1	0.82
UA7915C	FAIRCHILD	COM	-14.4	-15.6	1.5	0.86
MC7915C	MOTOROLA	COM	-14.4	-15.6	1.5	0.95
LM320L-15	NATIONAL	COM	-14.4	-15.6	0.1	1.00
LM79L15AC	NATIONAL	COM	-14.4	-15.6	0.1	1.00
LM7915C	NATIONAL	COM	-14.4	-15.6	1.5	1.08
LM79M15C	NATIONAL	COM	-14.4	-15.6	0.5	1.19
LM320ML-15	NATIONAL	COM	-14.4	-15.6	0.25	1.20
LM320MP-15	NATIONAL	COM	-14.4	-15.6	0.5	1.85
LM320T-15	NATIONAL	COM	-14.5	-15.5	1.5	2.49
LM320K-15	NATIONAL	COM	-14.4	-15.6	1.5	3.55

PART NUMBER	MANUFACTURER	TEMP RANGE	OUTPUT VOLTAGE (V) MIN	MAX	MAXIMUM OUTPUT CURRENT (A)	UNIT PRICE ($)
UA79M15	FAIRCHILD	MIL	-14.4	-15.6	0.5	3.80
LM320H-15	NATIONAL	COM	-14.6	-15.4	0.5	5.50
UA7915	FAIRCHILD	MIL	-14.4	-15.6	1.5	5.70
LM120H-15	NATIONAL	MIL	-14.7	-15.3	0.5	14.20
LM120K-15	NATIONAL	MIL	-14.7	-15.3	1.5	14.95

Table 7-17. Adjustable Negative Voltage Regulators.

PART NUMBER	MANUFACTURER	TEMP RANGE	OUTPUT VOLTAGE (V) MIN	MAX	MAXIMUM OUTPUT CURRENT (A)	UNIT PRICE ($)
UA79MGC	FAIRCHILD	COM	-2.23	-30	0.5	0.95
LM337LZ	NATIONAL	COM	-1.2	-37	0.1	1.05
UA79GC	FAIRCHILD	COM	-2.23	-30	1.5	1.14
LM337M	NATIONAL	COM	-1.2	-37	0.5	1.15
LM337T	NATIONAL	COM	-1.2	-37	1.5	1.95
LM304	NATIONAL	COM	-0.035	-30	0.02	3.50
MC1463G	MOTOROLA	COM	-3.8	-32	0.2	4.32
LM337H	NATIONAL	COM	-1.2	-37	0.5	4.50
LM337K	NATIONAL	COM	-1.2	-37	1.5	5.20
UA79G	FAIRCHILD	MIL	-2.23	-30	1.5	5.70
LM204	NATIONAL	IND	-0.015	-40	0.02	6.90
MC1463R	MOTOROLA	COM	-3.8	-32	0.5	7.70
LM337HVH	NATIONAL	COM	-1.2	-47	0.5	7.95
MC1563G	MOTOROLA	MIL	-3.6	-37	0.2	8.08
LM104	NATIONAL	MIL	-0.015	-40	0.02	9.65
LM337HVK	NATIONAL	COM	-1.2	-47	1.5	10.00
LM237H	NATIONAL	IND	-1.2	-37	0.5	12.45
MC1563R	MOTOROLA	MIL	-3.6	-37	0.5	13.61
LM237K	NATIONAL	IND	-1.2	-37	1.5	13.85
UA79HGC	FAIRCHILD	COM	-2.11	-24	5	16.25
LM137H	NATIONAL	MIL	-1.2	-37	0.5	18.00
LM137K	NATIONAL	MIL	-1.2	-37	1.5	19.60
LM237HVK	NATIONAL	IND	-1.2	-47	1.5	19.85

PART NUMBER	MANUFACTURER	TEMP RANGE	OUTPUT VOLTAGE (V) MIN	MAX	MAXIMUM OUTPUT CURRENT (A)	UNIT PRICE ($)
LM137HVH	NATIONAL	MIL	-1.2	-47	0.5	28.85
LM137HVK	NATIONAL	MIL	-1.2	-47	1.5	33.00
UA79HG	FAIRCHILD	MIL	-2.11	-24	5	58.33

Table 7-18. Dual-Polarity 12-Volt Regulators.

PART NUMBER	MANUFACTURER	TEMP RANGE	OUTPUT VOLTAGE (V) MIN	MAX	MAXIMUM OUTPUT CURRENT (A)	UNIT PRICE ($)
NE5553	SIGNETICS	COM	±11.5	±12.5	0.2	1.50
SE5553	SIGNETICS	MIL	±11.5	±12.5	0.2	3.20
LM326	NATIONAL	COM	±11.5	±12.5	0.05	3.30
LM126	NATIONAL	MIL	±11.8	±12.2	0.05	10.45

Table 7-19. Dual-Polarity 15-Volt Regulators.

PART NUMBER	MANUFACTURER	TEMP RANGE	OUTPUT VOLTAGE (V) MIN	MAX	MAXIMUM OUTPUT CURRENT (A)	UNIT PRICE ($)
LM325	NATIONAL	COM	±14.5	±15.5	0.05	3.30
LM325A	NATIONAL	COM	±14.8	±15.2	0.05	5.20
MC1468	MOTOROLA	COM	±14.5	±15.5	0.05	5.33
MC1568	MOTOROLA	MIL	±14.8	±15.2	0.05	6.08
LM125	NATIONAL	MIL	±14.8	±15.2	0.05	10.45

Table 7-20. Switchmode Power-Supply Controller ICs, Commercial Temperature Range.

PART NUMBER	MANUFACTURER	NO. OF OUTPUTS	MAXIMUM OUTPUT CURRENT (A)	TYPICAL REFERENCE VOLTAGE (V)	UNIT PRICE ($)
NE5561	SIGNETICS	1	0.02	3.75	1.20
MC34060	MOTOROLA	1	0.2	5	1.50
TL594C	T I	2	0.2	5	2.07
TL497AC	T I	1	0.5	1.2	2.19
UA494C	FAIRCHILD	2	0.2	5	2.19
TL494C	T I	2	0.2	5	2.30
CA2524	RCA	2	0.05	5	2.40

PART NUMBER	MANUFACTURER	NO. OF OUTPUTS	MAXIMUM OUTPUT CURRENT (A)	TYPICAL REFERENCE VOLTAGE (V)	UNIT PRICE ($)
MC34063	MOTOROLA	1	1	1.25	2.75
UA78S40C	FAIRCHILD	1	1	1.245	2.79
TL495C	T I	2	0.2	5	2.88
TL593C	T I	2	0.2	5	2.88
CA3524	RCA	2	0.05	5	2.90
NE5560	SIGNETICS	1	0.04	3.72	2.99
LM3524	NATIONAL	2	0.05	5	3.20
TL595C	T I	2	0.2	5	3.22
TL493C	T I	2	0.2	5	4.60
MC3420	MOTOROLA	2	0.04	7.8	6.79
LH1605C	NATIONAL	1	5	2.5	13.60
SH1605	FAIRCHILD	1	5	2.5	15.10

Table 7-21. Switchmode Power-Supply Controller ICs, Industrial Temperature Range.

PART NUMBER	MANUFACTURER	NO. OF OUTPUTS	MAXIMUM OUTPUT CURRENT (A)	TYPICAL REFERENCE VOLTAGE (V)	UNIT PRICE ($)
TL494I	T I	2	0.2	5	3.96
TL594I	T I	2	0.2	5	4.03
MC33063	MOTOROLA	1	1	1.25	4.17
TL595I	T I	2	0.2	5	4.26
TL593I	T I	2	0.2	5	4.60
TL497AI	T I	1	0.5	1.2	5.18

Table 7-22. Switchmode Power-Supply Controller ICs, Military Temperature Range.

PART NUMBER	MANUFACTURER	NO. OF OUTPUTS	MAXIMUM OUTPUT CURRENT (A)	TYPICAL REFERENCE VOLTAGE (V)	UNIT PRICE ($)
SE5561	SIGNETICS	1	0.02	3.75	2.20
CA1524	RCA	2	0.05	5	2.60
SE5560	SIGNETICS	1	0.04	3.72	4.12
UA78S40M	FAIRCHILD	1	1	1.245	8.84
MC35063	MOTOROLA	1	1	1.25	9.65

PART NUMBER	MANUFACTURER	NO. OF OUTPUTS	MAXIMUM OUTPUT CURRENT (A)	TYPICAL REFERENCE VOLTAGE (V)	UNIT PRICE ($)
TL497AM	T I	1	0.5	1.2	10.00
TL494M	T I	2	0.2	5	18.00
TL594M	T I	2	0.2	5	18.00
MC3520	MOTOROLA	2	0.04	7.8	18.15
SH1605SM	FAIRCHILD	1	5	2.5	75.00

Chapter 8

Digital-to-Analog Converters

A DIGITAL-TO-ANALOG (D/A CONVERTER, OR DAC) CONVERTS a digital input code (in the form of a "high" or "low" voltage on each bit input pin) to an analog current or voltage. The number of discrete analog output current or voltage levels is equal to 2^2, where n is the number of digital input lines.

FUNCTIONAL DESCRIPTION

Figure 8-1 shows a schematic of an 8-bit current output, DAC08 type D/A converter. Using external resistors (see Fig. 8-2), a reference current of 2 mA is established in Q1. The Q1 base-emitter voltage plus the drop across the Q1 emitter resistor biases Q2-Q10 and the so-called R-2R ladder network. The R-2R network causes the Q2 current to be half the Q1 current, the Q3 current to be half the Q2 current, and so forth. Q10 is called the remainder current and is not summed with the other currents.

With the switches set as shown in Fig. 8-1, approximately 1.992 mA flows into the I_{OUT1} pin (connected externally to ground, a resistor to ground, or an op amp circuit) through the switches and into the collectors of transistors Q2 through Q9. Zero current flows into the I_{OUT2} pin. Assume that this condition corresponds to a high voltage (digital "1") on inputs B1-B8.

If B8 is set low (digital "0"), the Q9 collector is connected to

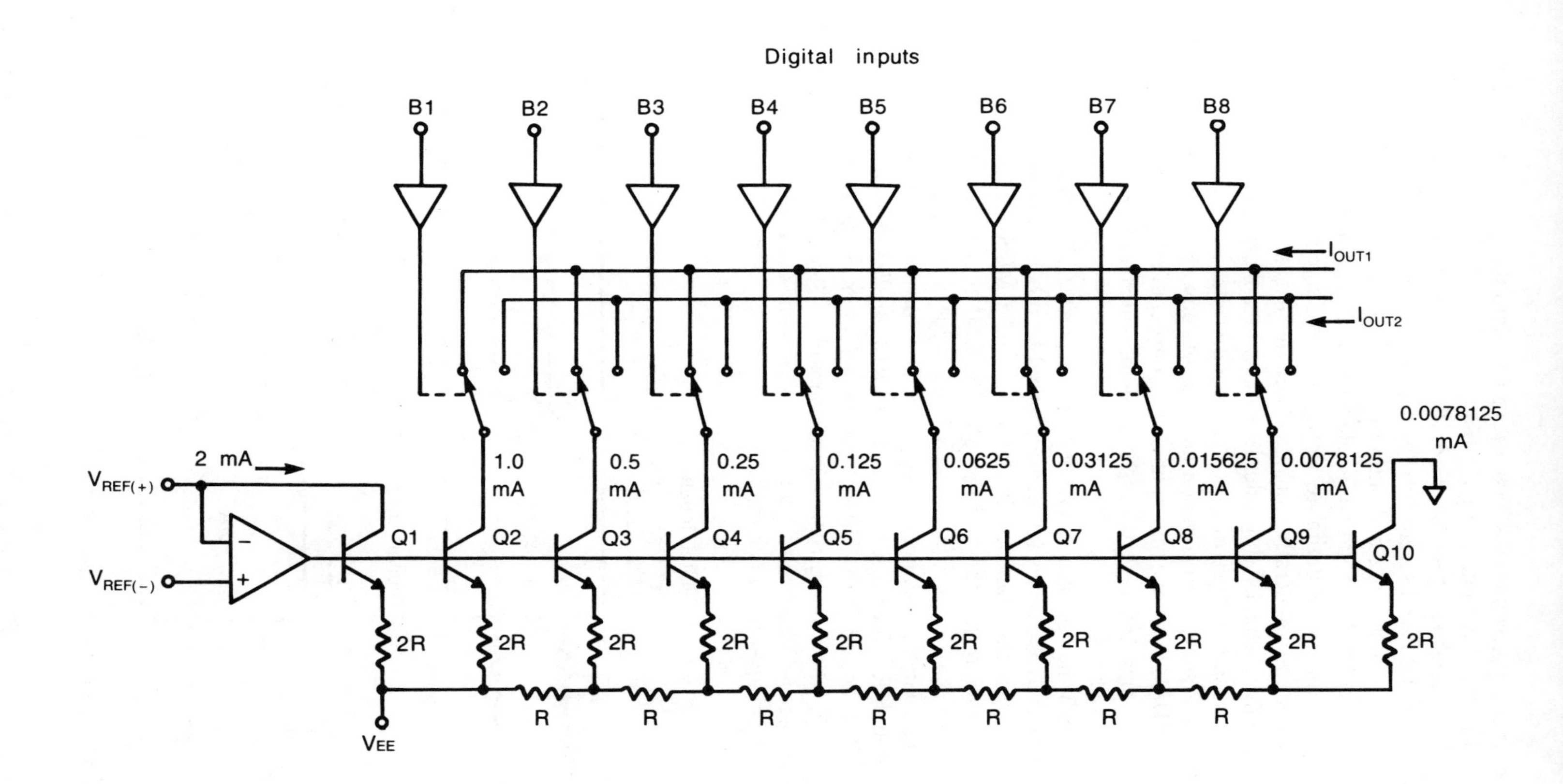

Fig. 8-1. 8-bit current-output, DAC08 type D/A converter.

I_{OUT2}, causing approximately 0.008 mA to flow into the I_{OUT2} pin and 1.984 mA to flow into the I_{OUT1} pin.

For this converter, there are 2^8, or 256, possible digital input combinations and 256 discrete current outputs. Table 8-1 shows a partial list of input codes and output currents. B8 is called the least significant bit (LSB) and B1 is called the most significant bit (MSB). (Note that in the digital IC world, the most significant bit in an 8-bit system is generally labeled D7 and the least significant bit is generally labeled D0. I wonder how many breadboards have been wired wrong because of this notation difference.)

Figure 8-2 shows two common methods of converting a current output DAC to a voltage output DAC. Some converters have an on-chip op amp and feedback resistor and provide both current and voltage outputs.

Figure 8-3 shows another current output type DAC. The Analog Devices AD7523 series (alternate sourced by other manufacturers) employs the standard R-2R ladder and CMOS analog switches. Assume that the resistance of the CMOS switches is negligible (or the 20 KΩ resistors have been reduced to compensate for the switch resistance). If V_{REF} equals +10 volts, 0.5 mA flows through R9, 0.25 mA flows through R10, 0.125 mA flows through R11, and so forth. These currents are summed and flow through R17. The total current is approximately 0.996 mA. The output voltage is −9.96 volts. Table 8-2 shows some of the possible output voltages for this circuit.

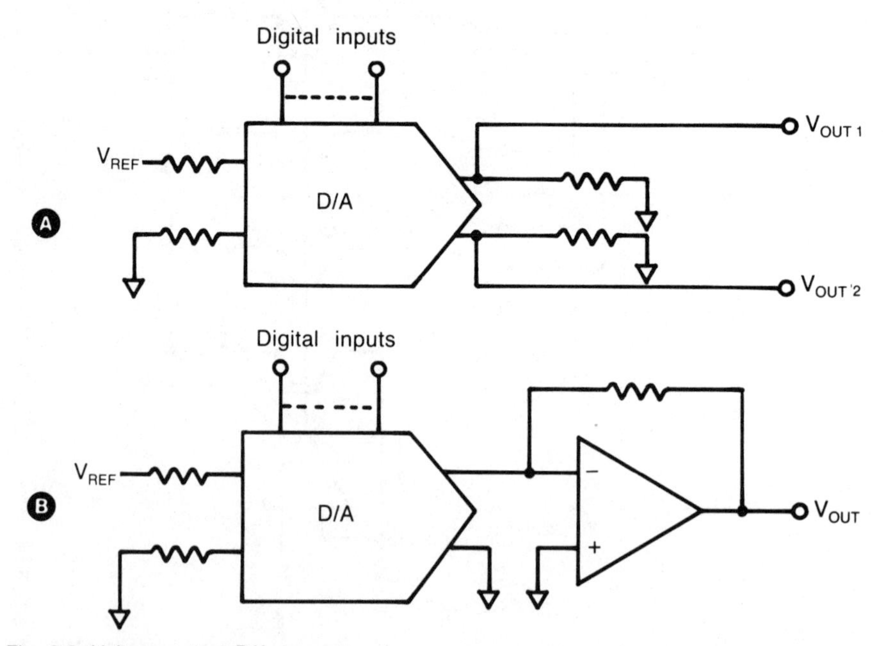

Fig. 8-2. Voltage-output D/A converters.

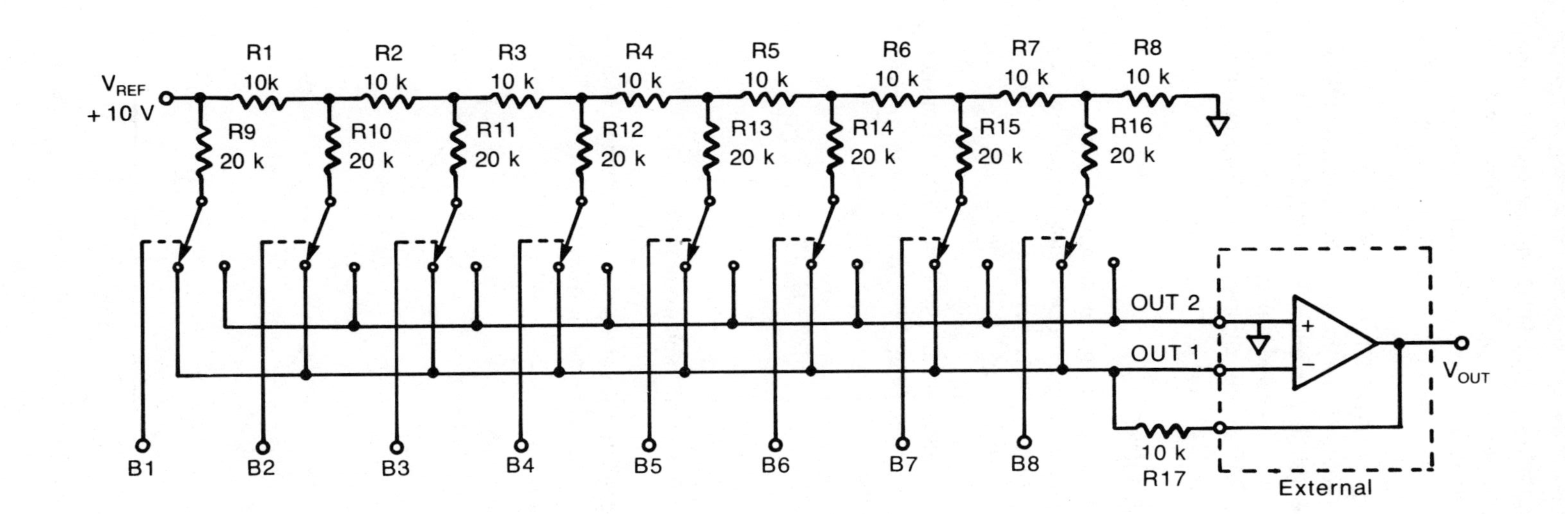

Fig. 8-3. Type AD7523 8-bit current-output D/A converter with external current-to-voltage op amp.

Note that the AD7523 (and other DACs of this configuration) will not work without an external op amp, in contrast with the DAC08 which will work with one external resistor to convert the current to voltage. However, in most applications, an op amp is required anyway in order to have some load driving capability.

Both the DAC08 and the AD7523 are called *multplying* DACs. This is because the output is proportional to the digital inputs *and* to an external reference (either current or voltage). This is equivalent to multiplication. Although some applications benefit from this feature, be aware that DAC specifications are usually guaranteed at only one value of V_{REF} (or I_{REF}).

PARAMETER DEFINITIONS

Figure 8-4 shows the ideal transfer function for a 3-bit D/A converter. For each digital input, there is one analog output level. For an ideal converter, the analog output levels are separated from each other by an amount equal to exactly 1 LSB, the current or voltage value of the least significant bit.

The value of 1 LSB is generally not specified in the data sheet. However, it may be calculated using the *Full Scale value* (given in the data sheet) as follows:

$$1 \text{ LSB} = (\text{F.S.})\ (2^n - 1)/(2^n)$$

where:

F.S. = Full Scale; and
n = number of bits of an n-bit converter

Full Scale is the ideal output voltage or current when all digital inputs are high or set to 1. Deviation from the ideal Full Scale

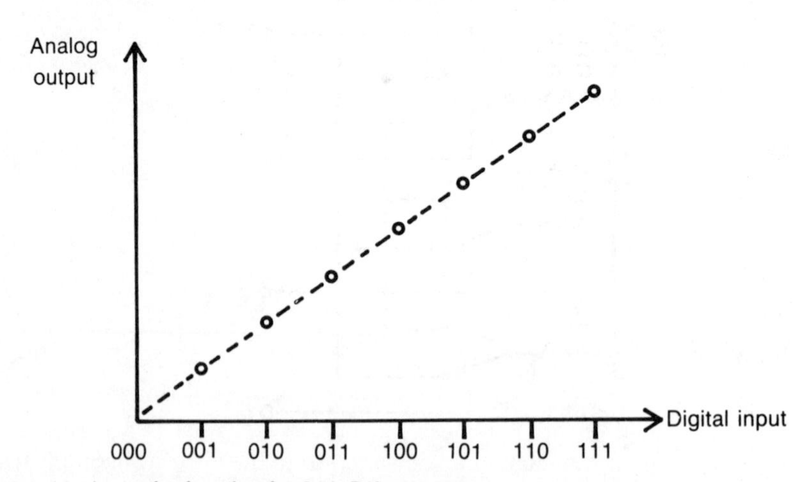

Fig. 8-4. Ideal transfer function for 3-bit D/A converter.

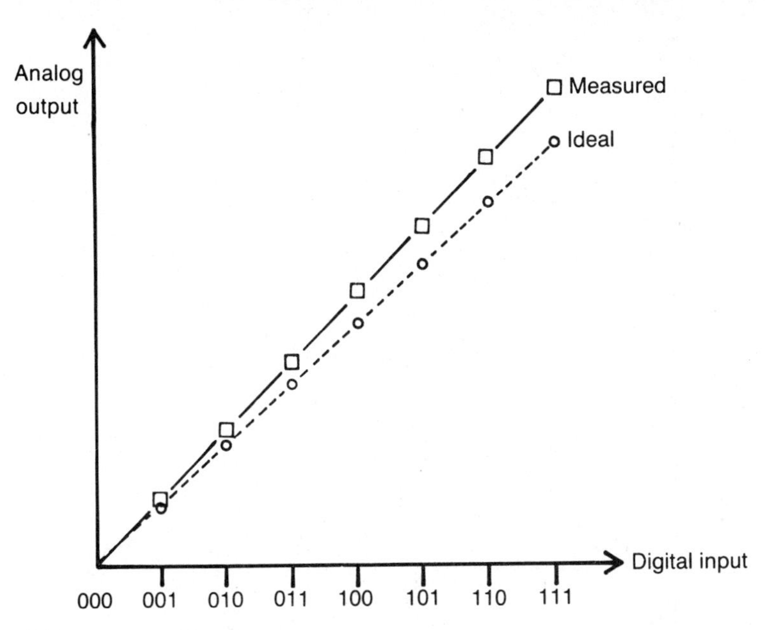

Fig. 8-5. D/A converter with gain error.

output voltage or current is called *gain error*. Gain error may be positive or negative, i.e., the measured Full Scale may be higher or lower than the manufacturer's design goal. Gain error is illustrated in Fig. 8-5.

Zero Scale is the ideal output voltage or current when all digital inputs are low or set to 0. For this condition, of course, the ideal output voltage or current is zero. Deviation from the ideal Zero Scale output voltage or current is called *offset error*. Offset may be positive or negative. Offset error is illustrated in Fig. 8-6. In current output DACs, offset error is due to leakage current. In voltage output DACs, offset error is primarily due to the op amp input offset voltage and secondarily due to input bias current flowing through the feedback resistor. Gain and offset errors may be externally adjusted to zero.

Resolution, Nonlinearity, and Monotonicity

Resolution is the number of bits of a given DAC. An 8-bit DAC, by definition, has 8-bit resolution. A 12-bit DAC has 12-bit resolution. A 12-bit DAC, however, may or may not be 12-bit accurate, depending on the amount of its nonlinearity, or how much its transfer function differs from the ideal.

Because no D/A converter is perfect, all D/A converters, by definition, are nonlinear. Nonlinearity, or relative accuracy, is defined as the maximum deviation from the ideal straight line transfer function after gain and offset errors have adjusted to zero.

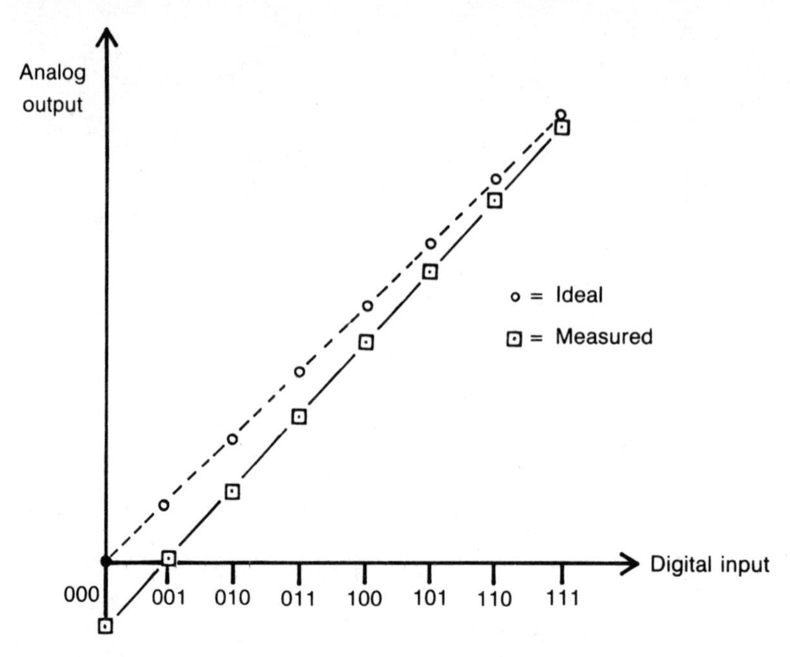

Fig. 8-6. D/A converter with offset error.

Nonlinearity is illustrated in Fig. 8-7.

Nonlinearity is specified in terms of ±LSBs or plus or minus percent of Full Scale (± % F.S.). Table 8-3 shows the relationship between LSBs and percent of Full Scale for different resolution DACs. A "true" n-bit DAC has a maximum nonlinearity of ± 1/2 LSB. An 8-bit DAC with 1/4 LSB nonlinearity is considered to have 9-bit accuracy. Also, a 12-bit DAC with 2 LSB nonlinearity is only 10-bit accurate. Resolution tells you the number of discrete output levels for a given DAC (2^n for an n-bit DAC), nonlinearity, or accuracy, tells you the maximum tolerance for any one of those output levels, except for zero and Full Scale.

In some applications, resolution is more important than accuracy. The transfer function may bow above the ideal straight line (as in Fig. 8-7) more than 1/2 LSB or sag more than 1/2 LSB below the ideal transfer function and still be satisfactory. In microprocessor control applications where there is feedback back to the microprocessor, the microprocessor can easily be programmed to "find" the right digital code to get the desired result, *as long as the D/A converter is monotonic.*

A DAC is *monotonic* if, for the entire analog output range (zero to full scale), the analog output stays the same or increases as the digital code input is incremented by 1 LSB. The transfer function shown in Fig. 8-7 is monotonic. The transfer function shown in Fig. 8-8 is nonmonotonic, because as the code is increased from 100 binary to 101 binary, the analog output decreases.

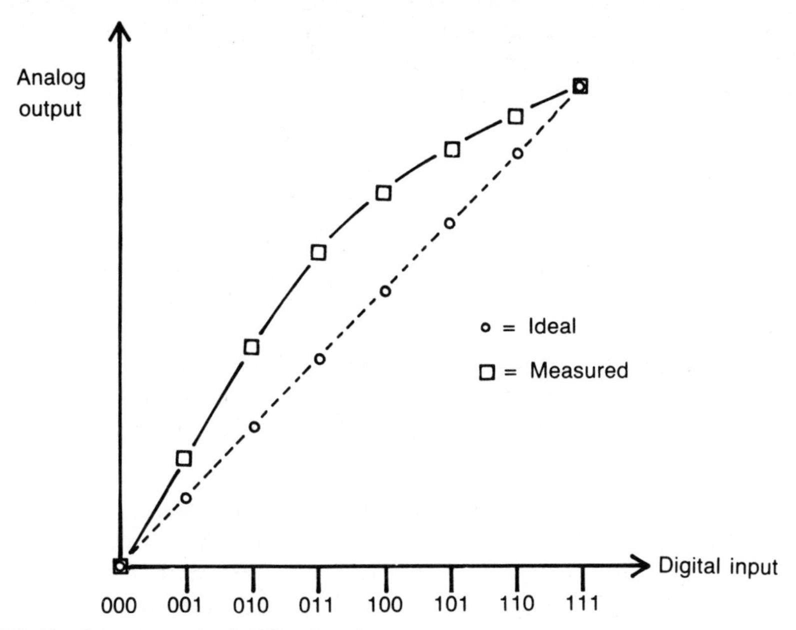

Fig. 8-7. Nonlinear, monotonic D/A converter.

Note that just because a DAC's nonlinearity is guaranteed to be less than ± 1 LSB that does not mean that the DAC is fully monotonic. The transfer function shown in Fig. 8-8 could be within the ± 1 LSB limits, but it is not monotonic. Consequently, manufac-

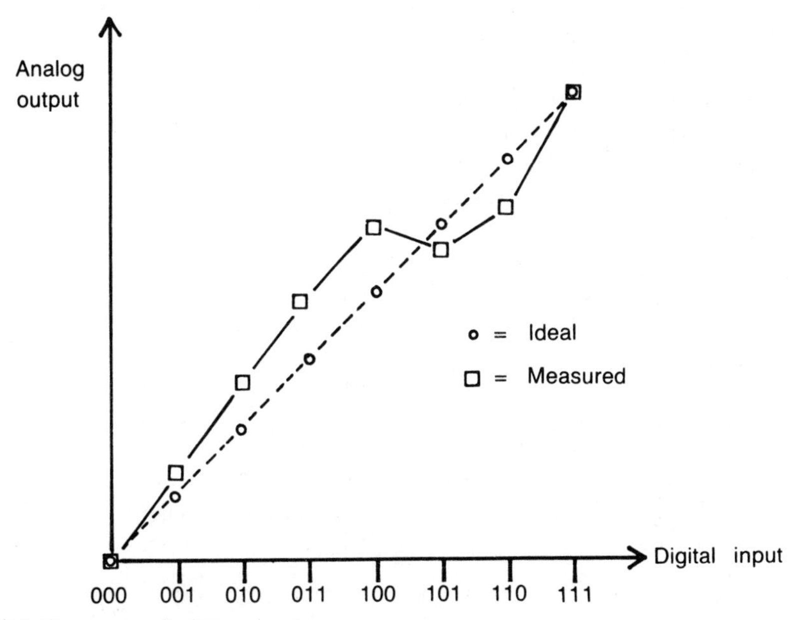

Fig. 8-8. Nonmonotonic D/A converter.

turers will either state in the data sheet that the part is guaranteed to be monotonic (for a certain number of bits and/or over a specified temperature range) or specify a parameter called *differential nonlinearity*.

Differential Nonlinearity

Differential nonlinearity is defined as the change (in equivalent LSBs, or fraction of LSB) in the analog output minus 1 LSB when the digital input code is changed 1 LSB. Stated another way, when the digital input code is changed by 1 LSB, the analog output is also supposed to change by 1 LSB. The difference between the actual change and 1 LSB is called the differential nonlinearity. Differential nonlinearity (DNL) is illustrated in Fig. 8-9. Let's walk through this graph point by point.

(1) The digital code is changed from 000 to 001. The measured analog output is equal to the ideal value. Therefore, the nonlinearity (NL), the amount of deviation from the ideal value at that point, is zero. The DNL is also zero because the digital code was changed by 1 LSB and the analog output changed by exactly 1 LSB.

(2) The digital code is changed from 001 to 010. The measured analog output is equivalent to 1/2 LSB below the ideal value. The NL is −1/2 LSB. The DNL is also −1/2 LSB because the digital code was changed by 1 LSB and the analog output changed by 1/2 LSB. DNL equals the analog output change minus 1 LSB, or 1/2 LSB minus 1 LSB, or −1/2 LSB.

(3) The digital code is changed from 010 to 011. The measured analog output is equal to the ideal value. NL equals 0 LSB. DNL equals +1 1/2 LSB (the equivalent change in analog output) minus 1 LSB, or +1/2 LSB.

(4) The digital code is changed from 011 to 100. The measured analog output is equivalent to 1/2 LSB above the ideal value. NL equals +1/2 LSB. DNL equals +1 1/2 LSB minus 1 LSB, or +1/2 LSB.

(5) The digital code is changed from 100 to 101. The measured analog output is 1 LSB below the ideal. NL is −1 LSB. DNL is −1/2 LSB minus 1 LSB, or −1 1/2 LSB. The transfer function is nonmonotonic at this point.

(6) The digital code is changed from 101 to 110. The measured analog output is 1/2 LSB above the ideal. NL is +1/2 LSB. DNL is +2 1/2 LSB minus 1 LSB, or +1 1/2 LSB.

(7) The digital code is changed from 110 to 111. The measured analog output is equal to the ideal. NL equals 0 LSB. DNL equals +1/2 LSB minus 1 LSB, or −1/2 LSB.

You can see that it is possible for a DAC to have a nonlinearity of ± 1 LSB and yet be nonmonotonic. However, if the differential

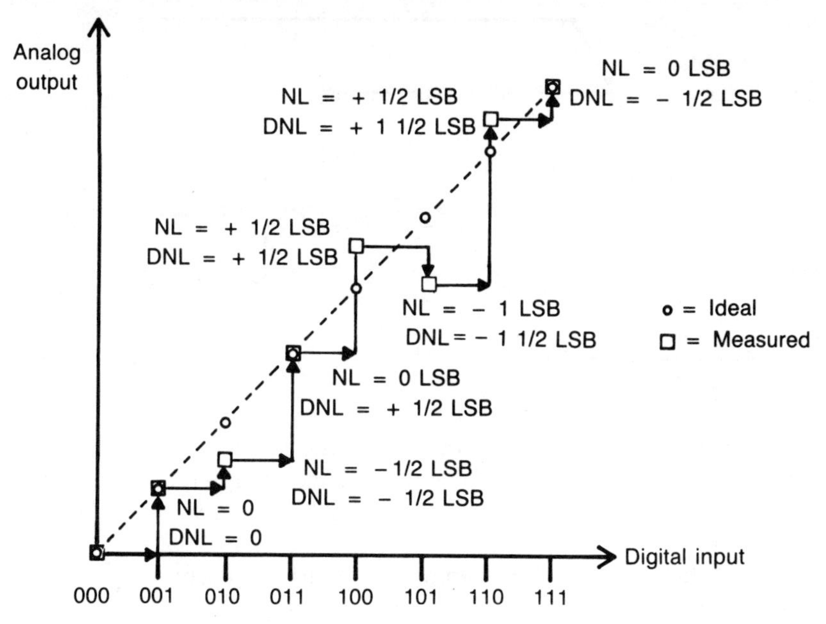

Fig. 8-9. Nonlinearity (NL) and differential nonlinearity (DNL).

nonlinearity is less than or equal to ±1 LSB, the DAC will be monotonic. In Fig. 8-9, if the analog output would have at least stayed the same when the digital code changed from 100 to 101 then the DNL would have been −1 LSB and the DAC would have been monotonic.

Gain Temperature Coefficient

The gain temperature coefficient is determined by measuring the full scale analog output voltage or current at room temperature (25 ° C) and at the temperature extremes for which the DAC is specified. The change in output voltage or current is divided by the change in temperature. For any given part, the gain drift may be positive or negative with respect to the room temperature value. Therefore, the gain temperature coefficient for a part type (and for DACs in general) is expressed in plus or minus parts-per-million per degrees Centigrade (±PPM/°C).

The importance of this specification can easily be overlooked. Let's say that the gain tempco is specified as ±80 PPM/°C for an 8-bit DAC with an external voltage reference. This translates into 0.008 %/°C, or 0.36% of full scale at 70° C. This is nearly equal to 1 LSB (0.39%) for an 8-bit DAC. This is before the tempco of the external voltage reference is considered. To make sure that your design would work for the worst case condition, you would have to add the tempco of the external reference to the DAC gain tempco. The statistical maximum overall gain tempco (the toler-

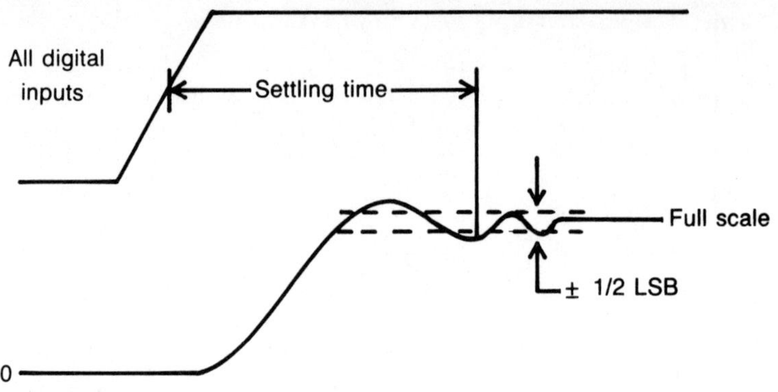

Fig. 8-10. Settling time measurement.

ance for a significant sample size, if tested) is the square root of the sum of the square of the DAC gain tempco and the reference tempco:

$$\text{Tempco(Total)} = \pm\sqrt{(\text{DAC Tempco})^2 + (\text{REF Tempco})^2}$$

The important thing to keep in mind is that PPM/°C translates into a percent change in the Full Scale output at the maximum temperature. Table 8-4 lists percent changes in Full Scale for different tempcos and the temperature extremes of the standard commercial, industrial, and military operating temperature ranges. You can use Table 8-3 to translate this percent change in Full Scale output to number of LSBs for a given DAC resolution.

Settling Time

Settling time is the time it takes for the analog output to change from zero to full scale within a specified tolerance, typically a percentage equivalent to ± 1/2 LSB. The measurement time starting point is the 50 percent point of the rise time of the digital inputs as they are all switched simultaneously from the low to high state. The settling time measurement waveforms are shown in Fig. 8-10. Settling time is critical in real-time control applications, particularly in video systems.

SELECTION GUIDES

Selection guides are given in Tables 8-5 through 8-24. Additional information is provided in Appendix E.

Table 8-1. Output Current Levels for D/A Converter Shown in Fig. 8-1.

B1	B2	B3	B4	B5	B6	B7	B8	IOUT1(mA)	IOUT2(mA)
1	1	1	1	1	1	1	1	1.992	0.000
1	1	1	1	1	1	1	0	1.984	0.008
1	1	1	1	1	1	0	1	1.976	0.016
1	1	1	1	1	1	0	0	1.967	0.023
1	0	0	0	0	0	0	0	1.000	0.992
0	0	0	0	0	0	0	1	0.008	1.984
0	0	0	0	0	0	0	0	0.000	1.992

Table 8-2. Output Voltage Levels for D/A Converter Shown in Fig. 8-3.

B1	B2	B3	B4	B5	B6	B7	B8	VOUT
1	1	1	1	1	1	1	1	-9.961
1	1	1	1	1	1	1	0	-9.922
1	1	1	1	1	1	0	1	-9.883
1	1	1	1	1	1	0	0	-9.844
1	0	0	0	0	0	0	0	-5.000
0	0	0	0	0	0	0	1	-0.039
0	0	0	0	0	0	0	0	0.000

Table 8-3. Nonlinearity in Percent of Full Scale and LSBs versus Resolution.

NON-LINEARITY IN LSBS	NON-LINEARITY IN PERCENT OF FULL SCALE VS. DAC RESOLUTION					
	6-BIT	8-BIT	10-BIT	12-BIT	14-BIT	16-BIT
1/8	0.19531	0.04883	0.01221	0.00305	0.00076	0.00019
1/4	0.39063	0.09766	0.02441	0.00610	0.00153	0.00038
1/2	0.78125	0.19531	0.04883	0.01221	0.00305	0.00076
1	1.56250	0.39063	0.09766	0.02441	0.00610	0.00153
2	3.12500	0.78125	0.19531	0.04883	0.01221	0.00305
3	4.68750	1.17188	0.29297	0.07324	0.01831	0.00458
4	6.25000	1.56250	0.39063	0.09766	0.02441	0.00610
5	7.81250	1.95313	0.48828	0.12207	0.03052	0.00763
8	12.5000	3.12500	0.78125	0.19531	0.04883	0.01221
16	25.0000	6.25000	1.56250	0.39063	0.09766	0.02441
32	50.0000	12.5000	3.12500	0.78125	0.19531	0.04883

Table 8-4. Gain Tempco versus Percent of Full Scale at Various Temperatures.

GAIN TEMPCO (±PPM/°C)	± PERCENT OF FULL SCALE AT -55°C	-40°C	-25°C	+70°C	+85°C	+125°C
5	0.04	0.0325	0.025	0.0225	0.03	0.05
10	0.08	0.065	0.05	0.045	0.06	0.1
20	0.16	0.13	0.1	0.09	0.12	0.2
30	0.24	0.195	0.15	0.135	0.18	0.3
40	0.32	0.26	0.2	0.18	0.24	0.4
50	0.4	0.325	0.25	0.225	0.30	0.5
60	0.48	0.39	0.3	0.27	0.36	0.6
80	0.64	0.52	0.4	0.36	0.48	0.8
100	0.8	0.65	0.5	0.45	0.6	1
120	0.96	0.78	0.6	0.54	0.72	1.2
160	1.28	1.04	0.8	0.72	0.96	1.6

Table 8-5. 6-Bit D/A Converters.

PART NUMBER	MANUFACTURER	TEMP RANGE	OUTPUT TYPE	MAXIMUM NON-LINEARITY (± LSB)	MAXIMUM GAIN TEMPCO (PPM/°C)	UNIT PRICE ($)	NOTES
MC1406L	MOTOROLA	COM	I	0.5	80 TYP	7.29	
DAC01D	P M I	COM	V	0.5	160	9.75	1
MC1506L	MOTOROLA	MIL	I	0.5	80 TYP	11.05	
DAC01C	P M I	COM	V	0.25	160	22.50	1
DAC01F	P M I	MIL	V	0.25	80	30.00	1
DAC01B	P M I	MIL	V	0.25	120	37.50	1
DAC01	P M I	MIL	V	0.25	80	45.00	1
DAC01A	P M I	MIL	V	0.125	80	75.00	1

1=INTERNAL REFERENCE 2=INTERNAL LATCHES

Table 8-6. 8-Bit Voltage-Output D/A Converters.

PART NUMBER	MANUFACTURER	TEMP RANGE	MAXIMUM NON-LINEARITY (± LSB)	MAXIMUM GAIN TEMPCO (PPM/°C)	UNIT PRICE ($)	NOTES
NE5018	SIGNETICS	COM	0.5	20 TYP	7.00	1,2
AD558J	ANALOG DEVICES	COM	0.5	60	8.95	1,2

PART NUMBER	MANUFACTURER	TEMP RANGE	MAXIMUM NON-LINEARITY (± LSB)	MAXIMUM GAIN TEMPCO (PPM/°C)	UNIT PRICE ($)	NOTES
MC6890	MOTOROLA	COM	0.5	50	10.90	1,2
NE5019	SIGNETICS	COM	0.25	20 TYP	11.00	1,2
AD558K	ANALOG DEVICES	COM	0.25	30	13.50	1,2
SE5018	SIGNETICS	MIL	0.5	20 TYP	15.00	1,2
MC6890A	MOTOROLA	MIL	0.5	50	19.45	1,2
AD558S	ANALOG DEVICES	MIL	0.5	60	19.50	1,2
AD558T	ANALOG DEVICES	MIL	0.25	30	26.95	1,2
SE5019	SIGNETICS	MIL	0.25	20 TYP	27.00	1,2
DAC82KG	BURR-BROWN	IND	0.5	50	33.90	1

1=INTERNAL REFERENCE 2=INTERNAL LATCHES

Table 8-7. 8-Bit Current-Output D/A Converters, Commercial Temperature Range.

PART NUMBER	MANUFACTURER	MAXIMUM NON-LINEARITY (± LSB)	MAXIMUM GAIN TEMPCO (PPM/°C)	UNIT PRICE ($)	NOTES
MC1408P6	MOTOROLA	2	20 TYP	1.46	
UA802A	FAIRCHILD	0.5	20 TYP	1.50	
UA802B	FAIRCHILD	1	20 TYP	1.50	
UA802C	FAIRCHILD	1	20 TYP	1.50	
UA801E	FAIRCHILD	0.5	50	1.52	
UA801C	FAIRCHILD	1	50	1.54	
DAC0806LC	NATIONAL	2	20 TYP	1.55	
MC1408P7	MOTOROLA	1	20 TYP	1.72	
DAC0807LC	NATIONAL	1	20 TYP	1.75	
MC1408P8	MOTOROLA	0.5	20 TYP	1.89	
DAC0801LC	NATIONAL	1	50	2.30	
DAC08C	P M I	1	80	2.33	
DAC0808LC	NATIONAL	0.5	20 TYP	2.40	
DAC0800LC	NATIONAL	0.5	50	2.80	
DAC08E	P M I	0.5	50	2.93	
AD1408-7	ANALOG DEVICES	1	20 TYP	3.50	

PART NUMBER	MANUFACTURER	MAXIMUM NON-LINEARITY (± LSB)	MAXIMUM GAIN TEMPCO (PPM/°C)	UNIT PRICE ($)	NOTES
DAC08H	P M I	0.25	50	3.68	
DAC0832LCN	NATIONAL	0.5	6	3.75	2
DAC0802LC	NATIONAL	0.25	50	4.15	
AD7523J	ANALOG DEVICES	0.5	120	4.00	
AD1408-8	ANALOG DEVICES	0.5	20	4.30	
NE5118	SIGNETICS	0.5	20 TYP	5.50	
DAC0831LCN	NATIONAL	0.25	6	5.63	2
AD7523K	ANALOG DEVICES	0.25	120	6.00	
DAC0830LCN	NATIONAL	0.125	6	6.35	2
HA-5618B-5	HARRIS	0.5	--	6.46	
HA-5618A-5	HARRIS	0.25	--	7.03	
AD7524J	ANALOG DEVICES	0.5	10	7.50	2
DAC808G	P M I	1	80	7.50	2
AD7523L	ANALOG DEVICES	0.125	120	8.00	
AD1408-9	ANALOG DEVICES	0.25	20	9.50	
AD7524K	ANALOG DEVICES	0.25	10	10.00	2
NE5119	SIGNETICS	0.25	20 TYP	10.50	
AD7524L	ANALOG DEVICES	0.125	10	12.50	2
MC10318CL6	MOTOROLA	2	10	19.45	
MC10318CL7	MOTOROLA	1	10	28.45	
MC10318L	MOTOROLA	0.5	10	42.00	
MC10318L9	MOTOROLA	0.25	10	70.60	

1=INTERNAL REFERENCE 2=INTERNAL LATCHES

Table 8-8. 8-Bit Current-Output D/A Converters, Industrial Temperature Range.

PART NUMBER	MANUFACTURER	MAXIMUM NON-LINEARITY (± LSB)	MAXIMUM GAIN TEMPCO (PPM/'C)	UNIT PRICE ($)	NOTES
DAC0832LCD	NATIONAL	0.5	6	5.63	2
DAC0830LCD	NATIONAL	0.125	6	8.60	2
DAC808F	P M I	0.5	80	9.00	2
DAC888F	P M I	0.5	80	9.00	2

PART NUMBER	MANUFACTURER	MAXIMUM NON-LINEARITY (± LSB)	MAXIMUM GAIN TEMPCO (PPM/°C)	UNIT PRICE ($)	NOTES
DAC808E	P M I	0.25	50	13.50	2
DAC888E	P M I	0.25	50	13.50	2
AD7524A	ANALOG DEVICES	0.5	10	13.50	2
AD7524B	ANALOG DEVICES	0.25	10	16.00	2
AD7524C	ANALOG DEVICES	0.125	10	18.50	2
DAC90BG	BURR-BROWN	0.5	50	19.40	1
DAC82KG	BURR-BROWN	0.5	50	33.90	1

1=INTERNAL REFERENCE 2=INTERNAL LATCHES

Table 8-9. 8-Bit Current-Output D/A Converters, Military Temperature Range.

PART NUMBER	MANUFACTURER	MAXIMUM NON-LINEARITY (± LSB)	MAXIMUM GAIN TEMPCO (PPM/°C)	UNIT PRICE ($)	NOTES
UA802	FAIRCHILD	0.5	20 TYP	2.85	
UA801	FAIRCHILD	0.5	50	3.80	
DAC0808L	NATIONAL	0.5	20 TYP	6.00	
DAC0800L	NATIONAL	0.5	50	7.10	
DAC08	P M I	0.5	80	7.13	
DAC0802L	NATIONAL	0.25	50	8.25	
DAC0830LD	NATIONAL	0.125	6	11.25	2
MC1508L8	MOTOROLA	0.5	20 TYP	11.46	
HI-5618B-2	HARRIS	0.5	--	13.37	
DAC08A	P M I	0.25	80	13.50	
AD1508-8	ANALOG DEVICES	0.5	20	14.60	
DAC808B	P M I	0.5	80	14.93	2
DAC888B	P M I	0.5	80	14.93	2
SE5118	SIGNETICS	0.5	20 TYP	15.50	
HI-5618A-2	HARRIS	0.25	--	15.56	
DAC808A	P M I	0.25	50	21.00	2
DAC888A	P M I	0.25	50	21.00	2
AD1508-9	ANALOG DEVICES	0.25	20	21.30	
SE5119	SIGNETICS	0.25	20 TYP	24.00	

PART NUMBER	MANUFACTURER	MAXIMUM NON-LINEARITY (± LSB)	MAXIMUM GAIN TEMPCO (PPM/°C)	UNIT PRICE ($)	NOTES
DAC90SG	BURR-BROWN	0.5	50	26.50	1
AD7524S	ANALOG DEVICES	0.5	10	32.25	2
AD9768S	ANALOG DEVICES	0.5	70	34.60	1
AD7524T	ANALOG DEVICES	0.25	10	38.50	2
AD7524U	ANALOG DEVICES	0.125	10	44.75	2

1=INTERNAL REFERENCE 2=INTERNAL LATCHES

Table 8-10. Dual 8-Bit D/A Converters with Internal Latches.

PART NUMBER	MANUFACTURER	TEMP RANGE	OUTPUT TYPE	MAXIMUM NON-LINEARITY (± LSB)	MAXIMUM GAIN TEMPCO (PPM/°C)	UNIT PRICE ($)
AD7528J	ANALOG DEVICES	COM	I	1	35	9.80
AD7528K	ANALOG DEVICES	COM	I	0.5	35	13.05
AD7528L	ANALOG DEVICES	COM	I	0.5	35	16.40
AD7528A	ANALOG DEVICES	IND	I	1	35	16.95
AD7528B	ANALOG DEVICES	IND	I	0.5	35	19.80
AD7528C	ANALOG DEVICES	IND	I	0.5	35	23.10
AD7528S	ANALOG DEVICES	MIL	I	1	35	32.90
AD7528T	ANALOG DEVICES	MIL	I	0.5	35	35.90
AD7528U	ANALOG DEVICES	MIL	I	0.5	35	41.90

1=INTERNAL REFERENCE 2=INTERNAL LATCHES

Table 8-11. 8-Bit Plus Sign D/A Converters with Internal Reference.

PART NUMBER	MANUFACTURER	TEMP RANGE	OUTPUT TYPE	MAXIMUM NON-LINEARITY (± LSB)	MAXIMUM GAIN TEMPCO (PPM/°C)	UNIT PRICE ($)
DAC208F	P M I	COM	V	1	60	12.75
DAC208E	P M I	COM	V	0.5	40	18.00
DAC208B	P M I	MIL	V	1	60	25.50
DAC208A	P M I	MIL	V	0.5	40	42.00

Table 8-12. 10-Bit Voltage-Output D/A Converters with Internal Reference.

PART NUMBER	MANUFACTURER	TEMP RANGE	MAXIMUM NON-LINEARITY (± LSB)	MAXIMUM GAIN TEMPCO (PPM/°C)	UNIT PRICE ($)
DAC03D	P M I	COM	4	60 TYP	11.93
NE5020	SIGNETICS	COM	1	20 TYP	12.00
DAC03C	P M I	COM	2	60 TYP	14.93
DAC03B	P M I	COM	1	60 TYP	21.00
DAC06G	P M I	COM	4	100	22.50
DAC03A	P M I	COM	0.5	60 TYP	27.00
DAC06F	P M I	COM	2	100	33.00
DAC06E	P M I	COM	1	100	67.50
DAC06C	P M I	MIL	4	120	90.00
DAC06B	P M I	MIL	2	90	127.50

Table 8-13. 10-Bit Current-Output D/A Converters, Commercial Temperature Range.

PART NUMBER	MANUFACTURER	MAXIMUM NON-LINEARITY (± LSB)	MAXIMUM GAIN TEMPCO (PPM/°C)	UNIT PRICE ($)	NOTES
AD7533J	ANALOG DEVICES	2	10	5.90	
DAC1022LCN	NATIONAL	2	10	6.00	
AD7533K	ANALOG DEVICES	1	10	7.35	
DAC1002LCN	NATIONAL	2	10	7.45	2
DAC1006LCN	NATIONAL	2	10	7.45	2
DAC1021LCN	NATIONAL	1	10	7.50	
DAC10G	P M I	1	50	8.25	
AD7533L	ANALOG DEVICES	0.5	10	8.85	
DAC1020LCN	NATIONAL	0.5	10	9.00	
DAC1001LCN	NATIONAL	1	10	8.95	2
DAC1007LCN	NATIONAL	1	10	8.95	2
AD7530JN	ANALOG DEVICES	2	10	9.75	
MC3410C	MOTOROLA	1	60	10.40	
DAC1000LCN	NATIONAL	0.5	10	10.45	2
DAC1006LCN	NATIONAL	0.5	10	10.45	2
DAC10F	P M I	0.5	25	12.75	

PART NUMBER	MANUFACTURER	MAXIMUM NON-LINEARITY (± LSB)	TYPICAL SETTLING TIME (uS)	UNIT PRICE ($)	NOTES
NE5410	SIGNETICS	0.5	40	13.45	
AD7530KN	ANALOG DEVICES	1	10	13.50	
MC3410	MOTOROLA	0.5	60	14.92	
AD7522JN	ANALOG DEVICES	2	10	15.00	
AD7520JN	ANALOG DEVICES	2	10	15.75	
HI-5610-5	HARRIS	0.5	5 TYP	16.04	
AD561J	ANALOG DEVICES	0.5	80 TYP	17.40	1
AD7522KN	ANALOG DEVICES	1	10	17.50	
DAC100DDQ3	P M I	3	120	19.93	1
DAC100DDQ4	P M I	3	120	19.93	1
AD7530LN	ANALOG DEVICES	0.5	10	18.50	
AD7522LN	ANALOG DEVICES	0.5	10	20.00	
AD7520KN	ANALOG DEVICES	1	10	20.50	
DAC100CCQ3	P M I	2	60	21.60	1
DAC100CCQ4	P M I	2	60	21.60	1
AD7520LN	ANALOG DEVICES	0.5	10	27.00	
AD561K	ANALOG DEVICES	0.25	30	28.75	1
DAC100BCQ3	P M I	1	60	30.60	1
DAC100BCQ4	P M I	1	60	30.60	1
DAC100ACQ3	P M I	0.5	60	46.80	1
DAC100ACQ4	P M I	0.5	60	46.80	1

1=INTERNAL REFERENCE 2=INTERNAL LATCHES

Table 8-14. 10-Bit Current-Output D/A Converters, Industrial Temperature Range.

PART NUMBER	MANUFACTURER	MAXIMUM NON-LINEARITY (± LSB)	MAXIMUM GAIN TEMPCO (PPM/°C)	UNIT PRICE ($)	NOTES
DAC1022LCD	NATIONAL	2	10	8.00	
AD7533A	ANALOG DEVICES	2	10	9.00	
DAC1021LCD	NATIONAL	1	10	9.15	
AD7533B	ANALOG DEVICES	1	10	10.25	
DAC1020LCD	NATIONAL	0.5	10	10.70	

PART NUMBER	MANUFACTURER	MAXIMUM NON-LINEARITY (± LSB)	MAXIMUM GAIN TEMPCO (PPM/°C)	UNIT PRICE ($)	NOTES
AD7533C	ANALOG DEVICES	0.5	10	12.55	
DAC1001LCD	NATIONAL	1	10	12.90	2
DAC1002LCD	NATIONAL	2	10	12.90	2
DAC1007LCD	NATIONAL	1	10	12.90	2
DAC1008LCD	NATIONAL	2	10	12.90	2
AD7530JD	ANALOG DEVICES	2	10	15.00	
DAC1000LCD	NATIONAL	0.5	10	16.10	2
DAC1006LCD	NATIONAL	0.5	10	16.10	2
AD7530KD	ANALOG DEVICES	1	10	18.50	
AD7520JD	ANALOG DEVICES	2	10	20.75	
AD7522JD	ANALOG DEVICES	2	10	26.50	2
AD7530LD	ANALOG DEVICES	0.5	10	28.50	
AD7522KD	ANALOG DEVICES	1	10	29.00	2
AD7520KD	ANALOG DEVICES	1	10	29.50	
DAC100DDQ7	P M I	3	120	30.00	1
AD7522LD	ANALOG DEVICES	0.5	10	31.50	2
DAC100DDQ8	P M I	3	120	36.00	1
AD7520LD	ANALOG DEVICES	0.5	10	39.50	
DAC100CCQ7	P M I	2	60	45.00	1
DAC100CCQ8	P M I	2	60	54.00	1
DAC100BCQ7	P M I	1	60	65.55	1
DAC100BCQ8	P M I	1	60	76.50	1
DAC100ACQ7	P M I	0.5	60	78.75	1
DAC100ACQ8	P M I	0.5	60	93.00	1
DAC100BBQ7	P M I	1	30	100.50	1
DAC100BBQ8	P M I	1	30	120.00	1
DAC100ABQ7	P M I	0.5	30	130.50	1
DAC100ABQ8	P M I	0.5	30	156.00	1
DAC100AAQ7	P M I	0.5	15	202.50	1
DAC100AAQ8	P M I	0.5	15	243.00	1

1=INTERNAL REFERENCE 2=INTERNAL LATCHES

Table 8-15. 10-Bit Current-Output D/A Converters, Military Temperature Range.

PART NUMBER	MANUFACTURER	MAXIMUM NON-LINEARITY (± LSB)	TYPICAL SETTLING TIME (uS)	UNIT PRICE ($)	NOTES
DAC10C	P M I	1	50	18.00	
DAC1022LD	NATIONAL	2	10	23.30	
AD7533S	ANALOG DEVICES	2	10	23.55	
SE5410	SIGNETICS	0.5	40	27.00	
DAC1021LD	NATIONAL	1	10	27.15	
AD7533T	ANALOG DEVICES	1	10	28.15	
DAC10B	P M I	0.5	25	30.00	
DAC1020LD	NATIONAL	0.5	10	30.00	
DAC1008LD	NATIONAL	2	10	31.15	2
AD7533U	ANALOG DEVICES	0.5	10	32.05	
DAC1007LD	NATIONAL	1	10	34.65	2
AD561S	ANALOG DEVICES	0.5	60 TYP	37.90	1
DAC1006LD	NATIONAL	0.5	10	39.25	2
DAC1002LD	NATIONAL	2	10	40.00	2
AD7520S	ANALOG DEVICES	2	10	42.00	
DAC1001LD	NATIONAL	1	10	43.15	2
DAC1000LD	NATIONAL	0.5	10	46.65	2
DAC100DDQ5	P M I	3	120	52.50	1
HI-5610-2	HARRIS	0.5	5 TYP	61.24	
AD7522S	ANALOG DEVICES	2	10	64.00	2
DAC100CCQ5	P M I	2	60	67.50	1
AD7520T	ANALOG DEVICES	1	10	70.00	
AD7522T	ANALOG DEVICES	1	10	70.25	2
DAC100DDQ6	P M I	3	120	70.50	1
AD7522U	ANALOG DEVICES	0.5	10	76.50	2
AD561T	ANALOG DEVICES	0.25	30	77.60	1
DAC100CCQ6	P M I	2	60	90.00	1
AD7520U	ANALOG DEVICES	0.5	10	97.00	
DAC100BCQ5	P M I	1	60	97.50	1
DAC100BCQ6	P M I	1	60	117.00	1

PART NUMBER	MANUFACTURER	MAXIMUM NON-LINEARITY (± LSB)	MAXIMUM GAIN TEMPCO (PPM/'C)	UNIT PRICE ($)	NOTES
DAC100ACQ5	P M I	0.5	60	150.00	1
DAC100BBQ5	P M I	1	30	160.80	1
DAC100ACQ6	P M I	0.5	60	180.00	1
DAC100BBQ6	P M I	1	30	192.90	1

1=INTERNAL REFERENCE 2=ON-CHIIP LATCHES

Table 8-16. 10-Bit Plus Sign D/A Converters with Internal Reference.

PART NUMBER	MANUFACTURER	TEMP RANGE	OUTPUT TYPE	MAXIMUM NON-LINEARITY (± LSB)	MAXIMUM GAIN TEMPCO (PPM/°C)	UNIT PRICE ($)
DAC210G	P M I	COM	V	1	30 TYP	14.25
DAC210F	P M I	COM	V	0.5	60	21.75
DAC05G	P M I	COM	V	5	100	22.50
DAC210E	P M I	COM	V	0.5	40	28.50
DAC02D	P M I	COM	V	4	150	30.00
DAC210B	P M I	COM	V	0.5	60	45.00
DAC02C	P M I	COM	V	2	60	45.00
DAC02B	P M I	COM	V	1	60	67.50
DAC05E	P M I	COM	V	2	100	67.50
DAC210A	P M I	COM	V	0.5	40	82.50
DAC05C	P M I	COM	V	5	120	90.00
DAC02A	P M I	COM	V	1	60	112.50
DAC05A	P M I	COM	V	2	60	202.50

Table 8-17. 12-Bit Voltage-Output D/A Converters.

PART NUMBER	MANUFACTURER	TEMP RANGE	MAXIMUM NON-LINEARITY (± LSB)	MAXIMUM GAIN TEMPCO (PPM/°C)	UNIT PRICE ($)	NOTES
AD7240J	ANALOG DEVICES	COM	1.25	6	16.00	
AD7240K	ANALOG DEVICES	COM	1	6	17.65	
AD7240A	ANALOG DEVICES	IND	1.25	6	18.50	
AD7240B	ANALOG DEVICES	IND	1	6	20.15	

PART NUMBER	MANUFACTURER	TEMP RANGE	MAXIMUM NON-LINEARITY (± LSB)	MAXIMUM GAIN TEMPCO (PPM/°C)	UNIT PRICE ($)	NOTES
DAC800-CBI-V	BURR-BROWN	COM	0.5	30	29.95	1
DAC80-CBI-V	BURR-BROWN	COM	0.5	30	36.50	1
DAC850-CBI-V	BURR-BROWN	IND	0.5	20	47.00	1
AD7240S	ANALOG DEVICES	MIL	1.25	6	65.90	
DAC851-CBI-V	BURR-BROWN	MIL	0.5	25	69.00	1
AD7240T	ANALOG DEVICES	MIL	1	6	73.90	
DAC85C-CBI-V	BURR-BROWN	COM	0.5	20	79.00	1
DAC87U-CBI-V	BURR-BROWN	MIL	0.5	60	95.00	1
AD3860K	ANALOG DEVICES	COM	0.5	10	104.00	1,2
DAC85-CBI-V	BURR-BROWN	IND	0.5	20	107.00	1
DAC87-CBI-V	BURR-BROWN	MIL	0.5	20	120.00	1
DAC85LD-CBI-V	BURR-BROWN	IND	0.5	10	142.75	1
AD3860S	ANALOG DEVICES	MIL	0.5	10	162.75	1,2

1=INTERNAL REFERENCE   2=INTERNAL LATCHES

Table 8-18. 12-Bit Current-Output D/A Converters, Commercial Temperature Range.

PART NUMBER	MANUFACTURER	MAXIMUM NON-LINEARITY (± LSB)	MAXIMUM GAIN TEMPCO (PPM/°C)	UNIT PRICE ($)	NOTES
DAC1222LCN	NATIONAL	8	10	6.75	
DAC1221LCN	NATIONAL	4	10	8.25	
DAC1220LCN	NATIONAL	2	10	9.75	
AD7531JN	ANALOG DEVICES	4	10	9.75	
AD7541AJ	ANALOG DEVICES	1	5	12.85	
AD7545J	ANALOG DEVICES	2	5	13.20	2
DAC312F	P M I	1	30	13.43	
AD7531KN	ANALOG DEVICES	2	10	13.50	
AD6012N	ANALOG DEVICES	2	40	13.95	
AD7541AK	ANALOG DEVICES	0.5	5	14.30	
Am6012DC	AMD	2	40	14.95	
HI-7541J	HARRIS	1	5	14.97	
AD7548J	ANALOG DEVICES	1	5	15.64	2

PART NUMBER	MANUFACTURER	MAXIMUM NON-LINEARITY (± LSB)	MAXIMUM GAIN TEMPCO (PPM/°C)	UNIT PRICE ($)	NOTES
HI-7541K	HARRIS	0.5	5	16.31	
AD7545K	ANALOG DEVICES	1	5	17.25	2
HI-562A-5	HARRIS	0.5	10 TYP	17.69	
AD7545L	ANALOG DEVICES	0.5	5	18.90	2
AD7548K	ANALOG DEVICES	0.5	5	18.97	2
AD566AJ	ANALOG DEVICES	0.5	10	19.95	
AD7521JN	ANALOG DEVICES	4	10	20.75	
AD7542J	ANALOG DEVICES	1	5	21.40	2
AD7543J	ANALOG DEVICES	1	5	21.40	2
AD567J	ANALOG DEVICES	0.5	50	22.50	1,2
AD7542K	ANALOG DEVICES	0.5	5	23.60	2
AD7543K	ANALOG DEVICES	0.5	5	23.60	2
AD565AJ	ANALOG DEVICES	0.5	50	23.95	1
DAC800-CBI-I	BURR-BROWN	0.5	30	23.95	1
AD7521KN	ANALOG DEVICES	2	10	25.50	
DAC312E	P M I	0.5	30	26.93	
MC3412	MOTOROLA	0.5	30	26.95	1
AD7541J	ANALOG DEVICES	1	10	27.50	
AD7545GL	ANALOG DEVICES	0.5	5	27.65	2
UA565J	FAIRCHILD	0.5	30	28.50	1
AD7531LN	ANALOG DEVICES	1	10	29.00	
AD566AK	ANALOG DEVICES	0.25	3	29.95	
Am6012ADC	AMD	2	20	29.95	
AD7541K	ANALOG DEVICES	0.5	10	30.00	
AD7542GK	ANALOG DEVICES	0.5	5	31.15	2
AD7543GK	ANALOG DEVICES	0.5	5	31.15	2
AD7521LN	ANALOG DEVICES	1	10	32.00	
UA565K	FAIRCHILD	0.25	20	34.20	1
AD567K	ANALOG DEVICES	0.25	20	34.50	1,2
DAC80-CBI-I	BURR-BROWN	0.5	30	34.50	1
AD565AK	ANALOG DEVICES	0.25	20	37.50	1

PART NUMBER	MANUFACTURER	MAXIMUM NON-LINEARITY (± LSB)	MAXIMUM GAIN TEMPCO (PPM/°C)	UNIT PRICE ($)	NOTES
AD563J	ANALOG DEVICES	0.5	50	53.30	1
AD562K	ANALOG DEVICES	0.5	5	73.50	
DAC85C-CBI-I	BURR-BROWN	0.5	20	77.00	1
AD563K	ANALOG DEVICES	0.25	20	82.10	1
DAC60-10	BURR-BROWN	1	30	158.00	1
DAC60-12	BURR-BROWN	0.5	30	172.00	1

1=INTERNAL REFERENCE 2=INTERNAL LATCHES

Table 8-19. 12-Bit Current-Output D/A Converters, Industrial Temperature Range.

PART NUMBER	MANUFACTURER	MAXIMUM NON-LINEARITY (± LSB)	MAXIMUM GAIN TEMPCO (PPM/°C)	UNIT PRICE ($)	NOTES
DAC1222LCD	NATIONAL	8	10	8.25	
DAC1221LCD	NATIONAL	4	10	9.30	
DAC1220LCD	NATIONAL	2	10	10.80	
DAC1210LCD	NATIONAL	2	6	11.90	2
DAC1232LCD	NATIONAL	2	6	11.90	2
DAC1219LCD	NATIONAL	1	6	13.40	
AD7541AA	ANALOG DEVICES	1	10	14.30	
DAC1209LCD	NATIONAL	1	6	14.60	2
DAC1231LCD	NATIONAL	1	6	14.60	2
DAC1218LCD	NATIONAL	0.5	6	14.90	
AD7531JD	ANALOG DEVICES	4	10	15.00	
AD7545A	ANALOG DEVICES	2	5	15.70	2
AD7541AB	ANALOG DEVICES	0.5	10	15.75	
DAC1208LCD	NATIONAL	0.5	6	17.15	2
DAC1230LCD	NATIONAL	0.5	6	17.15	2
AD7531KD	ANALOG DEVICES	2	10	18.50	
AD7548A	ANALOG DEVICES	1	5	18.64	2
AD7545B	ANALOG DEVICES	1	5	19.70	2
AD7545C	ANALOG DEVICES	0.5	5	21.35	2

PART NUMBER	MANUFACTURER	MAXIMUM NON-LINEARITY (± LSB)	MAXIMUM GAIN TEMPCO (PPM/°C)	UNIT PRICE ($)	NOTES
AD7548B	ANALOG DEVICES	0.5	5	21.78	2
HI-7541A	HARRIS	1	5	23.87	
AD7542A	ANALOG DEVICES	1	5	24.40	2
AD7543A	ANALOG DEVICES	1	5	24.40	2
AD7521JD	ANALOG DEVICES	4	10	25.75	
AD7542B	ANALOG DEVICES	0.5	5	26.60	2
AD7543B	ANALOG DEVICES	0.5	5	26.60	2
HI-7541B	HARRIS	0.5	5	28.56	
AD7545GC	ANALOG DEVICES	0.5	5	31.70	2
AD7542GB	ANALOG DEVICES	0.5	5	33.50	2
AD7543GB	ANALOG DEVICES	0.5	5	33.50	2
AD7521KD	ANALOG DEVICES	2	10	34.50	
AD7541A	ANALOG DEVICES	1	10	37.50	
DAC850-CBI-I	BURR-BROWN	0.5	20	39.00	1
AD7531LD	ANALOG DEVICES	1	10	39.50	
AD7541B	ANALOG DEVICES	0.5	10	40.00	
AD7521LD	ANALOG DEVICES	1	10	44.50	
AD562A	ANALOG DEVICES	0.5	5	96.40	
DAC85-CBI-I	BURR-BROWN	0.5	20	104.00	1
DAC63BG	BURR-BROWN	0.5	40	108.00	1
DAC63CG	BURR-BROWN	0.5	30	119.00	1

1=INTERNAL REFERENCE 2=INTERNAL LATCHES

Table 8-20. 12-Bit Current-Output D/A Converters, Military Temperature Range.

PART NUMBER	MANUFACTURER	MAXIMUM NON-LINEARITY (± LSB)	MAXIMUM GAIN TEMPCO (PPM/°C)	UNIT PRICE ($)	NOTES
DAC1222LD	NATIONAL	8	10	27.85	
DAC1221LD	NATIONAL	4	10	28.75	
DAC1220LD	NATIONAL	2	10	32.15	
DAC312B	P M I	1	40	40.43	

PART NUMBER	MANUFACTURER	MAXIMUM NON-LINEARITY (± LSB)	MAXIMUM GAIN TEMPCO (PPM/°C)	UNIT PRICE ($)	NOTES
AD7541AS	ANALOG DEVICES	1	5	41.85	
AD7545S	ANALOG DEVICES	2	5	44.55	2
Am6012DM	AMD	2	40	44.95	
AD7521S	ANALOG DEVICES	4	10	47.00	
AD7541AT	ANALOG DEVICES	0.5	5	48.80	
AD7548S	ANALOG DEVICES	1	5	56.10	2
AD7545T	ANALOG DEVICES	1	5	59.15	2
AD7542S	ANALOG DEVICES	1	5	63.20	2
AD7543S	ANALOG DEVICES	1	5	63.20	2
AD7548T	ANALOG DEVICES	0.5	5	63.77	2
AD7545U	ANALOG DEVICES	0.5	5	64.10	2
DAC851-CBI-I	BURR-BROWN	0.5	25	69.00	1
AD7542T	ANALOG DEVICES	0.5	5	71.80	2
AD7543T	ANALOG DEVICES	0.5	5	71.80	2
AD7521T	ANALOG DEVICES	2	10	75.00	
HI-562A-2	HARRIS	0.25	10 TYP	75.18	
AD566AS	ANALOG DEVICES	0.5	10	82.00	
AD7541S	ANALOG DEVICES	1	10	88.00	
AD567S	ANALOG DEVICES	0.5	30	89.00	1,2
UA565S	FAIRCHILD	0.5	30	95.00	1
AD7545GU	ANALOG DEVICES	0.5	5	95.04	2
MC3512	MOTOROLA	0.5	30	95.25	1
HI-7541S	HARRIS	1	5	96.00	
AD7542GT	ANALOG DEVICES	0.5	5	96.85	2
AD7543GT	ANALOG DEVICES	0.5	5	96.85	2
AD565AS	ANALOG DEVICES	0.5	30	97.00	1
AD7541T	ANALOG DEVICES	0.5	10	98.00	
AD7521U	ANALOG DEVICES	1	10	102.00	
HI-7541T	HARRIS	0.5	5	103.50	
AD566AT	ANALOG DEVICES	0.25	3	129.00	
AD565AT	ANALOG DEVICES	0.25	15	142.00	1

PART NUMBER	MANUFACTURER	MAXIMUM NON-LINEARITY (± LSB)	MAXIMUM GAIN TEMPCO (PPM/°C)	UNIT PRICE ($)	NOTES
UA565T	FAIRCHILD	0.25	20	142.50	1
AD562S	ANALOG DEVICES	0.25	5	209.70	
AD563S	ANALOG DEVICES	0.25	30	230.60	1
DAC10HT-1	BURR-BROWN	1	25	273.00	
AD563T	ANALOG DEVICES	0.25	10	293.60	1
DAC10HT	BURR-BROWN	0.5	10	295.00	

1=INTERNAL REFERENCE 2=INTERNAL LATCHES

Table 8-21. Quad 12-Bit D/A Converters with Internal Reference and Latches.

PART NUMBER	MANUFACTURER	TEMP RANGE	OUTPUT TYPE	MAXIMUM NON-LINEARITY (± LSB)	MAXIMUM GAIN TEMPCO (PPM/°C)	UNIT PRICE ($)
AD390J	ANALOG DEVICES	COM	V	0.75	40	148.00
AD390K	ANALOG DEVICES	COM	V	0.5	20	175.00
AD390S	ANALOG DEVICES	MIL	V	0.75	40	392.00
AD390T	ANALOG DEVICES	MIL	V	0.5	20	460.00

Table 8-22. 14-Bit D/A Converters with Internal Latches.

PART NUMBER	MANUFACTURER	TEMP RANGE	OUTPUT TYPE	MAXIMUM NON-LINEARITY (± LSB)	MAXIMUM GAIN TEMPCO (PPM/°C)	UNIT PRICE ($)
ICL7134BJC	INTERSIL	COM	I	2	8	21.40
ICL7134UJC	INTERSIL	COM	I	2	8	21.40
ICL7134BKC	INTERSIL	COM	I	1	8	32.95
ICL7134UKC	INTERSIL	COM	I	1	8	32.95
ICL7134BJI	INTERSIL	IND	I	2	8	47.85
ICL7134UJI	INTERSIL	IND	I	2	8	47.85
ICL7134BLC	INTERSIL	COM	I	0.5	8	49.45
ICL7134ULC	INTERSIL	COM	I	0.5	8	49.45
ICL7134BKI	INTERSIL	IND	I	1	8	61.05
ICL7134UKI	INTERSIL	IND	I	1	8	61.05

ICL7134BLI	INTERSIL	IND	I	0.5	8	74.25
ICL7134ULI	INTERSIL	IND	I	0.5	8	74.25
ICL7134BJM	INTERSIL	MIL	I	2	8	148.50
ICL7134UJM	INTERSIL	MIL	I	2	8	148.50
ICL7134BKM	INTERSIL	MIL	I	1	8	193.50
ICL7134UKM	INTERSIL	MIL	I	1	8	193.50

Table 8-23. 16-Bit Voltage-Output D/A Converters.

PART NUMBER	MANUFACTURER	TEMP RANGE	MAXIMUM NON-LINEARITY (± LSB)	MAXIMUM GAIN TEMPCO (PPM/°C)	UNIT PRICE ($)	NOTES
AD7546J	ANALOG DEVICES	COM	32	2	28.90	2
AD7546A	ANALOG DEVICES	IND	32	2	39.35	2
AD7546K	ANALOG DEVICES	COM	8	2	44.60	2
DAC701KH	BURR-BROWN	COM	2	25	48.00	1
DAC703KH	BURR-BROWN	COM	2	25	48.00	1
AD7546B	ANALOG DEVICES	IND	8	2	55.10	2
DAC71-COB-V	BURR-BROWN	COM	2	15	58.00	1
DAC71-CSB-V	BURR-BROWN	COM	2	15	58.00	1
DAC701BH	BURR-BROWN	IND	2	15	58.00	1
DAC703BH	BURR-BROWN	IND	2	15	58.00	1
DAC72C-COB-V	BURR-BROWN	COM	2	15	77.00	1
DAC72C-CSB-V	BURR-BROWN	COM	2	15	77.00	1
DAC72-COB-V	BURR-BROWN	IND	2	15	91.00	1
DAC72-CSB-V	BURR-BROWN	IND	2	15	91.00	1
DAC736J	BURR-BROWN	COM	1	10	220.00	1,2
DAC73J	BURR-BROWN	COM	1	10	242.00	1,2
DAC736K	BURR-BROWN	COM	0.5	10	260.00	1,2
DAC73K	BURR-BROWN	COM	0.5	10	286.00	1,2
DAC74	BURR-BROWN	COM	0.5	5	1495.00	1,2

1=INTERNAL REFERENCE 2=INTERNAL LATCHES

Table 8-24. 16-Bit Current-Output D/A Converters.

PART NUMBER	MANUFACTURER	TEMP RANGE	MAXIMUM NON-LINEARITY (± LSB)	MAXIMUM GAIN TEMPCO (PPM/°C)	UNIT PRICE ($)	NOTES
ICL7145JC	INTERSIL	COM	4	1 TYP	43.50	2
ICL7145KC	INTERSIL	COM	2	1 TYP	52.50	2
DAC71-COB-I	BURR-BROWN	COM	2	45	55.00	1
DAC71-CSB-I	BURR-BROWN	COM	2	45	55.00	1
ICL7145JI	INTERSIL	IND	4	1 TYP	55.50	2
ICL7145KI	INTERSIL	IND	2	1 TYP	67.50	2
DAC72C-COB-I	BURR-BROWN	COM	2	45	69.00	1
DAC72C-CSB-I	BURR-BROWN	COM	2	45	69.00	1
DAC72-COB-I	BURR-BROWN	IND	2	35	79.00	1
DAC72-CSB-I	BURR-BROWN	IND	2	35	79.00	1
DAC70C-COB-I	BURR-BROWN	COM	3.3	14	124.25	1
DAC70C-CSB-I	BURR-BROWN	COM	3.3	14	124.25	1
DAC70-COB-I	BURR-BROWN	IND	2	7	177.50	1
DAC70-CSB-I	BURR-BROWN	IND	2	7	177.50	1
DAC736J	BURR-BROWN	COM	1	10	220.00	1,2
DAC73J	BURR-BROWN	COM	1	10	242.00	1,2
DAC736K	BURR-BROWN	COM	0.5	10	260.00	1,2
DAC73K	BURR-BROWN	COM	0.5	10	286.00	1,2

1=INTERNAL REFERENCE   2=INTERNAL REFERENCES

Chapter 9

Analog-to-Digital Converters

AN ANALOG-TO-DIGITAL CONVERTER (A/D CONVERTER, OR ADC) converts an analog input voltage to a digital output code in the form of high or low voltages on the digital output lines. This chapter explains three types of A/D converters, defines the key parameters found in the data sheets, and presents selection guides.

TYPES OF A/D CONVERTERS

The three most common types of A/D converters are:

- ☐ successive approximation
- ☐ parallel, or flash
- ☐ integrating

The names of these types of A/D converters correspond to the methods of conversions used.

The Successive Approximation A/D Converter

Most A/D converters use the successive approximation method. A simplified block diagram of a successive approximation A/D converter is shown in Fig. 9-1. This type of converter consists of a successive approximation register (SAR), a clock, a D/A converter, a voltage reference, a comparator, and a set of output buffers.

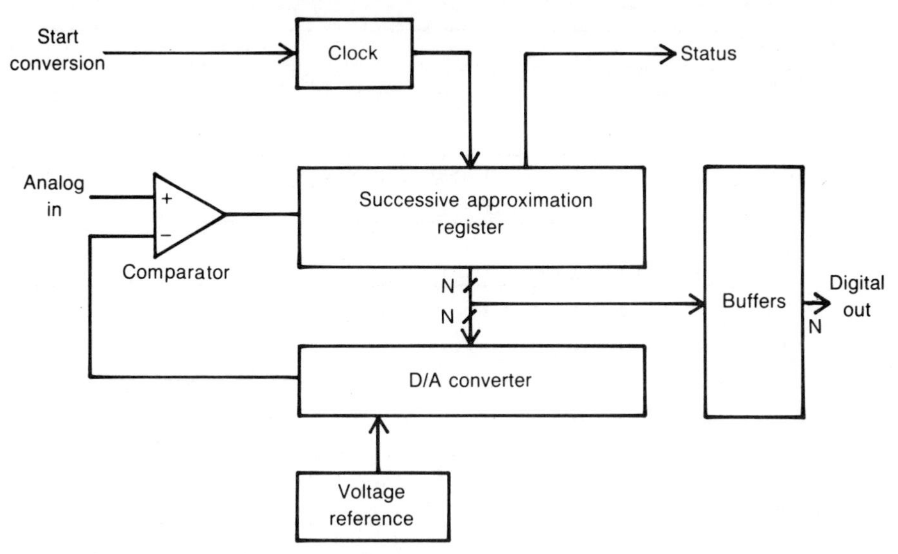

Fig. 9-1. Successive approximation A/D converter block diagram.

To see how this type of converter works, let's walk through the timing diagram shown in Fig. 9-2. This diagram is for an 8-bit converter. Let's assume that the analog voltage to be converted is +8.2 volts and that the voltage reference for the DAC is +10 volts (meaning that Full Scale is 9.961 volts). The bit values for bits B1 through B8 are given in Table 9-1, and D/A input and output values for this conversion are listed in Table 9-2. Now let's see what happens.

At time t_0, the START CONVERSION goes high for, let's say, 0.25 microseconds (μs). This enables the internal clock which runs, let's say, at 500 kHz, with a corresponding period (the time between the t numbers on the diagram) of 2 μs. Also at time t_0, the STATUS line goes high to indicate that conversion is in progress; and all the outputs of the SAR go high except BIT 1, which goes low. 01111111 is the first trial digital code and corresponds to a DAC output voltage of 4.961 volts (see Table 9-2).

At time t_1, the SAR reads the output of the comparator. Because the analog input is greater than 4.961 volts, the comparator output is high. Therefore, the SAR sets BIT 1 high and leaves this bit high for the remainder of the conversion. Also, at time t_1, the SAR sets BIT 2 low. Now the trial code is 10111111, corresponding to 7.461 volts.

At time t_2, the SAR again reads the output of the comparator. Because the analog input is greater than 7.461 volts, the comparator output is high, and the SAR sets BIT 2 high. BIT 3 is set low. The new trial code is 11011111. The DAC output is 8.711 volts.

Time t_3. The analog input is less than 8.711. The comparator

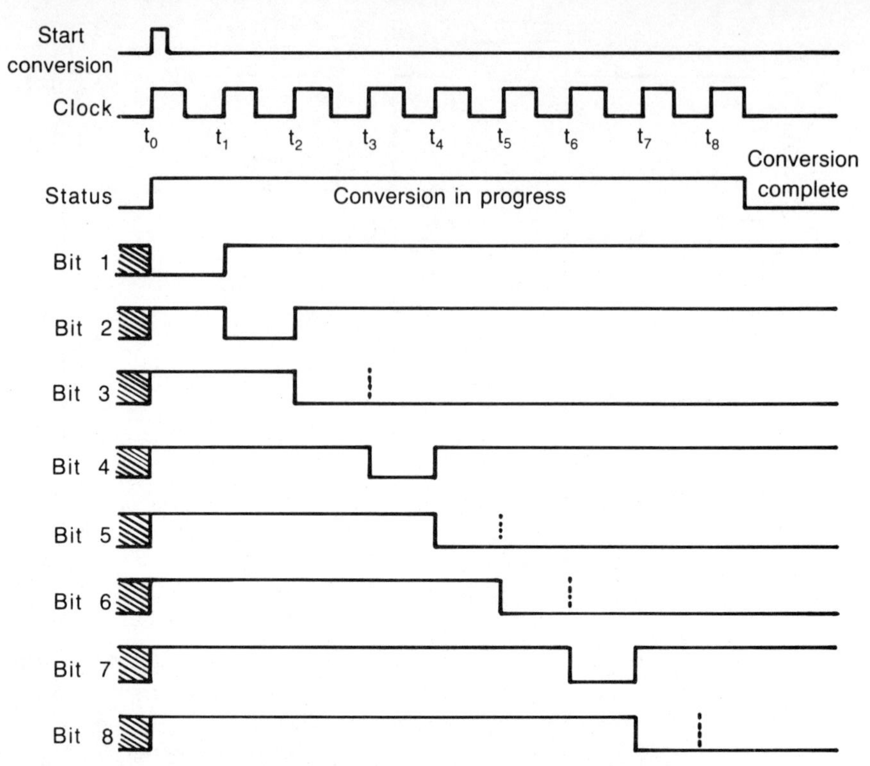

Fig. 9-2. Example timing diagram for 8-bit successive approximation A/D converter.

is low. The SAR leaves BIT 3 low. BIT 4 is set low. The trial code is 11001111. The DAC output is 8.086.

You get the idea. The SAR sequentially tests each bit to see if it should be set high or left low, and thus digitally *successively approximates* the value of the analog input. The final digital value is 11010010, corresponding to 8.203 volts, which is the closest possible approximation to 8.2 volts. The conversion takes 8 clock cycles or 16 μs. At the end of the conversion the STATUS line goes low, indicating that conversion is complete. If the analog input had been 8.21 volts (and assuming a perfect reference and D/A), then the final digital code would have been 11010011, or 8.242 volts.

The Flash Converter

A simplified block diagram of a flash A/D converter is shown in Fig. 9-3. To convert an analog input to n bits, 2^n comparators compare the input with 2^n different reference levels derived from a resistor chain.

A 6-bit flash converter, for example, uses 2^6 comparators. A set of 65 resistors (63 R value resistors and 2 R/2 value resistors) across a voltage reference divide the reference into 64 voltage levels which are connected to one side of the 64 comparators. The analog input is connected to the other side. When the LATCH ENABLE goes high, the comparator outputs, either high or low, are latched. The encoder section converts the 64 output lines to a 6-bit binary code.

This method is faster than the successive approximation method because the conversion for all digital output lines is done in parallel instead of serially. The speed is limited only by the propagation delay time of the comparator and encoder, which is typically less than 100 nanoseconds (ns) total. The RCA CA3300 6-bit flash converter typical conversion time is 70 ns. The Analog Devices AD9000 6-bit flash converter typical conversion time is 13 ns.

The disadvantage of this method is that the required number of comparators doubles for every additional bit of resolution. A 7-bit converter needs 128 comparators, an 8-bit converter needs 256 com-

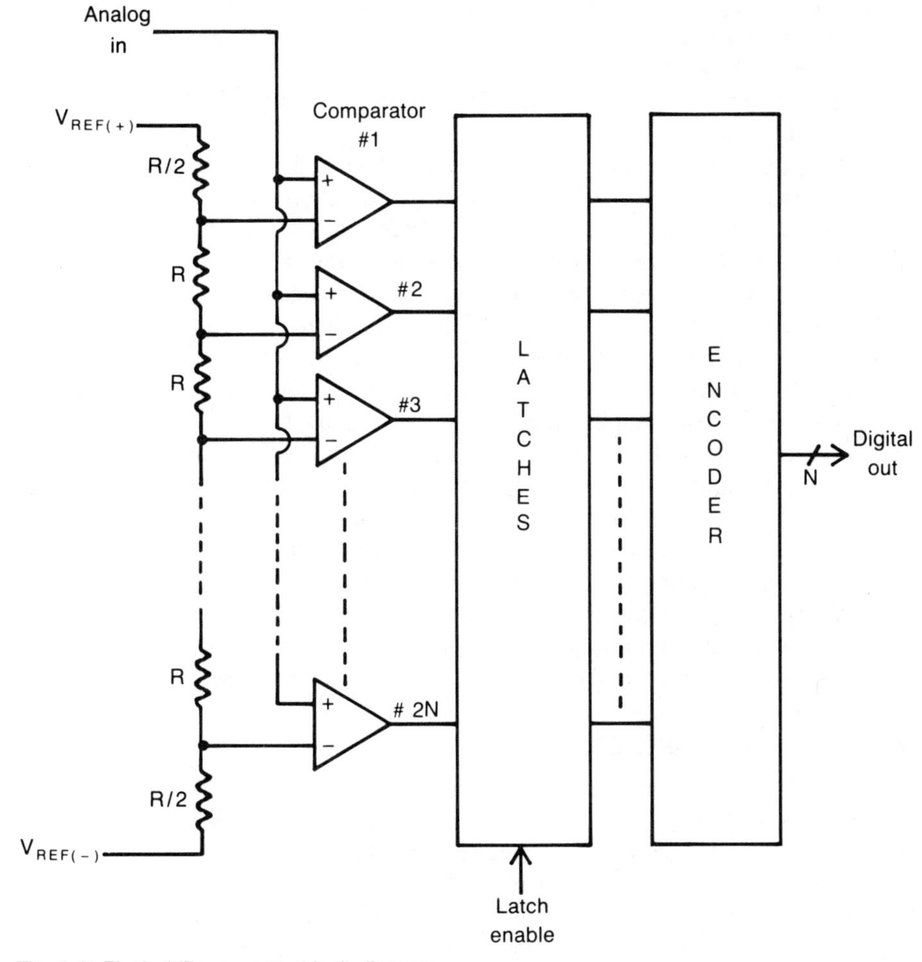

Fig. 9-3. Flash A/D converter block diagram.

parators, and so forth. This results in large chip sizes, low yield, and high cost. At present, the largest flash converter in production is the 8-bit converter, with 9-bit converters now being sampled by at least one manufacturer.

Flash converters are also called video A/D converters because they are mainly used in video or image processing applications.

Integrating A/D Converters

The third type of A/D converter is the *integrating* A/D converter. A block diagram of a simple integrating A/D converter is shown in Fig. 9-4.

The integrating A/D converter contains an integrator (the op amp, resistor R, and capacitor C), a comparator, some analog switches, and a digital processor section. The digital processor contains a clock, a counter, and the control logic for the switches. The capacitor is charged for a fixed time by the unknown analog input voltage and then discharged to zero by a known reference voltage. The discharge time, measured by the digital processor section, is then proportional to the input voltage. This is known as dual-slope conversion.

The integrating method is slow, taking milliseconds as compared to microseconds for the successive approximation method. For example, the Intersil ICL7109, a 12-bit plus sign intergrating A/D converter can perform a maximum of 30 conversions per second. That's 33 milliseconds per conversion. The maximum clock rate is about 250 kHz, or 4 μs per count. The conversion is in three phases: an auto-zero phase to null offsets in the op amp and comparator (2048 counts); an integrating phase to charge the capacitor using the unknown analog input voltage (2048 counts); and a de-integrate phase to discharge the capacitor using the reference voltage (4096 counts). The last phase is fixed at 4096 counts regardless of the input voltage. For voltages less than full scale, the digital processor latches the count and then waits until a total of 4096 counts before beginning the next auto-zero phase. The maximum 4096 counts for de-integration corresponds to 12-bit resolution. A total of 8192 counts for all three phases. The total conversion time is 8192 times 4 microseconds, or approximately 33 milliseconds.

PARAMETER DEFINITIONS

Most A/D parameters have a one-to-one correspondence to D/A parameters. In fact, you could change "digital input" to "digital output" and "analog output" to "analog input" in most of the definitions and figures given in Chapter 8 on D/A converters and you would have parameter definitions and figures for A/D converters. For this reason, only brief definitions for A/Ds are given below.

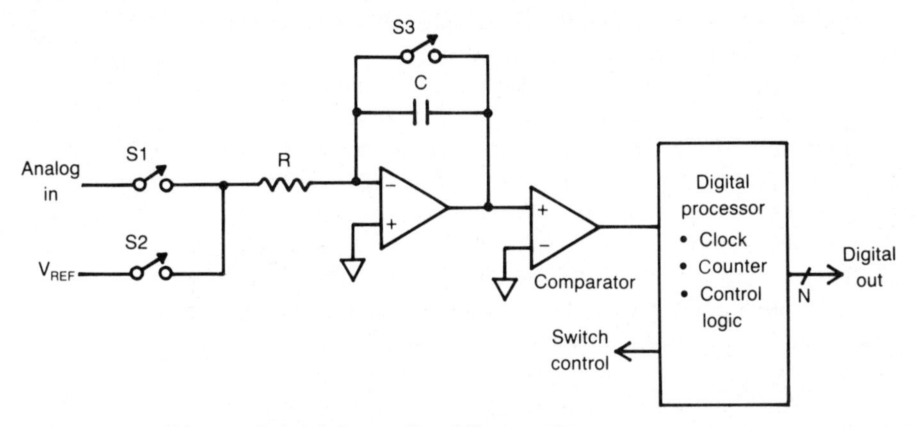

Fig. 9-4. Block diagram of simple integrating A/D converter.

If you skipped Chapter 8 or you are still a little fuzzy on D/A concepts and definitions, I recommend that you review the definitions and figures in Chapter 8 at this time.

Resolution is the number of digital output lines. An 8-bit A/D converter has 8 digital output lines, a 12-bit A/D converter has 12 digital output lines, and so forth. Resolution indirectly defines the smallest analog input "step size," or the difference between two analog input levels, than can cause the least significant bit (LSB) of the digital output to change from a 1 to 0, or vice versa.

Gain error is the amount that the analog input deviates from the ideal when the analog input is adjusted to produce a digital output corresponding to full scale (usually all 1's). *Offset error* is the amount that the analog input deviates from zero volts when the analog input is adjusted to a digital output of all 0's.

Relative accuracy, or *nonlinearity,* is the maximum amount, expressed in LSB's or fraction of LSB, that the analog input voltage may deviate from the ideal analog input voltage for any digital output code after the gain and offset errors have been adjusted to zero.

No missing codes. "No missing codes" is to A/D converters what "monotonicity" is the D/A converters. As the analog input is increased from zero to full scale, the digital output is supposed to increment one LSB at a time without skipping any codes. Manufacturers will either specify that there are no missing codes or they will specify a parameter called *differential nonlinearity.*

Differential nonlinearity (DNL) is the amount that the change in analog input deviates from the ideal change required to cause the digital output to change 1 LSB. For example, for an 8-bit A/D converter with a 10-volt range, it is supposed to take an analog input change of 39 millivolts (mV) to cause the output to change from 00000000 to 00000001. If it took 78 mV, then the DNL for this code change is +1 LSB. DNL must be measured for each and every code change. If, for the full range, the DNL is less than or equal to ±1 LSB then there will be no missing codes.

Note that this does not mean that the A/D relative accuracy, or nonlinearity, is less than or equal to ±1 LSB. It could be more or less than ±1 LSB. The nonlinearity specification tells you how much the A/D transfer function bows above or sags below the ideal straight-line transfer function. Differential nonlinearity tells you if there are any missing codes, meaning that for a successive approximation A/D, the internal DAC is nonmonotonic. A flash converter will have missing codes when the comparator input offset combined with adjacent resistor mismatch is greater than 1 LSB. Integrating A/D converters inherently do not have missing codes.

Gain temperature coefficient, expressed in PPM/°C or in maximum LSBs (or fraction) over the temperature range, is the amount that the full scale analog input voltage may change with temperature.

Conversion time. Conversion time for A/D converters corresponds to settling time for D/A converters and is the time required to convert the analog input voltage to a digital output code.

SELECTION GUIDES

Selection guides are given in Tables 9-3 through 9-11. Additional data is given in Appendix F.

Table 9-1. Bit Values for an 8-Bit A/D Converter with 10-Volt Range.

BIT	VALUE (VOLTS)	BIT	VALUE (VOLTS)
1 (MSB)	5.000	5	0.313
2	2.500	6	0.156
3	1.250	7	0.078
4	0.625	8 (LSB)	0.039

Table 9-2. D/A Input and Output Values for Example Shown in Fig. 9-2.

TIME	D/A INPUT	D/A OUTPUT (VOLTS)
0	01111111	4.961
1	10111111	7.461
2	11011111	8.711
3	11001111	8.086
4	11010111	8.398
5	11010011	8.242

TIME	D/A INPUT	D/A OUTPUT (VOLTS)
6	11010001	8.184
7	11010010	8.203
8	11010010	8.203

Table 9-3. 6-Bit A/D Converters.

PART NUMBER	MANUFACTURER	OPERATING TEMP (C) MIN	MAX	MAXIMUM NON-LINEARITY (+/- LSB)	TYPICAL CONVERSION TIME (uS)	UNIT PRICE ($)
NE5036	SIGNETICS	0	70	0.5	23	2.20
NE5037	SIGNETICS	0	70	0.5	9	2.40
CA3300	RCA	-40	85	0.8	0.07	21.25
AD9000J	ANALOG DEVICES	0	70	0.4 TYP	0.013	73.00
AD9000S	ANALOG DEVICES	-55	125	0.75	0.013	109.00

Table 9-4. 7-Bit A/D Converters.

PART NUMBER	MANUFACTURER	OPERATING TEMP (C) MIN	MAX	MAXIMUM NON-LINEARITY (+/- LSB)	TYPICAL CONVERSION TIME (uS)	UNIT PRICE ($)
MC10317L	MOTOROLA	0	70	0.2 TYP	0.07	73.00
MC10315L	MOTOROLA	0	70	0.2 TYP	0.07	73.00

Table 9-5. 8-Bit A/D Converters.

PART NUMBER	MANUFACTURER	OPERATING TEMP (C) MIN	MAX	MAXIMUM NON-LINEARITY (+/- LSB)	TYPICAL CONVERSION TIME (uS)	UNIT PRICE ($)
TLC541I	T I	-40	85	0.5	19	3.86
ADC0804LC	NATIONAL	0	70	1	110	4.05
ADC0833CCN	NATIONAL	0	70	1	80	4.80
TLC540I	T I	-40	85	0.5	10	5.08
ADC0805LC	NATIONAL	-40	85	1	110	5.15
ADC0809CC	NATIONAL	-40	85	1	100	5.37
TL533	T I	-40	85	0.5 TYP	300	6.10
TL521	T I	-40	85	0.5 TYP	100 MIN	6.10
ADC0803LC	NATIONAL	-40	85	0.5	110	6.20
TL522	T I	-40	85	0.25 TYP	200 MIN	6.39

PART NUMBER	MANUFACTURER	OPERATING TEMP (C) MIN	MAX	MAXIMUM NON-LINEARITY (+/- LSB)	TYPICAL CONVERSION TIME (uS)	UNIT PRICE ($)
TL531	T I	-40	85	0.5 TYP	300	6.79
TLC533AI	T I	-40	85	0.5	30	8.03
ADC0817CC	NATIONAL	-40	85	1	100	8.92
ADC0802LC	NATIONAL	-40	85	0.5	110	9.25
ADC0833BCN	NATIONAL	0	70	0.5	80	9.75
NE5034	SIGNETICS	0	70	0.5	17	10.00
ADC0808CC	NATIONAL	-40	85	0.5	100	10.12
TL532	T I	-40	85	0.5	300	10.35
TL520	T I	-40	85	0.25 TYP	70 MIN	10.81
TLC532AI	T I	-40	85	0.5	15	11.20
AD673J	ANALOG DEVICES	0	70	0.5	20	12.35
AD7574J	ANALOG DEVICES	0	70	0.75	15	12.50
TL530	T I	-40	85	0.5	300	12.65
ADC0833CCJ	NATIONAL	-40	85	1	80	12.75
ADC0816CC	NATIONAL	-40	85	0.5	100	14.92
AD7574K	ANALOG DEVICES	0	70	0.5	15	15.00
AD7574A	ANALOG DEVICES	-25	85	0.75	15	15.50
ADC0800PC	NATIONAL	0	70	1	50	16.60
ADC0801LC	NATIONAL	-40	85	0.25	110	17.00
AD7574B	ANALOG DEVICES	-25	85	0.5	15	18.00
Am6148	AMD	0	70	0.5	1	19.40
ADC0833BCJ	NATIONAL	-40	85	0.5	80	21.75
Am6108	AMD	0	70	0.5	1	22.45
AD570J	ANALOG DEVICES	0	70	0.5	25	28.60
AD7574S	ANALOG DEVICES	-55	125	0.75	15	30.00
AD7574T	ANALOG DEVICES	-55	125	0.5	15	35.00
ADC0800P	NATIONAL	-55	125	1	50	35.05
ADC0802L	NATIONAL	-55	125	0.5	110	41.60
ADC0801L	NATIONAL	-55	125	0.25	110	50.25
AD673S	ANALOG DEVICES	-55	125	0.5	20	61.30
AD570S	ANALOG DEVICES	-55	125	0.5	25	61.30

PART NUMBER	MANUFACTURER	OPERATING TEMP (C) MIN	MAX	MAXIMUM NON-LINEARITY (+/- LSB)	TYPICAL CONVERSION TIME (uS)	UNIT PRICE ($)
ADC0808C	NATIONAL	-55	125	0.5	100	62.00
ADC82	BURR-BROWN	-25	85	0.5	2.8 MAX	69.00
ADC0816C	NATIONAL	-55	125	0.5	100	74.00
ADC60-08	BURR-BROWN	0	70	0.5	0.88 MAX	285.00

Table 9-6. 10-Bit A/D Converters.

PART NUMBER	MANUFACTURER	OPERATING TEMP (C) MIN	MAX	MAXIMUM NON-LINEARITY (+/- LSB)	TYPICAL CONVERSION TIME (uS)	UNIT PRICE ($)
AD573J	ANALOG DEVICES	0	70	1	20	21.70
AD573K	ANALOG DEVICES	0	70	0.5	20	29.20
ADC1021CCD	NATIONAL	-40	85	1	200	31.40
UA571J	FAIRCHILD	0	70	1	25	42.47
AD571J	ANALOG DEVICES	0	70	1	25	45.30
UA571K	FAIRCHILD	0	70	0.5	25	49.40
AD571K	ANALOG DEVICES	0	70	0.5	25	52.20
ADC80AG-10	BURR-BROWN	-25	85	0.5	21 MAX	85.00
UA571S	FAIRCHILD	-55	125	0.5	25	96.62
AD571S	ANALOG DEVICES	-55	125	1	25	103.40
AD573S	ANALOG DEVICES	-55	125	1	20	103.40
ADC84KG-10	BURR-BROWN	0	70	0.5	6 MAX	105.00
ADC85C-10	BURR-BROWN	0	70	0.5	6 MAX	119.00
ADC85-10	BURR-BROWN	-25	85	0.5	6 MAX	143.00
AD579J	ANALOG DEVICES	0	70	0.5	2.2 MAX	145.50
AD579K	ANALOG DEVICES	0	70	0.5	1.8 MAX	177.00
AD579T	ANALOG DEVICES	-55	125	0.5	1.8 MAX	253.50
ADC60-10	BURR-BROWN	0	70	0.5	1.88 MAX	316.00

Table 9-7. 10-Bit Plus Sign A/D Converters.

PART NUMBER	MANUFACTURER	OPERATING TEMP (C) MIN	MAX	MAXIMUM NON-LINEARITY (+/- LSB)	TYPICAL CONVERSION TIME (uS)	UNIT PRICE ($)
AD7571J	ANALOG DEVICES	0	70	1	80	32.20
AD7571K	ANALOG DEVICES	0	70	1	80	37.90

PART NUMBER	MANUFACTURER	OPERATING TEMP (C) MIN	MAX	MAXIMUM NON-LINEARITY (+/- LSB)	TYPICAL CONVERSION TIME (uS)	UNIT PRICE ($)
AD7571A	ANALOG DEVICES	-25	85	1	80	40.45
AD7571B	ANALOG DEVICES	-25	85	0.75	80	46.10
AD7571S	ANALOG DEVICES	-55	125	1	80	121.30
AD7571T	ANALOG DEVICES	-55	125	0.75	80	138.35

Table 9-8. 12-Bit A/D Converters.

PART NUMBER	MANUFACTURER	OPERATING TEMP (C) MIN	MAX	MAXIMUM NON-LINEARITY (+/- LSB)	TYPICAL CONVERSION TIME (uS)	UNIT PRICE ($)
ADC1211HC	NATIONAL	-25	85	2	100	45.00
HI-574AJD-5	HARRIS	0	70	1	25	46.07
AD574AJ	ANALOG DEVICES	0	70	1	25	49.50
ADC1210HC	NATIONAL	-25	85	0.5	100	54.65
HI-574AKD-5	HARRIS	0	70	0.5	25	59.82
AD574AK	ANALOG DEVICES	0	70	0.5	25	65.00
Am6112	AMD	0	70	1 TYP	3.3	67.50
ADC80AG-12	BURR-BROWN	-25	85	0.5	25 MAX	87.00
ADC1211H	NATIONAL	-55	125	2	100	90.00
AD574AL	ANALOG DEVICES	0	70	0.5	25	95.00
ADC84KG-12	BURR-BROWN	0	70	0.5	10 MAX	119.00
HI-574ASD-2	HARRIS	-55	125	1	25	123.76
ADC1210H	NATIONAL	-55	125	0.5	100	125.45
ADC85C-12	BURR-BROWN	0	70	0.5	10 MAX	132.00
HI-5712-5	HARRIS	0	75	0.5	9	135.76
AD578J	ANALOG DEVICES	0	70	0.5	6 MAX	141.75
AD574AS	ANALOG DEVICES	-55	125	1	25	145.00
AD572AD	ANALOG DEVICES	-25	85	0.5	25 MAX	160.00
AD578K	ANALOG DEVICES	0	70	0.5	4.5 MAX	166.50
ADC85-12	BURR-BROWN	-25	85	0.5	10 MAX	172.00
ADC87U	BURR-BROWN	-55	125	0.5	7.5	172.00
HI-574ATD-2	HARRIS	-55	125	0.5	25	178.75

PART NUMBER	MANUFACTURER	OPERATING TEMP (C) MIN	MAX	MAXIMUM NON-LINEARITY (+/- LSB)	TYPICAL CONVERSION TIME (uS)	UNIT PRICE ($)
AD572BD	ANALOG DEVICES	-25	85	0.5	25 MAX	191.00
AD5240K	ANALOG DEVICES	0	70	0.5	5 MAX	193.00
AD574AT	ANALOG DEVICES	-55	125	0.5	25	195.00
AD5211B	ANALOG DEVICES	-25	85	0.5	13 MAX	200.00
AD5212B	ANALOG DEVICES	-25	85	0.5	13 MAX	200.00
AD5214B	ANALOG DEVICES	-25	85	0.5	13 MAX	200.00
AD5215B	ANALOG DEVICES	-25	85	0.5	13 MAX	200.00
AD5205B	ANALOG DEVICES	-25	85	0.5	50 MAX	207.00
AD5204B	ANALOG DEVICES	-25	85	0.5	50 MAX	207.00
AD5202B	ANALOG DEVICES	-25	85	0.5	50 MAX	207.00
AD5201B	ANALOG DEVICES	-25	85	0.5	50 MAX	207.00
AD578L	ANALOG DEVICES	0	70	0.5	3 MAX	207.25
ADC87	BURR-BROWN	-55	125	0.5	7.5	240.00
AD5240B	ANALOG DEVICES	-25	85	0.5	5 MAX	254.00
HI-5712A-5	HARRIS	0	75	0.5	9	273.00
AD574AU	ANALOG DEVICES	-55	125	0.5	25	285.00
HI-5712-2	HARRIS	-55	125	0.5	9	286.50
AD5204T	ANALOG DEVICES	-55	125	0.5	50 MAX	300.00
AD5205T	ANALOG DEVICES	-55	125	0.5	50 MAX	300.00
AD5201T	ANALOG DEVICES	-55	125	0.5	50 MAX	300.00
AD5202T	ANALOG DEVICES	-55	125	0.5	50 MAX	300.00
ADC60-12	BURR-BROWN	0	70	1	1.88 MAX	326.00
AD5212T	ANALOG DEVICES	-55	125	0.5	13 MAX	330.00
AD5215T	ANALOG DEVICES	-55	125	0.5	13 MAX	330.00
AD5211T	ANALOG DEVICES	-55	125	0.5	13 MAX	330.00
AD5214T	ANALOG DEVICES	-55	125	0.5	13 MAX	330.00
AD572SD	ANALOG DEVICES	-55	125	0.5	25 MAX	383.00
HI-5712A-2	HARRIS	-55	125	0.5	9	436.90
ADC10HT-1	BURR-BROWN	-55	200	2	30	448.00
ADC10HT	BURR-BROWN	-55	200	0.5	30	485.00

Table 9-9. 12-Bit Plus Sign A/D Converters.

PART NUMBER	MANUFACTURER	OPERATING TEMP (C) MIN	MAX	MAXIMUM NON-LINEARITY (+/- LSB)	TYPICAL CONVERSION TIME (uS)	UNIT PRICE ($)
ICL7109C	INTERSIL	0	70	1 COUNT	33333	15.00
AD7552K	ANALOG DEVICES	0	70	1	160000	16.10
ICL7109I	INTERSIL	-20	85	1 COUNT	33333	29.70
ICL7109M	INTERSIL	-55	125	1 COUNT	33333	61.15

Table 9-10. 14-Bit A/D Converters.

PART NUMBER	MANUFACTURER	OPERATING TEMP (C) MIN	MAX	MAXIMUM NON-LINEARITY (+/- LSB)	MAXIMUM CONVERSION TIME (uS)	UNIT PRICE ($)
ICL7115JC	INTERSIL	0	70	2	40	58.50
ICL7115JI	INTERSIL	-25	85	2	40	88.50
ICL7115KC	INTERSIL	0	70	1	40	102.00
ICL7115KI	INTERSIL	-25	85	1	40	132.00
ICL7115LC	INTERSIL	0	70	0.5	40	148.50
ICL7115LI	INTERSIL	-25	85	0.5	40	178.50

Table 9-11. 16-Bit A/D Converters.

PART NUMBER	MANUFACTURER	OPERATING TEMP (C) MIN	MAX	MAXIMUM NON-LINEARITY (+/- LSB)	TYPICAL CONVERSION TIME (uS)	UNIT PRICE ($)
ADC71J	BURR-BROWN	0	70	4	50 MAX	129.00
ADC71K	BURR-BROWN	0	70	2	50 MAX	161.00
ADC72J	BURR-BROWN	0	70	4	50 MAX	173.00
ADC72A	BURR-BROWN	-25	85	4	50 MAX	207.00
ADC72K	BURR-BROWN	0	70	2	50 MAX	216.00
ADC76J	BURR-BROWN	0	70	4	15 MAX	229.00
ADC72B	BURR-BROWN	-25	85	2	50 MAX	258.00
ADC76K	BURR-BROWN	0	70	2	15 MAX	265.00
ADC73J	BURR-BROWN	0	70	1	150	285.00
ADC731J	BURR-BROWN	0	70	1	150	335.00
ADC73K	BURR-BROWN	0	70	0.5	150	345.00
ADC100-BOB	BURR-BROWN	0	70	4	200000 MAX	387.00
ADC731K	BURR-BROWN	0	70	0.5	150	395.00

Chapter 10

Sample-And-Hold Amplifiers

A SAMPLE-AND-HOLD AMPLIFIER (S/H AMPLIFIER, OR SHA) IS a special-purpose, dual-mode amplifier that is designed to follow, or track, an analog input signal and then, upon receipt of an external "hold" command, hold the analog level that is present at the input.

Figure 10-1 shows a block diagram of a typical data acquisition system in which an S/H amplifier is placed between an analog multiplexer and an A/D converter to ensure that the input to the A/D converter is constant during conversion. Also, an S/H amplifier may be used in applications where it is important to record an analog voltage level at a specific point in time or at a certain time relative to other events.

FUNCTIONAL DESCRIPTION

Figure 10-2 shows a block diagram of the Harris HA-2420/2425 and the Analog Devices AD583 sample-and-hold IC (both IC's are pin-for-pin compatible, both circuits are schematically almost identical). This IC consists of two op amps, a low-leakage bipolar analog switch, and a TTL-compatible logic circuit to control the switch. The first op amp is a high-gain amplifier. The second op amp is a high-impedance unity-gain amplifier. Generally, the inverting input of the first op amp is tied to the output of the second op amp

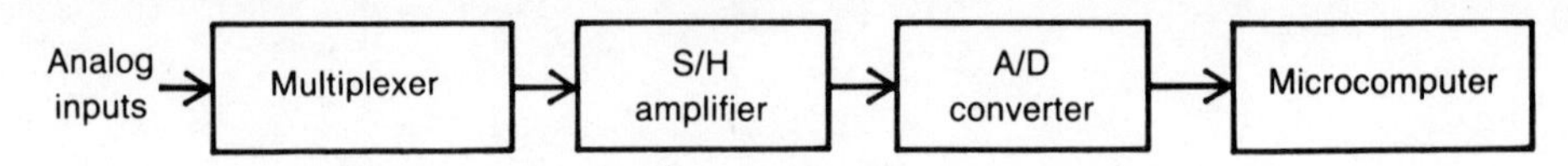

Fig. 10-1. Block diagram of a typical data acquisition system.

so that the overall gain of the S/H amplifier is equal to one. However, by making the inverting input of the first amplifier available, Harris and Analog Devices have added some versatility.

When the S/H control input is low, the switch is closed and the amplifier is in the sample mode. Assuming that the amplifier is wired to have a gain of one, then whatever analog voltage is present at the $V_{IN(+)}$ input appears across the external hold capacitor C_H and at the output V_{OUT}.

When the S/H control input goes high, the switch opens. This is the hold mode. The voltage across the capacitor remains the same as what it was just prior to the S/H control input going high, and any variation at the input will not be seen at the output. The hold capacitor is isolated from the external load by the high-impedance unity-gain amplifier and, thus, maintains its charge until the S/H control input goes low.

The above discussion assumes an ideal S/H amplifier. Now let's look at actual performance characteristics and the definitions of key data sheet parameters.

PARAMETER DEFINITIONS

Figure 10-3 shows the waveforms for the sample-hold operation performed on a trapezoidal-shaped analog input voltage. (Note that the shape of the analog input could be anything—dc, varying dc, sinusoidal, or any combination.) The S/H amplifier samples the incoming waveform until about half way up the leading edge, at which time it receives the hold command.

There is a finite time delay between the time that the S/H con-

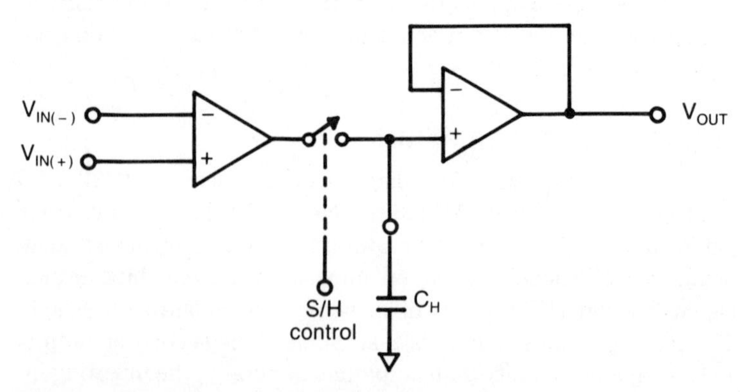

Fig. 10-2. Block diagram of a sample-and-hold IC.

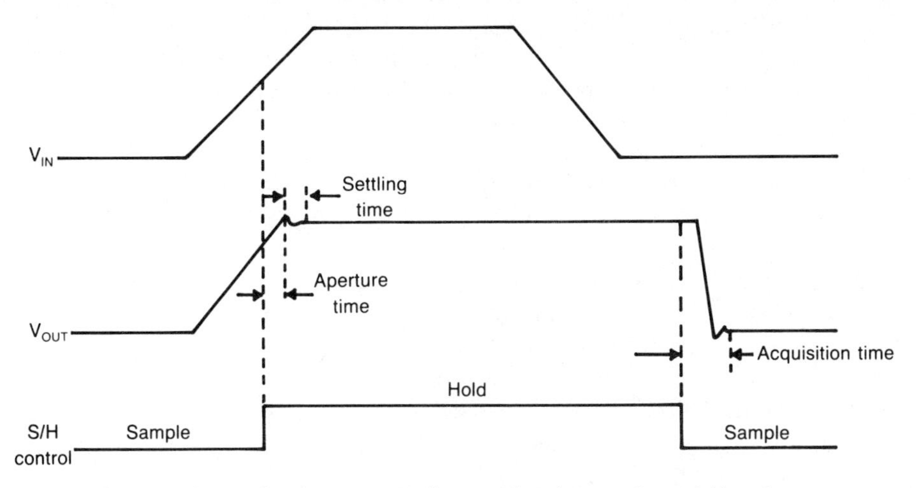

Fig. 10-3. S/H waveforms showing aperture time, setting time, and acquisition time.

trol input goes high and the time that the switch opens. This time delay is called *aperture time* and is on the order of 100-200 nanoseconds for general-purpose S/H amplifiers and on the order of 10-50 nanoseconds for high-speed S/H amplifiers. The shorter the aperture time, the smaller the error between the actual output level and the ideal level, i.e., the analog level you wanted to capture. For a very slow ramp (or slowly varying inputs), the error may be insignificant. For fast changing inputs, this error can theoretically be reduced to zero by having the S/H control input line go high early by an amount of time equal to the aperture time.

After aperture time comes *settling time*, the time it takes the output to settle to within some percentage (0.01%, or sometimes 0.1%) of the final value or to within 1 millivolt of the final value. Settling time is typically less than 1 microsecond. This parameter, unfortunately, is not always specified. This may or may not be important in some applications. In a data acquisition system, A/D conversion generally should not begin until after the aperture time and the settling time are complete. However, if, for example, the A/D conversion takes 25 microseconds and the sum of aperture time and settling time is less than 10-15 microseconds, then conversion could start at the same time that the S/H control line goes high (hold mode). This is alright for successive approximation converters because the low-order bits are done last.

Going from the hold mode to the sample mode, there is a time delay called the *acquisition time*. See Fig. 10-2. Acquisition time has three components: (1) the time required to close the switch, (2) the time required for the amplifier to slew to the new voltage level, and (3) the time required for the output to settle to within 1 millivolt or some percentage of the final value. The largest component is the amplifier slewing time, the time it takes the high gain input amplifier to charge or discharge the hold capacitor. Therefore, the larger the hold capacitor, the larger the acquisition time.

The acquisition time parameter in the data sheets is usually specified as the time required for the output to change 10 volts *for a specified value of the hold capacitor*. This is important, because a parameter called *droop rate* is also directly related to the value of the hold capacitor. Droop rate (which will be defined shortly), relatively speaking, is low when the hold capacitor value is *high*. Acquisition time is low when the hold capacitor is *low*. I have seen data sheets specify acquisition time for a low-value hold capacitor and then specify droop rate for a high-value hold capacitor. Usually, a set of curves showing both parameters as a function of hold capacitor value is included in the data sheet.

The next largest component of acquisition time is settling time. (Don't confuse this settling time with the settling time which follows aperture time. In the hold mode, settling time is separate from aperture time. In the sample mode, settling time is included in the acquisition time.) For this reason, some data sheets will show one value of acquisition time to settle to 0.1% and a slightly longer acquisition time to settle to 0.01% of final value.

Droop rate. In the hold mode, droop rate is the change in hold capacitor voltage divided by the change in time for a given hold capacitor. It is usually specified in millivolts-per-millisecond (mV/ms). Droop, the change in hold capacitor voltage, is due to leakage current (through the switch or other path) and the input bias current of the buffer amplifier and may be positive or negative. Since droop rate is dependent on the value of the external hold capacitor, some data sheets specify *droop current* instead. Assuming that the hold capacitor has negligible leakage, then droop rate may be calculated from droop current as follows:

$$\text{Droop Rate} = dV_C/dT = I/C$$

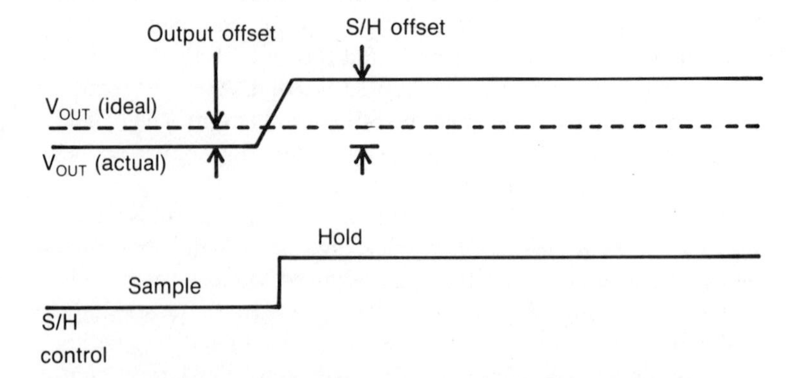

Fig. 10-4. S/H waveforms showing S/H offset and output offset.

where:

I is the droop current; and

C is the hold capacitor value

Sample-to-hold offset. Sample-to-hold offset is the change in output voltage when the mode is changed from sample to hold. See Fig. 10-4. For example, assume that the input voltage is zero (the usual test condition for this parameter). In the sample mode, ideally the output should be zero. It isn't, of course, mainly due to input offset voltage common to all operational amplifiers. Now if the circuit is switched from the sample mode to the hold mode, the output will typically shift 1-2 millivolts due to internal transient currents getting coupled to the hold capacitor. The sample-to-hold offset may be positive or negative and is sometimes called *hold step.*

SELECTION GUIDES

Selection guides are given in Tables 10-1 through 10-3. Additional data is given in Appendix G.

Table 10-1. Sample-and-Hold Amplifiers, Commercial Temperature Range.

PART NUMBER	MANUFACTURER	TYPICAL ACQUIS. TIME (uS)	TYPICAL APERTURE TIME (nS)	TYPICAL DROOP RATE (mV/mS)	UNIT PRICE ($)
NE5537	SIGNETICS	20	125	0.002	2.70
UA398	FAIRCHILD	20	125	0.003	3.33
SMP-11G	P M I	5	50	0.08	4.13
LF398	NATIONAL	20	200	0.003	4.25
LF398A	NATIONAL	20	200	0.003	6.50
SMP-81F	P M I	3.5	50	0.1	8.25
HA-2425-5	HARRIS	5	30	0.005	9.77
AD582K	ANALOG DEVICES	6	200	0.1 MAX	12.50
HA-5320-5	HARRIS	1.5	25	0.1	13.38
SMP-11F	P M I	5	50	0.07	14.25
SMP-10F	P M I	5	50	0.005	14.25
SMP-81E	P M I	3.5	50	0.1	16.13
AD583K	ANALOG DEVICES	5	50	0.005	23.90
SMP-11E	P M I	5	50	0.06	26.25
SMP-10E	P M I	5	50	0.005	26.25
SHC80KP	BURR-BROWN	10 MAX	40	0.5	51.00
AD346J	ANALOG DEVICES	1.6	30	0.1	69.00

PART NUMBER	MANUFACTURER	TYPICAL ACQUIS. TIME (uS)	TYPICAL APERTURE TIME (nS)	TYPICAL DROOP RATE (mV/mS)	UNIT PRICE ($)
ADSHC-85	ANALOG DEVICES	5 MAX	25	0.2	91.00
SHC85	BURR-BROWN	4.5 MAX	30	0.125	95.00
SHM60	BURR-BROWN	0.8	12	1	154.00
HTC-0300A	ANALOG DEVICES	0.15	6	1	191.00
HTC-0300	ANALOG DEVICES	0.17	6	5	223.00
HTS-0025	ANALOG DEVICES	0.03	5	0.2	288.00
HTS-0010K	ANALOG DEVICES	0.016	-2	0.1 MAX	369.00

Table 10-2. Sample-and-Hold Amplifiers, Industrial Temperature Range.

PART NUMBER	MANUFACTURER	TYPICAL ACQUIS. TIME (uS)	TYPICAL APERTURE TIME (nS)	TYPICAL DROOP RATE (mV/mS)	UNIT PRICE ($)
SHC298AM	BURR-BROWN	6	30	0.025	6.95
IH5110I	INTERSIL	4	120	0.0005	8.95
IH5111I	INTERSIL	4	120	0.0005	10.45
UA298	FAIRCHILD	20	125	0.003	10.91
IH5112I	INTERSIL	4	120	0.0005	11.25
IH5113I	INTERSIL	4	120	0.0005	12.40
IH5114I	INTERSIL	4	120	0.0005	12.75
IH5115I	INTERSIL	4	120	0.0005	13.90
LF298	NATIONAL	20	200	0.003	16.50
LH0043C	NATIONAL	30	20	0.002	22.25
LH0023C	NATIONAL	50	150	0.002	27.60
LH0053C	NATIONAL	8	10	0.01	42.15
HTC-0500A	ANALOG DEVICES	0.85	30	0.5	106.00

Table 10-3. Sample-and-Hold Amplifiers, Military Temperature Range.

PART NUMBER	MANUFACTURER	TYPICAL ACQUIS. TIME (uS)	TYPICAL APERTURE TIME (nS)	TYPICAL DROOP RATE (mV/mS)	UNIT PRICE ($)
SE5537	SIGNETICS	20	125	0.002	11.40
IH5110M	INTERSIL	4	120	0.0005	13.50
IH5112M	INTERSIL	4	120	0.0005	16.50

PART NUMBER	MANUFACTURER	TYPICAL ACQUIS. TIME (uS)	TYPICAL APERTURE TIME (nS)	TYPICAL DROOP RATE (mV/mS)	UNIT PRICE ($)
IH5111M	INTERSIL	4	120	0.0005	16.80
IH5113M	INTERSIL	4	120	0.0005	17.65
IH5114M	INTERSIL	4	120	0.0005	18.00
IH5115M	INTERSIL	4	120	0.0005	19.15
UA198	FAIRCHILD	20	125	0.003	24.93
SMP-11B	P M I	5	50	0.07	33.53
SMP-10B	P M I	5	50	0.005	33.53
LH0023	NATIONAL	50	150	0.001	34.20
LF198	NATIONAL	20	200	0.003	34.60
AD582S	ANALOG DEVICES	6	200	0.1 MAX	34.80
LH0043	NATIONAL	30	20	0.001	38.50
HA-2420-2	HARRIS	5	30	0.005	44.66
LF198A	NATIONAL	20	200	0.003	51.20
SMP-11A	P M I	5	50	0.06	52.50
SMP-10A	P M I	5	50	0.005	52.50
LH0053	NATIONAL	5	10	0.006	60.00
HA-5320-2	HARRIS	1.5	25	0.1	61.00
AD346S	ANALOG DEVICES	1.6	30	0.1	108.00
SHC85ET	BURR-BROWN	4.5 MAX	30	0.125	129.00
ADSHC-85ET	ANALOG DEVICES	5 MAX	25	0.2	138.00
HTC-0500S	ANALOG DEVICES	0.85	30	0.5	144.00
HTC-0300AM	ANALOG DEVICES	0.15	6	1	259.00
HTC-0300M	ANALOG DEVICES	0.17	6	5	305.00
HTS-0025M	ANALOG DEVICES	0.03	5	0.2	365.00
HTS-0010S	ANALOG DEVICES	0.016	-2	0.1 MAX	438.00

Chapter 11

Special Purpose Amplifiers

IN THIS CHAPTER I WILL DISCUSS FIVE TYPES OF SPECIAL-PURpose amplifiers:

- ☐ Unity-gain amplifiers
- ☐ Instrumentation amplifiers
- ☐ Operational transconductance amplifiers
- ☐ Norton current-mode operational amplifiers
- ☐ Video amplifiers

These amplifiers are "special purpose" only in the sense that they are less universal in their applications than the op amp. As a result (I suppose), there are fewer parts available. Nevertheless, these amplifiers are standard circuits having definite characteristics that make them especially suitable for particular types of applications.

UNITY-GAIN AMPLIFIERS

Figure 11-1 shows two types of unity-gain amplifiers—(A) the voltage follower and (B) the buffer amplifier. Both of these amplifiers are single-ended amplifiers and have high input impedance, low output impedance, and a voltage gain of 1. The buffer amplifier differs from the voltage follower in that the buffer amplifier

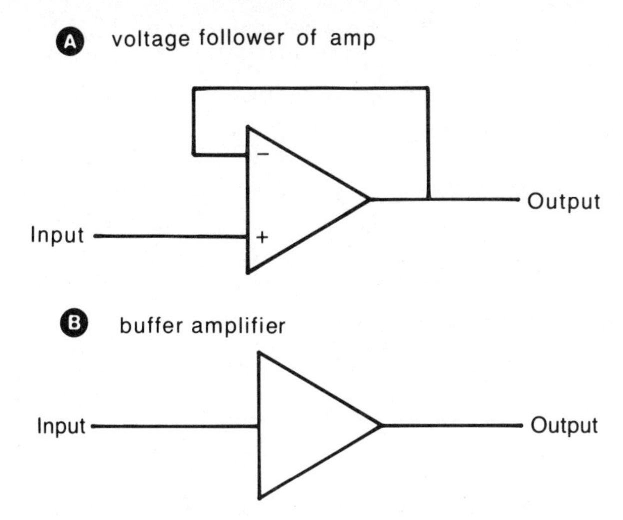

Fig. 11-1. Unity-gain amplifiers.

is designed to have a high output current capability, typically greater than 100 mA. For this reason, the buffer amplifier is often called a current booster. Table 11-1 lists available voltage follower ICs and buffer amplifiers in alphabetical order by manufacturer.

As Fig. 11-1 shows, the voltage follower IC is an op amp with the inverting input internally connected to the output. Figure 11-2 shows two voltage followers used as high-impedance buffers for a fixed-gain differential amplifier. Without these followers, the input impedance would be equal to the sum of R1 and R2 (R1 + R2 in parallel with R3 + R4 in the general case).

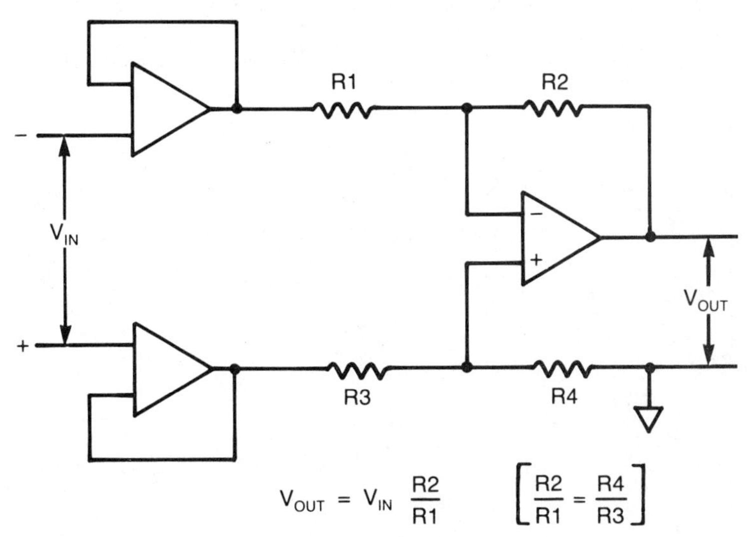

Fig. 11-2. Voltage followers used as high-impedance buffers for a differential amplifier.

Figure 11-3 shows the schematic of the National LM310 voltage follower IC. The base of Q8 is the inverting input and is internally wired to the output through a 5 kΩ resistor (R7).

National is the only manufacturer making voltage follower ICs—the LM102 series (102, 202, and 302 for military, industrial, and commercial temperature grade versions, respectively), the LM110 series and the LH2110 series (hybrid dual LM110). These are all old designs, and I doubt that National will be offering any new voltage follower ICs. When they were originally designed, these ICs were special op amps optimized for high slew rates at unity gain-10 V/μs for the LM102 series and 30 V/μs for the LM110

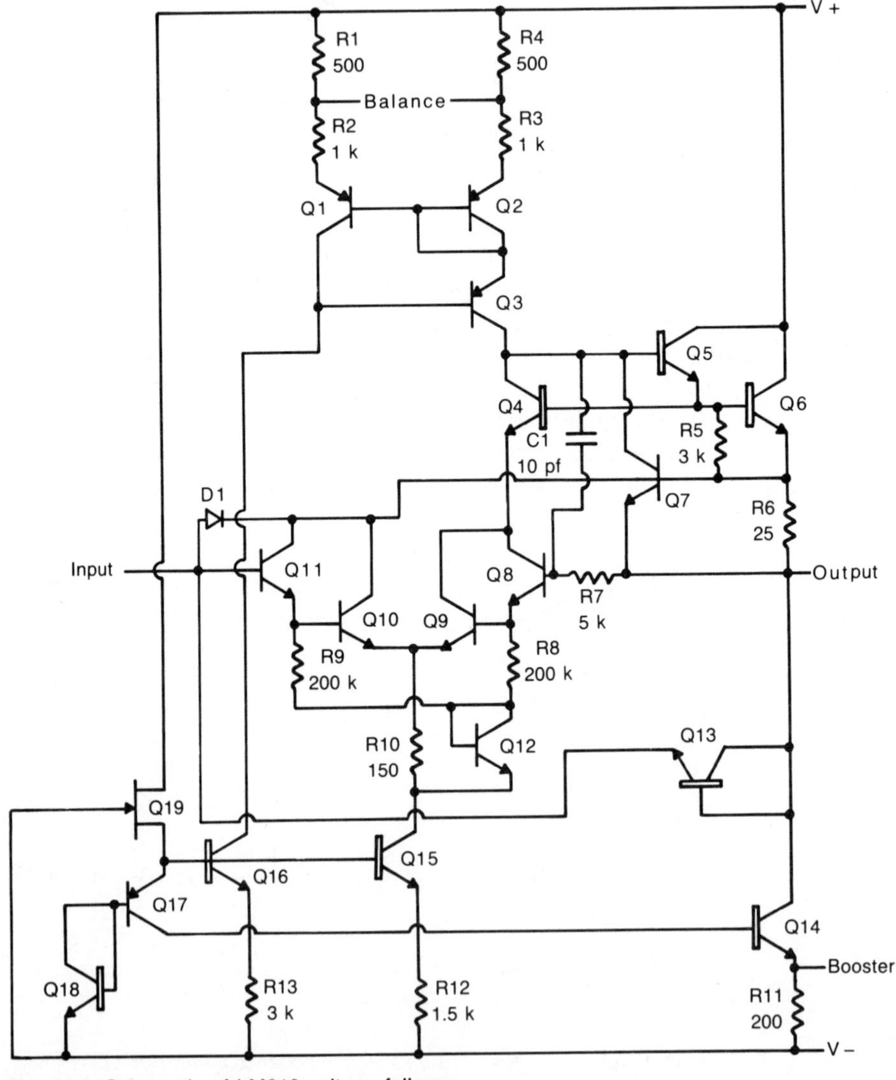

Fig. 11-3. Schematic of LM310 voltage follower.

and LH2110 series. That was before the advent of low-cost, high-slew-rate FET-input op amps.

Buffer Amplifiers

Figure 11-4 shows the buffer amplifier in its most common application, being used as a current booster in a closed-loop op-amp circuit. In this application, the buffer amplifier's voltage gain (typically 0.85 to 0.95) is not critical. However, in unity-gain applications, such as being used as a driver amplifier (driving a low-resistance load or high capacitance), the actual buffer amplifier gain should be taken into account.

Figure 11-5 shows the schematic of the National LH0033 buffer amplifier, alternated sourced by Analog Devices (the ADLH0033). Q1 operates as a source-follower driving a standard four-transistor output stage (see Chapter 3). Minimum dc input-to-output offset voltage is achieved as follows. The Q4-D1-R1 combination is a current source. If the Q5 and Q6 betas are about the same, then the input transistor Q1 operates at the same current as Q4. If Q1 and Q4 are matched, then Q1 will have the same gate-to-source voltage as Q4—one diode drop plus one 50-ohm IR drop. Because these are the same components that are connected to the Q1 source, then the voltage at the Q2 emitter approximately equals the input voltage. Finally, the output voltage is approximately equal to the Q2 emitter voltage, and thus equal to the input voltage. In practice, R1 and R2 are factory laser-trimmed such that the output offset voltage is less than 20 mV.

Figure 11-6 shows a schematic of the National LH0002 buffer amplifier, the lowest-cost buffer amplifier in Table 11-1 ($7.80 for the LH0002C). This circuit is simpler than the LH0033, using just four transistors wired as two stages of complementary emitter-followers. However, the LH0002 has a lower input resistance and lower slew rate than the FET-input LH0033.

The Harris HA2630/2635 offers the highest output current capability at the expense of lower slew rate and bandwidth. The

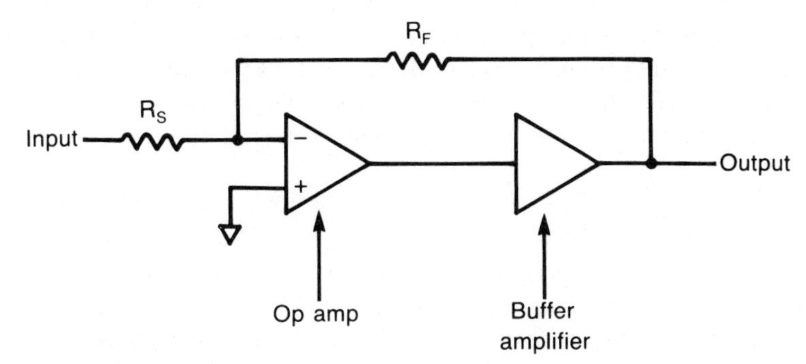

Fig. 11-4. Buffer amplifier used as an op-amp current booster.

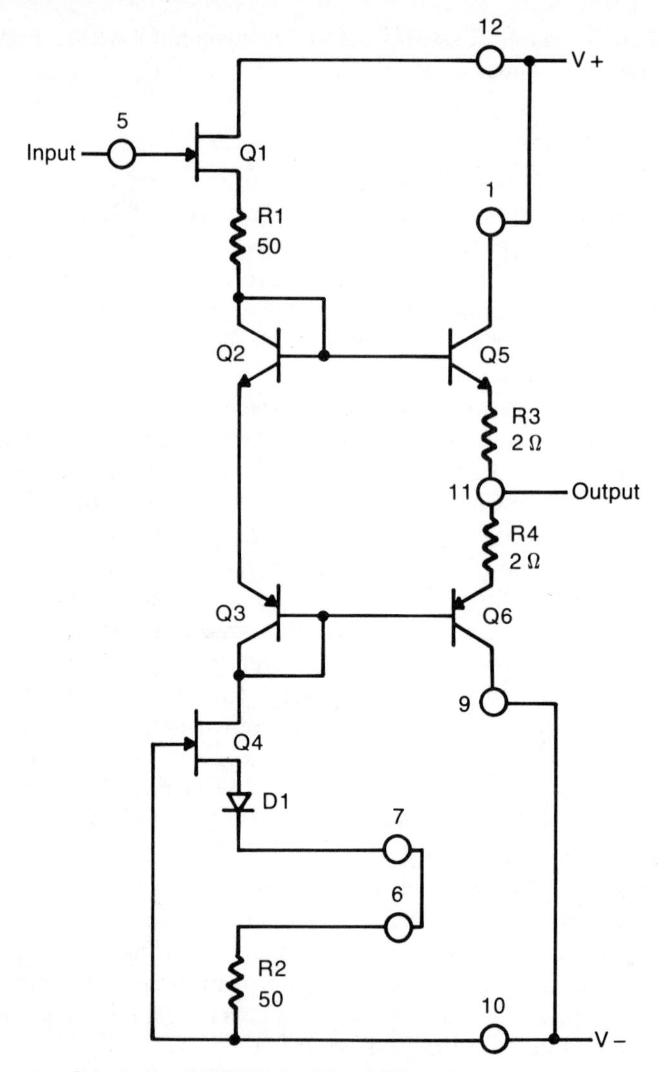

Fig. 11-5. Schematic of LH0033 buffer amplifier.

300 mA maximum output current shown in Table 11-1 is the guaranteed minimum (the minimum "maximum"). The typical maximum output current is 400 mA and the absolute maximum rating is 700 mA. The Motorola MC1438R/1538R has a typical maximum output current of 300 mA and an absolute maximum rating of 350 mA.

The National LH0063 offers the highest slew rate—6000 v/μs typical, driving a 1 kΩ load with an input of 10 V, and 2400 V/μS typical (2000 min), driving a 50-ohm load with an input of 10 V. 2400 V/μs is equivalent to 2.4 V/ns, or a 10-volt swing in just over 4 ns. National literally calls the LH0063 a "Damn Fast Buffer Amplifier." That's the data sheet title.

The PMI BUF-03 offers the highest minimum gain (for a buffer amplifier; National's voltage follower ICs have higher gain but virtually no drive capability). Table 11-1 shows the minimum gain for a 10 kΩ load and an input voltage of 10 V. For a 1 kΩ load the minimum gain drops to 0.9925 for the BUF-03A and BUF-03E and 0.9905 for the BUF-03B and BUF-03F. The minimum slew rate is 220 V/μs for the "A" and "E" versions and 180v/μs for the "B" and "F" versions.

INSTRUMENTATION AMPLIFIERS

Figure 11-7 shows the basic instrumentation amplifier. The instrumentation amplifier is a fixed-gain differential amplifier. The key purpose of the instrumentation amplifier is to amplify small signals riding on large commonmode voltages. The gain of the amplifier is usually programmed by one or two external resistors. Instrumentation amplifiers are designed to have the following characteristics:

- ☐ High input impedance
- ☐ High common-mode rejection
- ☐ Low offset
- ☐ Low offset drift

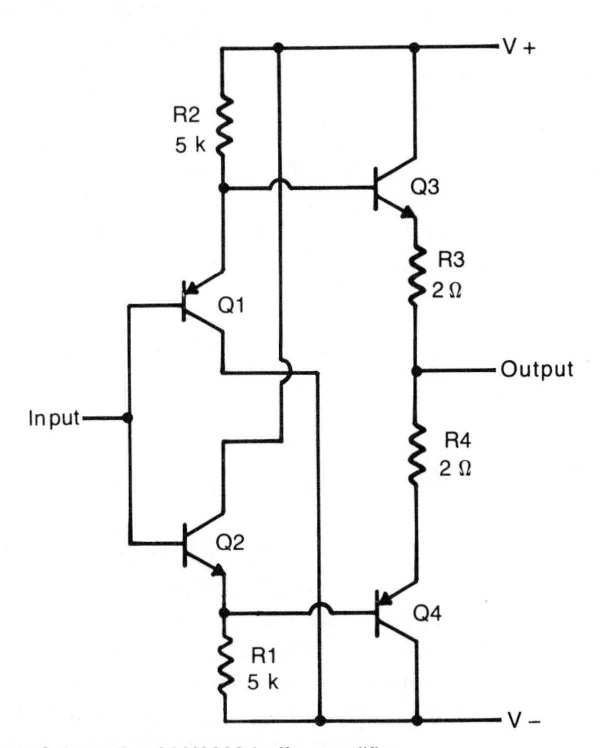

Fig. 11-6. Schematic of LH0002 buffer amplifier.

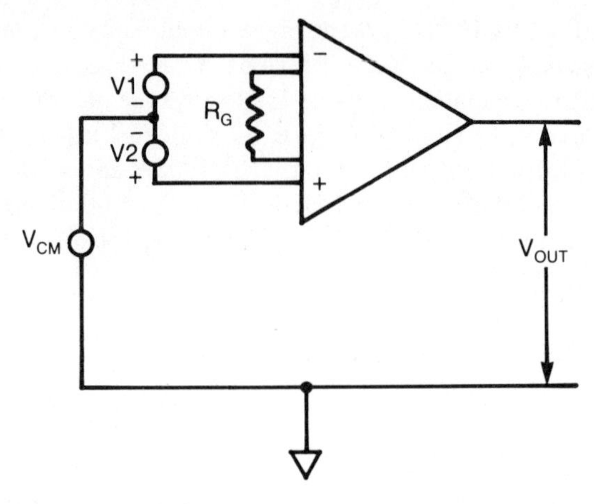

Fig. 11-7. Basic instrumentation amplifier.

- ☐ Low gain tolerance
- ☐ Low gain drift
- ☐ Low gain nonlinearity

Figure 11-8 shows the internal structure of a typical IC instrumentation amplifier. This is the traditional three-op-amp approach. Chances are that you can build this circuit yourself for less money using discrete op amps and film resistors (or resistor packs with very tight matching characteristics). The main advantage of the IC instrumentation amplifier are small size (one package) and guaranteed performance specifications.

In Fig. 11-8, amplifier A1 and A2 are noninverting amplifiers each with a gain of 1 + 20 k/(RG/2). A2 is wired as a unity-gain differential amplifier. Let's call the minus input V1, the plus input V2, and the corresponding gains G1 and G2. The equation for the output voltage is:

$$V_{OUT} = (V2\ G2) - (V1\ G1) = G1\ (V2 - V1) \qquad [G2 = G1]$$

The gain equation is:

$$\text{GAIN} = G1 = G2 = 1 + 20\ \text{k}/(\text{RG}/2) = 1 + 40\ \text{k}/\text{RG}$$

Figure 11-8 shows two lines called SENSE and REFERENCE. The SENSE line is sometimes internally wired to the output and is not available. The REFERENCE line is sometimes called the COMMON, or GROUND, but is always available. When the SENSE is available, it can be used with the REFERENCE line for both driving remote loads and driving low-impedance loads as shown in Fig. 11-9. These configurations eliminate errors due to voltage drops in the output

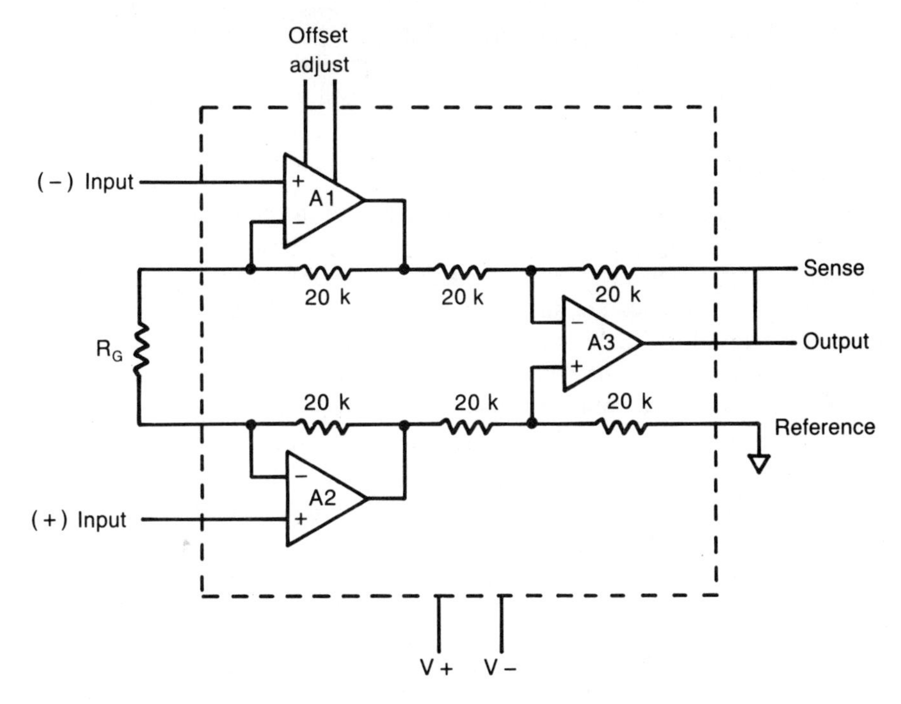

Fig. 11-8. Internal structure of typical instrumentation amplifier IC.

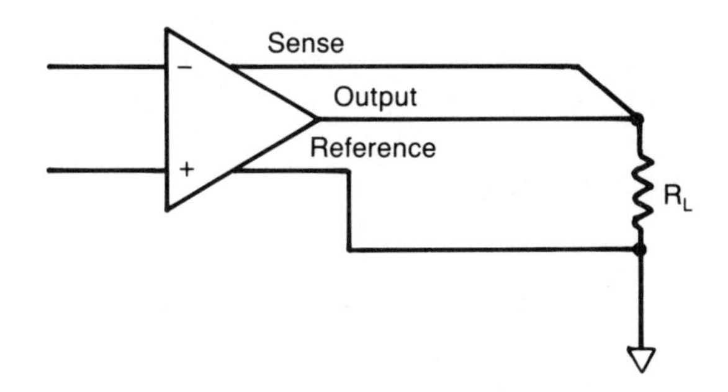

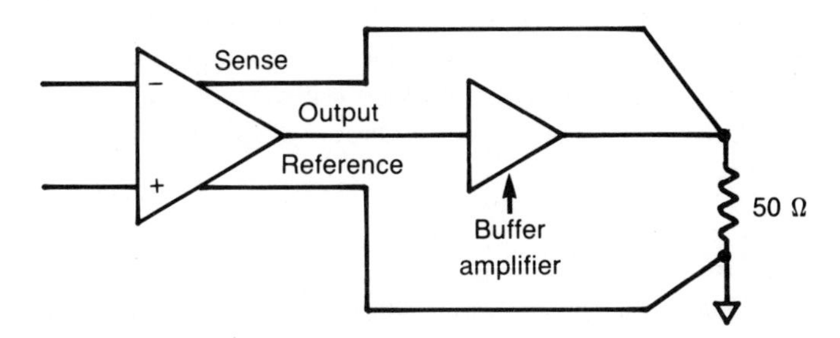

Fig. 11-9. Using the SENSE and REFERENCE lines.

and ground lines, and, in the case of adding the buffer amplifier inside the feedback loop, increase the drive capability of the instrumentation amplifier without degrading the gain accuracy.

Specifications

Commond-mode rejection, or CMR, is defined as the ability of a differential amplifier to reject, or *not amplify*, that portion of each input signal voltage which is common to each other. CMR is tested by tying both inputs together and applying a dc or low-frequency signal (less than 60 Hz) to the shorted inputs. Theoretically, no signal should be present at the output. CMR is expressed in dB for a given gain configuration and is equal to 20 LOG(V_{IN}/V_{OUT}).

Input offset voltage is the voltage at the input required to produce zero volts at the output. *Input offset drift* is the change in input offset voltage with temperature, expressed in microvolts per degrees Centigrade (μV/°C).

There are three specifications associated with gain accuracy: (1) gain equation error, (2) gain drift (temperature coefficient), and (3) gain nonlinearity.

Gain equation error is defined as the deviation from computed gain using the manufacturer's gain equation for that particular IC. The calculation assumes a nominal value for RG. Generally, the error is different for different values of RG, i.e., different gains. Therefore, gain equation error is specified as a percent of one or more specific gains and does not include the tolerance of the gain setting resistor RG. If the nominal gain in your application is critical, then you will have to use a precision trimpot in series with a precision resistor, both of which will have to have tight initial tolerances (in accordance with your requirements) and excellent temperature stability.

Gain drift is the variation in gain with respect to temperature and is expressed in PPM/°C. Gain drift, like gain equation error, is also a function of gain.

Gain nonlinearity is defined as the deviation from the ideal straight line transfer function of the amplifier for a specified gain. That is, if you plot input voltage versus output voltage for a given gain, the plot should be a perfectly straight line. The amount that the actual deviates from the ideal straight line is called nonlinearity (just like A/D and D/A converters, only both input and output are analog). Nonlinearity is expressed as a percent of full scale (% F.S.) at a specified gain.

Selection Guides

Table 11-2 lists available instrumentation amplifiers by manufacturer and highlights three specifications: input offset voltage, offset drift, and gain nonlinearity. Input offset voltage and drift are specified at gain $G = 1000$ and gain nonlinearity is specified at gain $G = 1$.

The best price/performance amplifier in the table is the Burr-Brown INA101, available in industrial and military temperature grades. The INA104 offers the same performance in commercial temperature grade. The price is higher because the 104 includes a fourth amplifier which can be used as an output active filter, level shifter, an additional gain block, or coaxial shield driver (used to improve common-mode rejection). Burr-Brown's older 36XX series amplifiers are not price/performance competitive with the newer 101 and 104 and are not likely to be used in new designs.

Analog Devices AD624 offers on-chip gain resistors with pin-programmable gains of 1, 100, 200, 500, and 1000. This device also offers an outstanding 4 nanovolts per root Hertz ($nV/\sqrt{Hz}$) typical input noise voltage at $f = 1$ kHz. (Burr-Brown INA101 and 104 are 13 and 17 $nV/\sqrt{Hz}$ respectively.)

Before Analog Devices came out with the AD624 (and AD524), the best low-noise instrumentation amplifier was the National LH0038, which offers 6 $nV/\sqrt{Hz}$ from 100 Hz to 10 kHz. Although the LH0038 offers a fourth coaxial shield amplifier, this part can not compete with the Analog Devices part with a commercial temp grade price of $59.15 (versus $23.30 for the AD624BD) and a military temp grade price of $118.60 (versus $38.90 for the AD624SD). Note that although these will be old price figures by the time you read this, I doubt that National's price for the LH0038 will be lower than the AD624. National is going to have to introduce new IC instrumentation amplifiers in order to compete with both Analog Devices and Burr-Brown in this marketplace.

OPERATIONAL TRANSCONDUCTANCE AMPLIFIERS

Figure 11-10 shows the schematic of the basic operational transconductance amplifier, or OTA. The OTA converts an input voltage to an output current according to the following equation:

$$I_{OUT} = V_{IN} \left(\tfrac{1}{2} I_{ABC}\right) / (kT/q)$$

where:

V_{IN} is the voltage difference between the (+) input and the (−) input;

I_{ABC} is the input stage operating current (Q10 current in Fig. 11-10); and

kT/q is approximately equal to 26 mV at room temperature (see Chapters 3 and 4).

The above equation is only valid for V_{IN} less than about 25

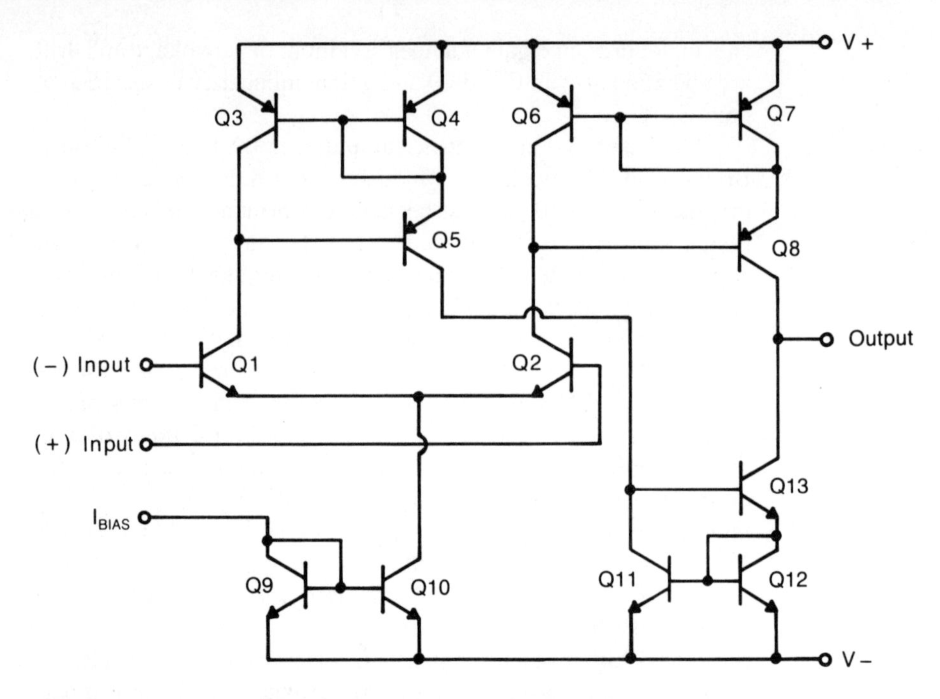

Fig. 11-10. Basic operational transconductance amplifier (OTA).

mV. Above this level the OTA transfer function begins to become nonlinear. Recall from Chapters 3 and 4 that a 60 mV differential input voltage will cause the collector currents of a differential pair to have a 10 to 1 ratio. 120 mV will cause a 100:1 ratio. At this level, one side is ON, the other is OFF.

The term $(\frac{1}{2}I_{ABC})/(kT/q)$ in the above equation is called the *transconductance* of the amplifier and is denoted by the symbol g_m. Transconductance, by definition, is output current divided by input voltage and is expressed in mhos (ohm spelled backwards). For an I_{ABC} of 500 μA, the transconductance at room temperature (27 °C) is equal to (0.5) (500 μA)/26 mV, or about 9.6 μA output current per mV of differential input voltage, or 9600 μmhos. The bias current I_{ABC} in Fig. 11-10 is set by an external resistor connected between the base-collector junction of Q9 and V+. This allows the OTA transconductance to be programmable—a useful feature for some applications. Most OTAs have an I_{ABC} current range of about 0.5 μA to 1 mA.

Linearizing Diodes

Figure 11-11 shows the schematic of the National LM13600 dual OTA (only one shown) with so-called "linearizing diodes" and a buffer (Darlington transistor). The purpose of the linearizing diodes is to make the transfer equation (presented above) linear and valid beyond a differential input voltage of 25 mV. Actually it is more accurate to say that these diodes, when biased, convert the OTA into a current amplifier.

Figure 11-12 shows the model of the OTA with linearizing diodes biased. The model assumes that Q4 and Q5 base currents are negligible. First, assume that the input current I_S equals zero. A current I_D is forced into the anode of the diodes and $I_D/2$ is pulled out of each of the cathodes of the diodes. Because the diodes are identical and the transistors are identical, then the voltage from the base of Q4 to the base of Q5 is zero. I4 and I5 are equal and I_{OUT} is zero.

Now inject a small input current I_S which must be less than $I_D/2$ (but may be positive or negative). Then the currents shall be distributed as shown in Fig. 11-12.

The transfer function is developed as follows:

1. $I_{OUT} = I5 - I4$
2. $I_{ABC} = I4 + I5$

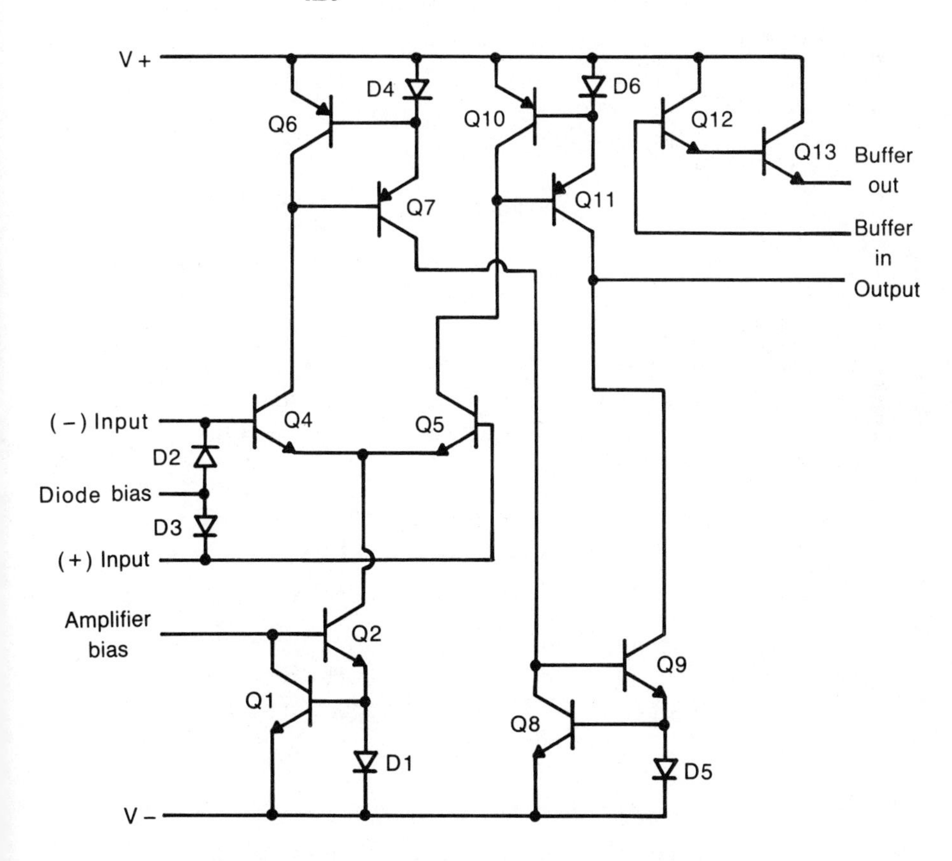

Fig. 11-11. Schematic of LM13600 dual OTA with linearizing diodes and buffers (one OTA shown).

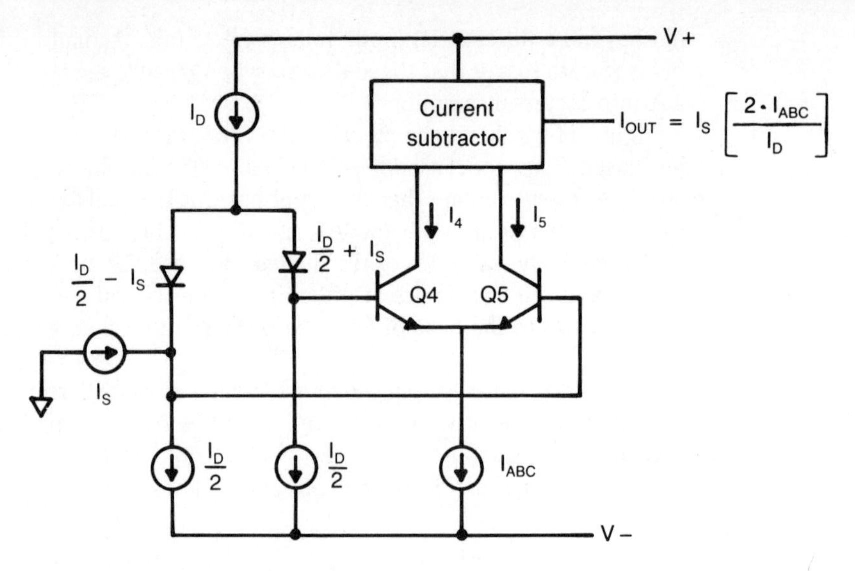

Fig. 11-12. Model of OTA with linearizing diodes.

3. If you divide each of the above equations by 2 and add, you get:

$$I5 = I_{ABC}/2 + I_{OUT}/2$$

4. Similarly, if you divide each of the equations (1 and 2) by 2 and subtract (1) from (2), you get:

$$I4 = I_{ABC}/2 - I_{OUT}/2$$

5. Because the difference in voltages across the diodes is exactly equal to the difference in voltages across the bases of the transistors, and because the diodes are identical geometries and the transistors are identical geometries, then the following is true:

$$V_{IN} = (kT/q) \ln \left[\frac{I_D/2 + I_S}{I_D/2 - I_S}\right] = (kT/q) \ln \left[\frac{I5}{I4}\right]$$

6. Because these terms are equal the arguments of the logarithms must be equal. Setting the arguments equal to each other and substituting the equations for I4 and I5 developed in steps 3 and 4 and rearranging terms, you get:

$$I_{OUT} = I_S \, (2 \, I_{ABC}/I_D)$$

The OTA is now a current amplifier. But you can make it back into an OTA by replacing the current source I_S by a voltage source and a resistor. The advantage of using the linearizing diodes is that the transfer function is both linear and independent of temperature (not counting the temperature coefficients of external current-setting resistors).

Note that the current source I_D and the two $I_D/2$ sources are not included in the National LM13600. They *are* included in the RCA CA3280. In the CA3280 all three current sources are simultaneously programmed by one external resistor.

Applications

The OTA may be used in any application where the output (or parameter to be manipulated) is a current instead of a voltage. One such example is shown in Fig. 11-13 where a dual OTA (National LM13600 or RCA CA3280) is used to create a voltage-controlled triangle and square wave generator. Note the OTA symbol—an op amp with a programmable current source hung on the output.

In this application the capacitor C is charged at a linear rate (I1/C) by a constant-current source from the OTA. When the voltage on the capacitor reaches some value (determined by the second OTA output current I2 and resistor R) the second OTA output current changes polarity, reversing the voltage across the resistor R and also switching the input stage of the first OTA. The first OTA output current then changes polarity and begins to discharge at a linear rate. The result is a triangle wave at the emitter of Q2 and a square wave at the emitter of Q4. Additionally, the frequency of oscillation is controlled by the voltage V_C which sets the value of I1.

Selection Guide

Table 11-3 lists available OTAs. Input offset voltage for OTAs is defined as that input voltage which is required to obtain zero output current.

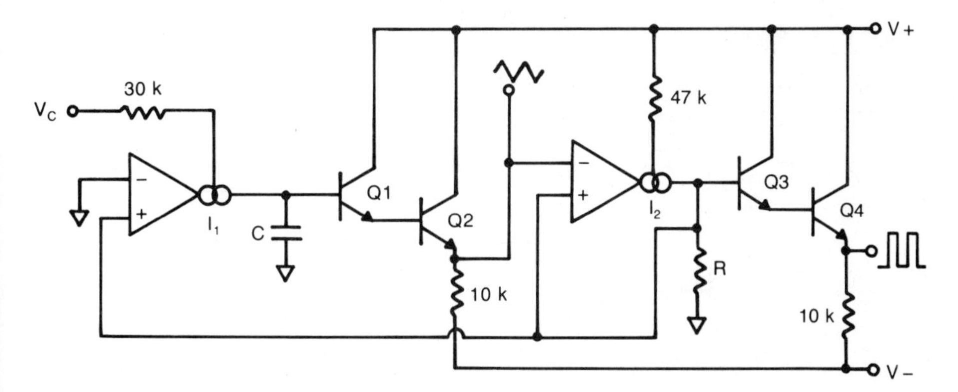

Fig. 11-13. Dual OTA used as a voltage-controlled triangle/squarewave generator.

The LM3080, CA3060 and CA3080 are basic OTAs—no linearizing diodes and no Darlington transistor buffer.

The LM13600, LM13700, and NE5517 have linearizing diodes and buffers. The LM13700 has a higher maximum buffer input current than the 13600—2 μA versus 0.5 μA when a 5 kΩ resistor is connected from the emitter of the output buffer transistor to V−.

The CA3094 has no linearizing diodes but does have a buffer transistor that can source up to 100 mA continuous, 300 mA peak.

The CA3280 has no buffer transistor but does have linearizing diodes with on-chip programmable diode-bias current sources. 3 dB bandwidth is 9 MHz (2 MHz is typical for most other OTAs); slew rate is 125 V/μs (compared to 50 Vμs for other OTAs); input noise is 8 nV/$\sqrt{\text{Hz}}$ at 1 kHz (not specified for most other OTAs). Perhaps the only problem with the CA3280 is that it is not second-sourced.

NORTON CURRENT-MODE OPERATIONAL AMPLIFIERS

Figure 11-14 shows a simplified schematic of the so-called Norton op amp. In the configuration shown, Q3 gets its base current from the Q4 output transistor through an external resistor R_F. The dc output voltage is equal to the Q3 V_{BE} plus the Q3 base current times the value of R_F.

Current flowing into the noninverting input $I_{IN(+)}$ flows into Q1 and is mirrored over to Q2. If the current flowing into the inverting input $I_{IN(-)}$ is equal to the noninverting current (and, therefore, the Q2 collector current), then zero current flows through the external feedback resistor R_F (not counting the Q3 base current). If

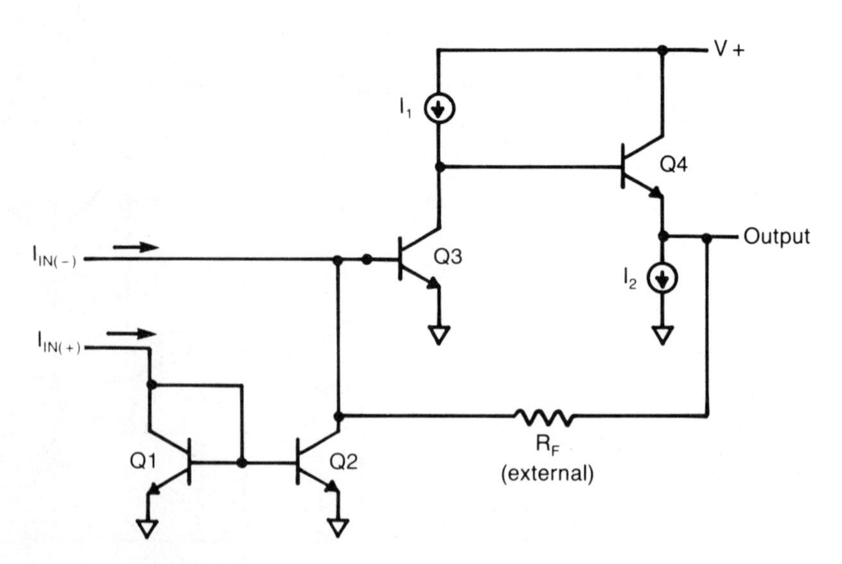

Fig. 11-14. The Norton op-amp.

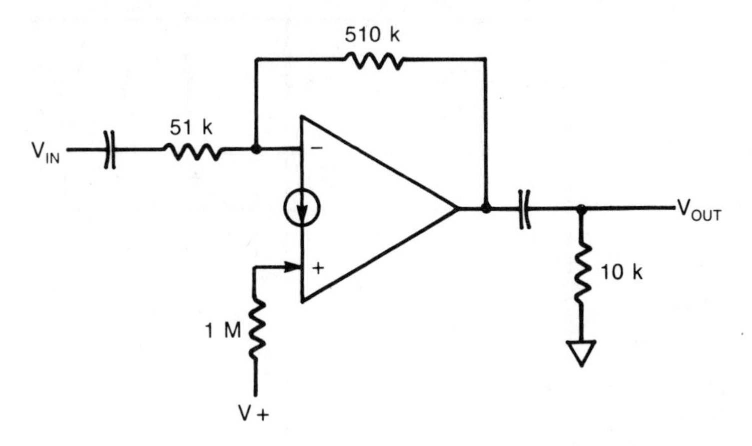

Fig. 11-15. Norton op-amp used as an inverting amplifier with a gain of 10.

$I_{IN(-)}$ is less than $I_{IN(+)}$ then the output voltage goes more positive. If $I_{IN(-)}$ is more than $I_{IN(+)}$ then the output voltage goes negative (less positive).

Figure 11-15 shows the Norton op amp used as an inverting amplifier with a gain of 10. (Notice the symbol for the Norton op amp—arrow into the plus input and a current source symbol connected from the minus to plus inputs.) In this configuration the plus input is biased by V+ through a 1 MΩ resistor. This will cause a nearly equal current to flow through the 510 kΩ resistor, thus biasing the output to a dc level approximately equal to V+/2 which allows for maximum output swing in both directions. Note that the Norton op amp, unlike the standard op amp, must be ac-coupled on both input and output in order to reference the input and output signals to ground.

Table 11-4 lists available Norton op amps, all of which are quad amplifiers except the LM359 which is a dual. Also, the LM359 is a high speed version of the standard Norton amplifier, having a typical unity-gain bandwidth of 30 MHz. When operating at a gain of 10, the LM359 has a 3 dB bandwidth of 25 MHz, making it suitable for general-purpose video-amplifier applications. To date, however, no one is second-sourcing the LM359.

VIDEO AMPLIFIERS

The standard video amplifier is undoubtably the μA733, the schematic diagram of which is shown in Fig. 11-16. R8, Q8, and R15 set up the amplifier bias currents. About 0.9 mA flows through Q8 when the supplies are +6 V and −6 V. The current in Q7 is approximately R15/R7 times 0.9 mA, or (0.9) (1400/300) mA, or 4.2 mA. This current value is one of the determining factors of the amplifier gain which is pin-selectable to be 10, 100, or 400.

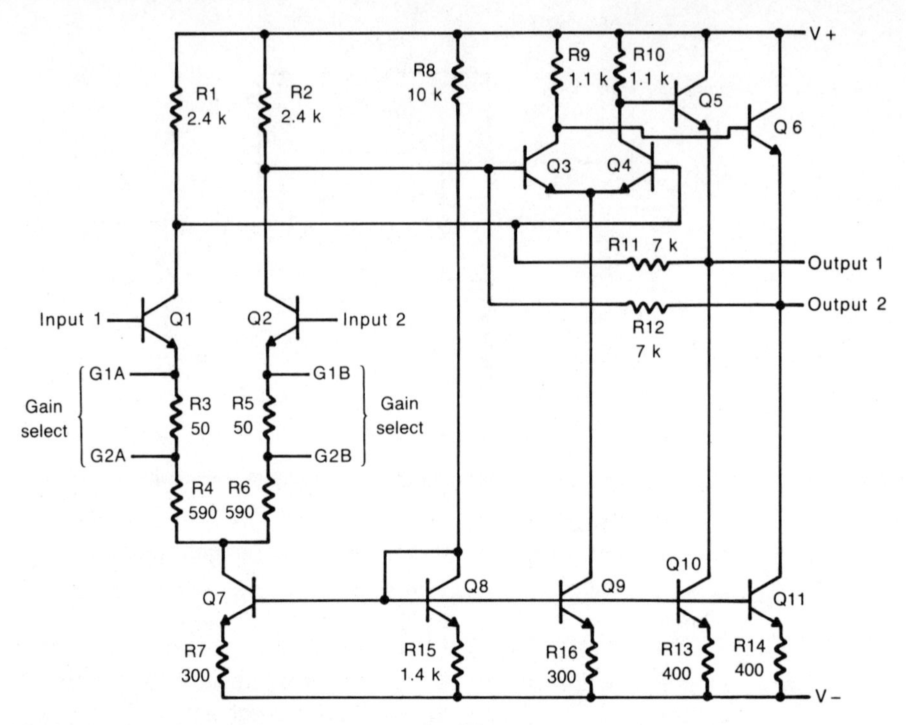

Fig. 11-16. Schematic of the μA733 video amplifier.

Recall from Chapter 3 that the gain of a differential stage is:

$$\text{GAIN} = R_L/(R_E + r_e)$$

where:

R_L is the collector load resistance;
R_E is the emitter resistance; and
r_e is the dynamic emitter resistance equal to $(kT/q)/I_E$

With I_E equal to about 2 mA (one-half the Q7 current) and kT/q equal to 26 mV at room temperature, then r_e is equal to 26mV/2mA, or 13 ohms. Also, the effective collector load resistance in Fig. 11-16 is 7000 ohms (R11, R12), not 2400 ohms (R1, R2). Therefore, with gain pins G_{1A} and G_{1B} shorted the gain is:

$$\text{GAIN 1} = 7000/(0 + 13) = 538$$

This is higher than the typical data sheet number of 400 for two reasons: (1) finite transistor betas reduce the gain about 7 percent; and (2) there is a physical emitter bulk resistance of 2-3 ohms that should be added to the 13 ohms. These two effects reduce the gain to about 6500/15, or 433. Using 3-ohms for bulk resistance yields a gain of 406.

With gain pins G_{1A} and G_{1B} open and G_{2A} and G_{2B} shorted the gain is:

$$\text{GAIN 2} = 6500/(50+15) = 100$$

With all gain pins open the gain is:

$$\text{GAIN 3} = 6500/(590+50+15) = 6500/655 = 9.9$$

Another popular video amplifier is the Signetics NE592 which is identical to the μA733 except that the 590-ohm gain setting resistors are not included. Also, the input stage current source arrangement is slightly different. See Fig. 11-17. This configuration allows only two pin-selectable gains—100 and 400. Also, if you want a gain of 400, or you want to program the gain with a resistor between G_{1A} and G_{1B}, Signetics offers the NE592 in an 8-pin mini-DIP

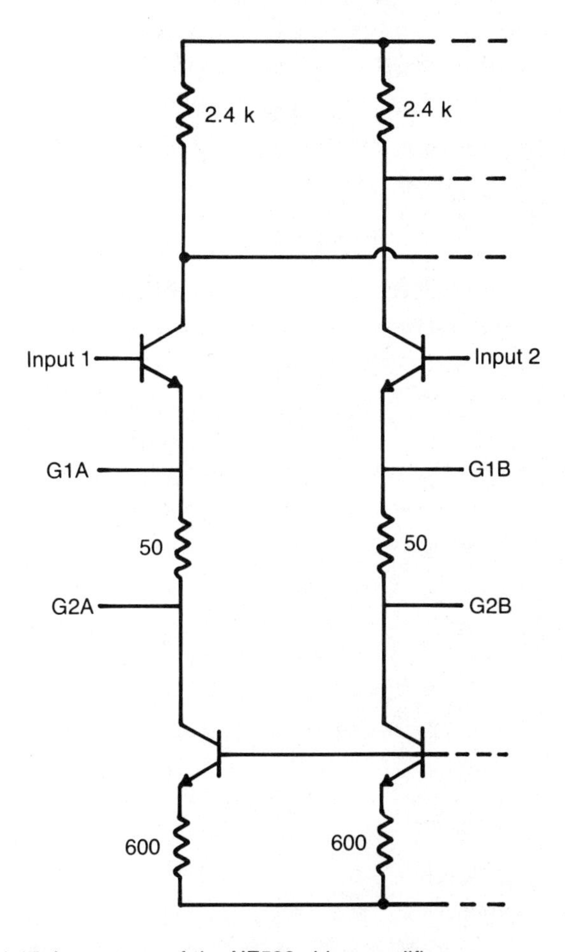

Fig. 11-17. Input stage of the NE593 video amplifier.

package. This package does not bring out gain pins G_{2A} and G_{2B}.

Table 11-5 lists available video amplifiers, most of which are 733s and 592s. Note that not all 733s or 592s are listed. For example, Texas Instruments makes a μA733. As I said in the Introduction, in the interest of keeping down the size of this book, I have listed only those alternate source ICs that the second-source manufacturer has labeled with his own standard prefix, indicating that the design is pin-compatible and probably, *but not necessarily*, meets most of the original specifications. On the other hand, TI's μA733, for example, is a true second-source and will meet all of the original Fairchild μA733 specifications.

RCA's CA3001 and CA3040 are old designs and probably should not be considered for new systems.

Table 11-1. Voltage Followers/Buffer Amplifiers.

PART NUMBER	MANUFACTURER	TEMP GRADE	MINIMUM GAIN (V/V)	TYP SLEW RATE (V/uS)	MAXIMUM OUTPUT CURRENT (A)	UNIT PRICE ($)
ADLH0033G	ANALOG DEVICES	MIL	0.96	1500	0.1	39.00
ADLH0033CG	ANALOG DEVICES	IND	0.96	1400	0.1	24.00
HOS-100AH	ANALOG DEVICES	IND	0.94	1400	0.1	21.00
HOS-100SH	ANALOG DEVICES	MIL	0.95	1500	0.1	33.00
3329/03	BURR-BROWN	IND	1 TYP	600 MIN	0.1	38.60
3553AM	BURR-BROWN	IND	0.92	2000 MIN	0.2	36.00
HA-2630	HARRIS	MIL	0.85	500	0.3	40.76
HA-2635	HARRIS	COM	0.85	500	0.3	18.97
MC1438R	MOTOROLA	COM	0.85	75	0.3	16.51
MC1538R	MOTOROLA	MIL	0.9	75	0.3	31.67
LM102H	NATIONAL	MIL	0.999	10	0.001	11.75
LM202H	NATIONAL	IND	0.999	10	0.001	9.00
LM302H	NATIONAL	COM	0.9985	10	0.001	4.50
LM110H	NATIONAL	MIL	0.999	30	0.001	4.35
LM210H	NATIONAL	IND	0.999	30	0.001	5.90
LM310N	NATIONAL	COM	0.999	30	0.001	2.40
LH0002H	NATIONAL	MIL	0.95	200	0.1	21.30
LH0002CN	NATIONAL	COM	0.95	200	0.1	7.80
LH0033G	NATIONAL	MIL	0.97	1500	0.1	39.20
LH0033CG	NATIONAL	IND	0.96	1400	0.1	25.50

PART NUMBER	MANUFACTURER	TEMP GRADE	MINIMUM GAIN (V/V)	TYP SLEW RATE (V/uS)	MAXIMUM OUTPUT CURRENT (A)	UNIT PRICE ($)
LH0063K	NATIONAL	MIL	0.94	6000	0.25	45.00
LH0063CK	NATIONAL	IND	0.94	6000	0.25	36.00
*LH2110D	NATIONAL	MIL	0.999	30	0.001	28.25
*LH2210D	NATIONAL	IND	0.999	30	0.001	24.60
*LH2310D	NATIONAL	COM	0.999	30	0.001	14.90
BUF-03AJ	P M I	MIL	0.996	250	0.07	35.25
BUF-03BJ	P M I	MIL	0.994	250	0.07	18.75
BUF-03EJ	P M I	COM	0.996	250	0.07	20.25
BUF-03FJ	P M I	COM	0.994	250	0.07	11.25

* DENOTES DUAL AMPLIFIER.

Table 11-2. Instrumentation Amplifiers.

PART NUMBER	MANUFACTURER	TEMP GRADE	MAXIMUM INPUT OFFSET (mV)	MAXIMUM OFFSET DRIFT (uV/°C)	MAXIMUM NON-LINEARITY (% F.S.)	UNIT PRICE ($)
AD521JD	ANALOG DEVICES	COM	3	15	0.2	14.90
AD521KD	ANALOG DEVICES	COM	1.5	5	0.2	23.10
AD521LD	ANALOG DEVICES	COM	1	2	0.1	33.80
AD521SD	ANALOG DEVICES	MIL	1.5	5	0.2	36.00
AD522AD	ANALOG DEVICES	IND	0.4	6	0.005	43.50
AD522BD	ANALOG DEVICES	IND	0.2	2	0.001	53.75
AD522SD	ANALOG DEVICES	MIL	0.2	6	0.001	72.50
AD524AD	ANALOG DEVICES	IND	0.25	2	0.01	14.85
AD524BD	ANALOG DEVICES	IND	0.1	0.75	0.005	19.40
AD524CD	ANALOG DEVICES	IND	0.05	0.5	0.003	26.90
AD524SD	ANALOG DEVICES	MIL	0.1	2	0.01	35.95
AD624AD	ANALOG DEVICES	IND	0.2	2	0.005	17.85
AD624BD	ANALOG DEVICES	IND	0.075	0.5	0.003	23.30
AD624CD	ANALOG DEVICES	IND	0.025	0.25	0.001	35.00
AD624SD	ANALOG DEVICES	MIL	0.075	2	0.005	38.90
INA101AM	BURR-BROWN	IND	0.05	2	0.005	14.00
INA101CM	BURR-BROWN	IND	0.025	0.75	0.002	18.40

PART NUMBER	MANUFACTURER	TEMP GRADE	MAXIMUM INPUT OFFSET (mV)	MAXIMUM OFFSET DRIFT (uV/°C)	MAXIMUM NON-LINEARITY (% F.S.)	UNIT PRICE ($)
INA101SM	BURR-BROWN	MIL	0.025	0.25	0.002	19.50
INA104HP	BURR-BROWN	COM	0.05	2	0.005	18.75
INA104JP	BURR-BROWN	COM	0.025	0.75	0.002	22.50
INA104KP	BURR-BROWN	COM	0.025	0.25	0.002	29.00
3626AP	BURR-BROWN	IND	0.4	6	0.02	32.00
3626BP	BURR-BROWN	IND	0.2	3	0.01	34.20
3626CP	BURR-BROWN	IND	0.2	1	0.01	41.35
3627AM	BURR-BROWN	IND	0.25	30	0.001	12.50
3627BM	BURR-BROWN	IND	0.25	20	0.001	16.75
3629AP	BURR-BROWN	IND	0.05	3	0.005	27.85
3629BP	BURR-BROWN	IND	0.025	1.5	0.003	35.50
3629CP	BURR-BROWN	IND	0.025	0.75	0.003	41.40
3629SM	BURR-BROWN	MIL	0.025	1.5	0.003	46.35
3630AM	BURR-BROWN	IND	0.05	2	0.005	44.00
3630BM	BURR-BROWN	IND	0.025	0.75	0.002	62.25
3630CM	BURR-BROWN	IND	0.025	0.25	0.002	95.00
3630SM	BURR-BROWN	MIL	0.025	0.75	0.002	95.00
LH0036G	NATIONAL	MIL	1	10 TYP	0.03 TYP	60.90
LH0036CG	NATIONAL	IND	2	10 TYP	0.03 TYP	33.45
LH0038D	NATIONAL	MIL	0.1	0.25	0.0001 TYP	118.60
LH0038CD	NATIONAL	IND	0.15	1	0.0001 TYP	59.15
LH0084D	NATIONAL	MIL	5	10 TYP	0.002 TYP	77.75
LH0084CD	NATIONAL	IND	10	10 TYP	0.002 TYP	50.25

Table 11-3. Operational Transconductance Amplifiers.

PART NUMBER	MANUFACTURER	TEMP GRADE	OTA'S PER PKG	MAXIMUM INPUT OFFSET (mV)	UNIT PRICE ($)
LM3080N	NATIONAL	COM	1	5	.97
LM3080AN	NATIONAL	COM	1	2	3.00
LM13600N	NATIONAL	COM	2	4	1.30
LM13600AN	NATIONAL	COM	2	1	1.50

PART NUMBER	MANUFACTURER	TEMP GRADE	OTA'S PER PKG	MAXIMUM INPUT OFFSET (mV)	UNIT PRICE ($)
LM13700N	NATIONAL	COM	2	4	1.30
LM13700AN	NATIONAL	COM	2	1	1.50
CA3060E	RCA	IND	3	5	2.56
CA3080E	RCA	COM	1	5	.80
CA3080AE	RCA	MIL	1	5	1.10
CA3094E	RCA	MIL	1	5	1.12
CA3094AE	RCA	MIL	1	5	1.56
CA3094BT	RCA	MIL	1	5	2.60
CA3280E	RCA	COM	2	3	1.84
CA3280AE	RCA	MIL	2	0.5	3.84
NE5517N	SIGNETICS	COM	2	5	1.30
NE5517AN	SIGNETICS	COM	2	2	1.50

Table 11-4. Norton Op Amps.

PART NUMBER	MANUFACTURER	TEMP GRADE	OP AMPS PER PKG	MAXIMUM SUPPLY VOLTAGE (V)	TYP BAND-WIDTH (MHz)	UNIT PRICE ($)
MC3301P	MOTOROLA	IND	4	28	4	.88
MC3401P	MOTOROLA	COM	4	18	4	.86
LM359N	NATIONAL	COM	2	22	30	1.78
LM2900N	NATIONAL	IND	4	32	2.5	1.45
LM3301N	NATIONAL	IND	4	28	2.5	.80
LM3401N	NATIONAL	COM	4	18	2.5	.76
LM3900N	NATIONAL	COM	4	32	2.5	.85
CA3401E	RCA	COM	4	18	5	.94

Table 11-5. Video Amplifiers.

PART NUMBER	MANUFACTURER	TEMP GRADE	TYP GAIN (V/V)	TYP 3dB BW (MHz)	UNIT PRICE ($)
UA733DM	FAIRCHILD	MIL	400	40	3.23
UA733PC	FAIRCHILD	COM	400	40	.76
MC1733CP	MOTOROLA	COM	400	40	.93
MC1733G	MOTOROLA	MIL	400	40	2.04

PART NUMBER	MANUFACTURER	TEMP GRADE	TYP GAIN (V/V)	TYP 3dB BW (MHz)	UNIT PRICE ($)
LM733H	NATIONAL	MIL	400	40	3.40
LM733CN	NATIONAL	COM	400	40	.80
CA3001	RCA	MIL	8.9	29	2.50
CA3040	RCA	MIL	70.8	55	4.20
NE592N	SIGNETICS	COM	400	40	.84
SE592N	SIGNETICS	MIL	400	40	1.42
TL592P	T I	COM	400	50	1.15

Chapter 12

Analog Math Blocks

THE PURPOSE OF THIS CHAPTER IS TO PRESENT PRINCIPLES and selection guides for the following types of linear ICs:

- ☐ Multipliers
- ☐ Dividers
- ☐ Logarithmic amplifiers
- ☐ Multifunction converters
- ☐ RMS-to-dc converters

Each of these types of ICs can perform one or more mathematical operations on an input voltage by making use of the fundamental logarithmic property of the semiconductor junction.

MULTIPLIERS

There are two basic types of analog multipliers—the variable-transconductance multiplier, and the log-antilog multiplier.

Figure 12-1 shows a simplified schematic of a 4-quadrant variable-transconductance multiplier. The multiplier is a 4-quadrant multiplier because each of the X and Y inputs may be either positive or negative. In general, the values of I, R_x, R_Y, R3, R4, and the gain G of the output amplifier are selected such that the transfer function is

$$V_{OUT} = (V_X V_Y)/10$$

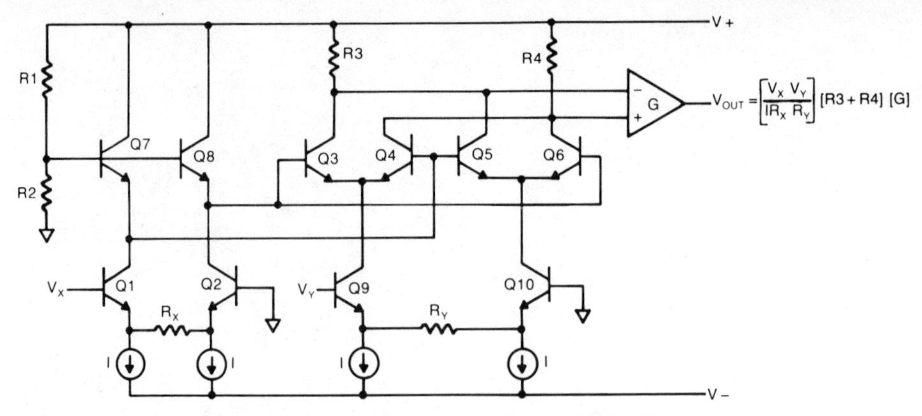

Fig. 12-1. Basic four-quadrant variable transconductance multiplier.

Figure 12-2 shows an example solution for a variable-transconductance multiplier with the X input equal to 2 volts and the Y input equal to 3 volts. For this example, I have set the four current sources equal to 1 mA and R_x and R_Y to 20 kΩ. This means that when V_X or V_Y equals 10 V (typical maximum) then the current through R_X or R_Y is 0.5 mA, or half the available current.

R1 and R2 have been selected such that 12 V is at the bases of Q7 and Q8, allowing the inputs to be able to go to 10 V. R3 and R4 have been selected so that the maximum drop across them is 3 V, making sure that Q3-Q6 never saturate. The gain G of the output amplifier is set to 10, making the overall transfer function $(V_X\ V_Y)/10$.

With V_X equal to 2 V, the current through R_X equals 100 μA; Q1 current is 1.1 mA; and Q2 current is 0.9 mA. (Warning: all circuit solutions in this chapter assume that transistor base current is negligible.) The ratio of Q1 current to Q2 current will force the emitter voltages of Q7 and Q8 to have a difference of

$$V = (kT/Q) \ln (I1/I2) = 0.026 \ln (1.1/0.9) = 5.2 \text{ mV}$$

With V_Y equal to 3 V, the current through R_Y equals 150 μA; Q9 current is 1.14 mA; and Q2 current is 0.85 mA. The ratio of Q3 current to Q4 current and the ratio of Q5 current to Q6 current is set by the 5.2 mV generated by Q7 and Q8 and is equal to

$$I3/I4 = I6/I5 = e^{(qV/kT)} = e^{(0.0052/0.026)} = 1.22$$

or

$$I3 = 1.22\ I4 \text{ and } I6 = 1.22\ I5$$

Because we know that

$$I3 + I4 = 1.15 \text{ mA and } I5 + I6 = 0.85 \text{ mA}$$

then we can substitute as follows:

$$\begin{aligned} 1.22\ I4 + I4 &= 1.15 \text{ mA} \\ 2.22\ I4 &= 1.15 \text{ mA} \\ I4 &= 1.15 \text{ mA}/2.22 = 518\ \mu A \\ I3 &= 1.15 \text{ mA} - I4 = 632\ \mu A \end{aligned}$$

$$\begin{aligned} 1.22\ I5 + I5 &= 0.85 \text{ mA} \\ 2.22\ I5 &= 0.85 \text{ mA} \\ I5 &= 0.85 \text{ mA}/2.22 = 383\ \mu A \\ I6 &= 0.85 \text{ mA} - I5 = 467\ \mu A \end{aligned}$$

I3 and I5 are summed through R3. I4 and I6 are summed through R4. The result is a voltage difference across the resistors equal to 60 mV which is multiplied by 10 by the output amplifier to get 0.6 V, or $(V_X\ V_Y)/10$.

If you take the time to write the equations for I3-I6 (the Q3-Q6 currents) in terms of V_X, R_X, V_Y, and R_Y, and substitute into

$$V_{OUT} = [(I4+I6)\ (R4) - (I3+I5)\ (R3)]\ [G]$$

you will get the equation shown in Fig. 12-1:

$$V_{OUT} = [(V_X\ V_Y)/(I\ R_X\ R_Y)]\ [R3+R4]\ [G]$$

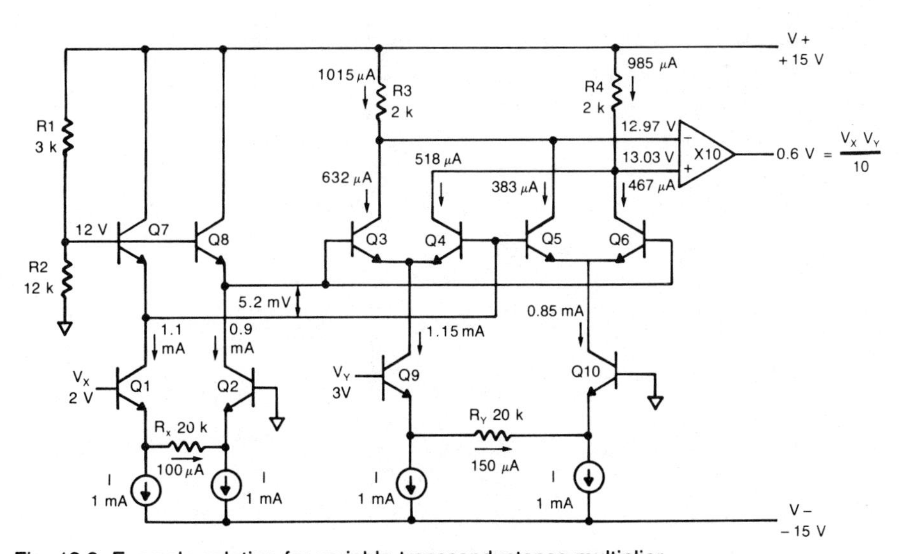

Fig. 12-2. Example solution for variable transconductance multiplier.

Log-Antilog Multiplier

Figure 12-3 shows the schematic of the one-quadrant log-antilog type multiplier. This multiplier is only a one-quadrant multiplier because the inputs must be positive voltages. This multiplier is based on the mathematical law that states that the antilog of the sum of the logarithms of two numbers is equivalent to multiplication:

$$\text{ANTILOG}\,[\,\text{LOG}(X) + \text{LOG}(Y)\,] = X\,Y$$

The voltage at the emitter of Q3 is proportional to the log of V_{REF}. The voltage at the base of Q1 is proportional to the ratio of the log of V_{REF} to the log of V_X. The voltage at the emitter of Q2 is proportional to the log of V_Y. The Q4 base-emitter voltage is proportional to the sum of the log of V_Y and the V_X/V_{REF} log ratio. Q4 converts the base-emitter voltage to a collector current proportional to the antilog of its base-emitter voltage. The op amp A4 converts this current to a voltage.

Figure 12-4 shows an example solution of the log-antilog multiplier when the X input equals 2 volts and the Y input equals 3 volts. This solution assumes matched transistors, neglible base current, and a transistor reverse saturation current I_S of 1×10^{-15} (or 1E-15). The exact value of I_S is not important as long as it is the same for all transistors. A different I_S will result in different base-emitter voltages, but the voltage at the base of Q4 and the voltage at the output will be the same.

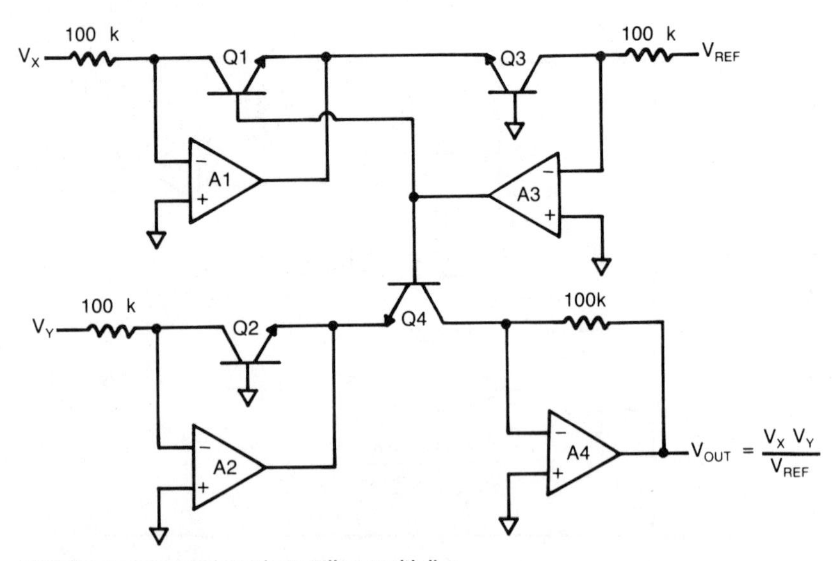

Fig. 12-3. Basic one-quadrant log-antilog multiplier.

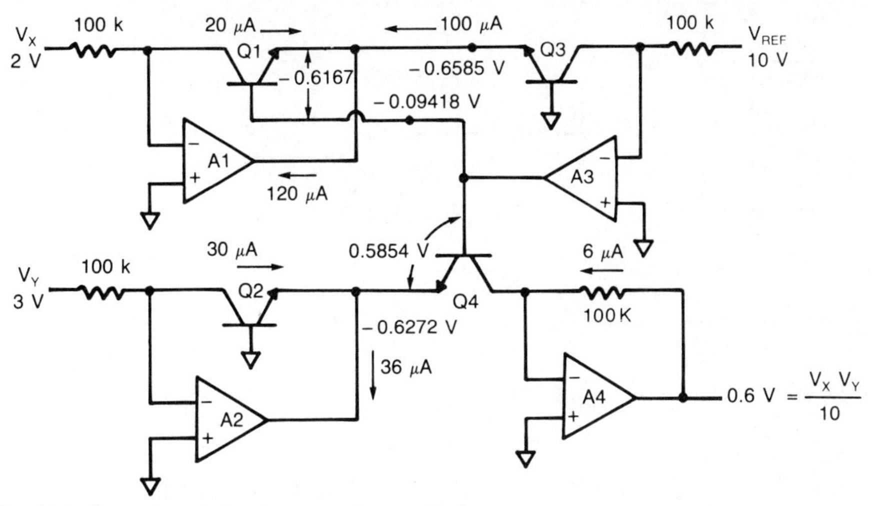

Fig. 12-4. Example solution for log-antilog multiplier.

The voltage at the base of Q4 can be solved without knowing I_S:

$$V_{B4} = (kT/q) \ln (I1/I3) = 0.026 \ln 0.2 = -0.0418 \text{ V}$$

The voltage at the emitter of Q2 and Q4 is

$$V_{E2} = V_{E4} = (kT/q) \ln (I2/I_S) = 0.026 \text{ in } [(30E-6)/(1E-15)] = -0.6272 \text{ V}$$

The Q4 base-emitter voltage is −0.0418 V minus −0.6272 V, or 0.5854 V, and the Q4 collector current is

$$I4 = I_S e^{(qV/kT)} = 1E-15\ e^{(0.5854/0.026)} = 6\ \mu A$$

The output voltage is 6 μA times 100 kΩ, or 0.6 mV, or $(V_X V_Y)/10$.

Selection Guides

Tables 12-1 through 12-3 list available multipliers by manufacturer for commercial, industrial, and military temperature ranges. Maximum multiply error is in percent of full scale at 25° C. Error drift is the temperature coefficient of the multiply error and is in percent of full scale per degree Centigrade.

Feedthrough is defined as the peak-to-peak voltage at the output when either the X or Y input is zero and the other input is being driven by a 10 V peak-to-peak 50 Hz sine wave. For

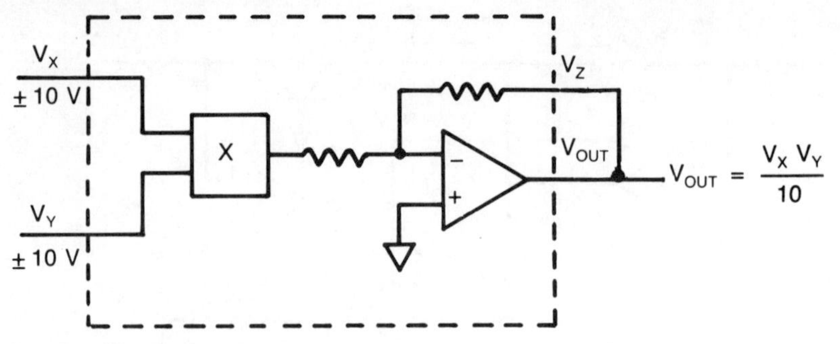

Fig. 12-5. Multiplier IC connected as a multiplier.

variable-transconductance type multipliers, the X input is worse than the Y input. The feedthrough numbers in Tables 12-1 through 12-3 are X-input numbers.

Many of the IC multipliers in Tables 12-1 through 12-3 have been designed to be easily used to perform other mathematical functions. Figures 12-5 through 12-8 show functional block diagrams of an IC multiplier connected as a multiplier, divider, squarer, and square rooter. To facilitate these functions, the basic transconductance multiplier shown in Fig. 12-1 is modified such that the output amplifier is normally open loop. The feedback resistor is connected to the output in the multiplier and squarer modes but is used as an input for the divider and square rooter modes. Note that in the divider mode the X input is limited to negative voltages. In the square rooter mode the Z input must be positive; the output is negative.

Table 12-4 shows typical divider, squarer, and square rooter errors for various IC multipliers. Notice that the divider error is a function of the denominator V_X.

DIVIDERS

Table 12-5 lists two ICs that have been specifically designed

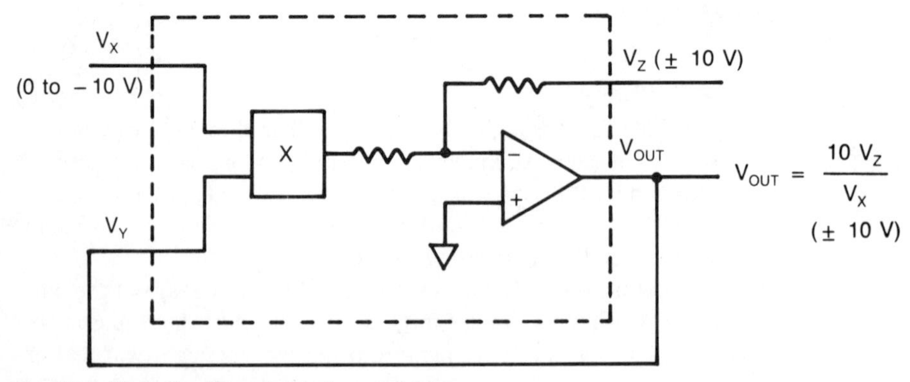

Fig. 12-6. Multiplier IC connected as a divider.

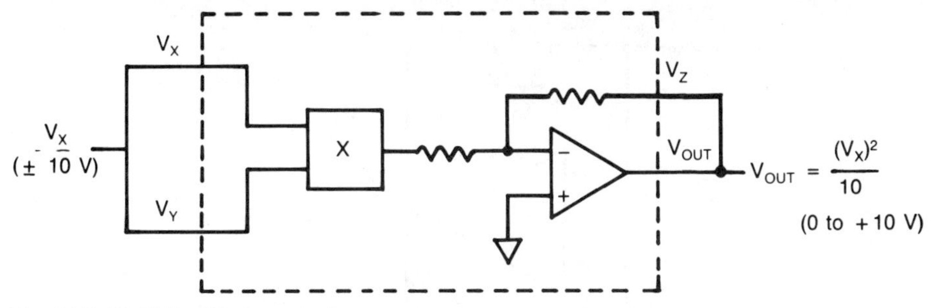

Fig. 12-7. Multiplier IC connected as a squarer.

to be dividers. The AD535 is a transconductance type with the output externally connected to the Y-input. The DIV100 is a log-antilog type. If you look back at Fig. 12-3 you can see that the divider function is achieved by making V_{REF} the denominator D, V_Y equal to 10 V, and V_X the numerator N. The transfer function becomes

$$V_{OUT} = 10\ N/D$$

The DIV100 has its own internal 10 V reference and added circuitry so that the numerator may be positive or negative. The denominator, however, must be positive. For the AD535, the numerator may also be positive or negative. However, the 535 may be connected in one configuration for a positive denominator and another for a negative denominator.

LOGARITHMIC AMPLIFIERS

The basic log amp is shown in Fig. 12-9. The voltage at the base of Q3 is proportional to the log of V_X/V_{REF}. The voltage is then scaled up by the op amp A1 and the resistors R1 and R2. The equation for V_{OUT} is

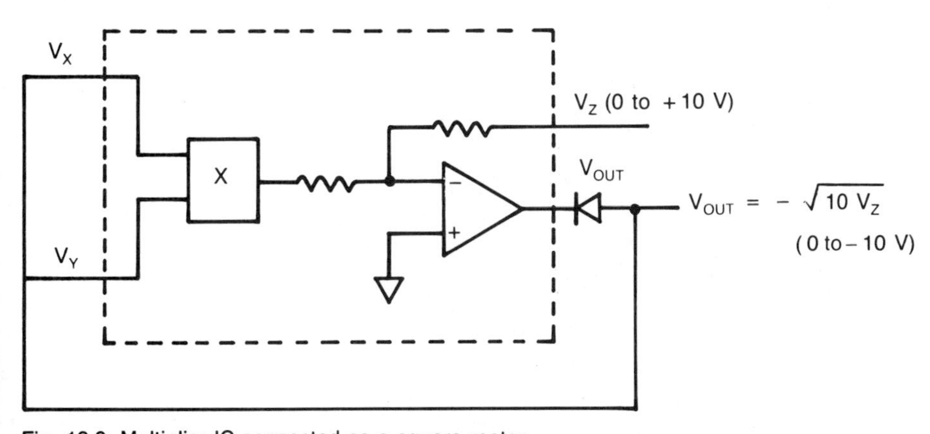

Fig. 12-8. Multiplier IC connected as a square rooter.

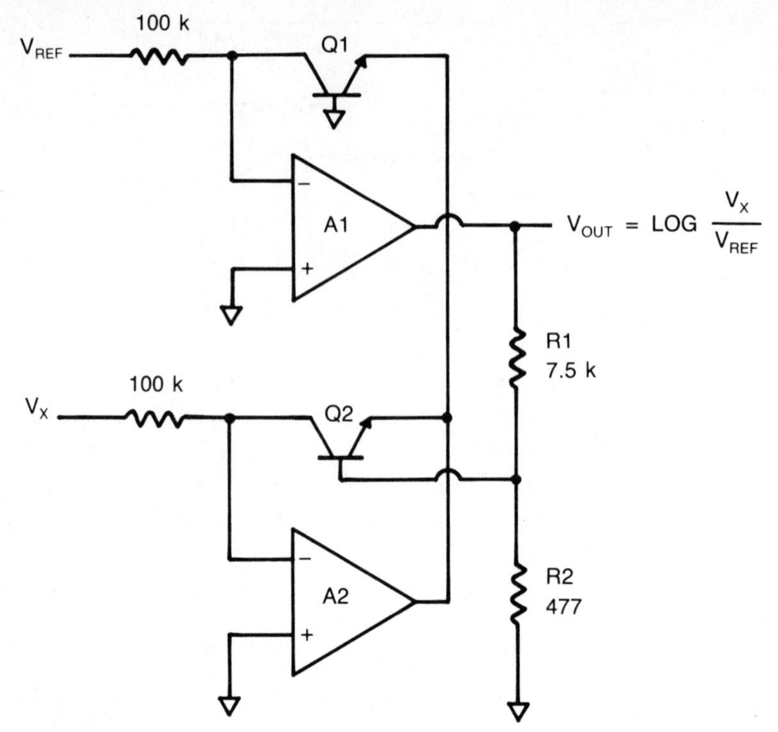

Fig. 12-9. Basic log amp.

$$V_{OUT} = [\,(R1+R2)/R2\,]\,[kT/q]\,\ln\,(V_X/V_{REF})$$

To convert this natural log function to a base 10 log function, we make use of the relation

$$\ln X = 2.3 \log_{10} X$$

Substituting, we get

$$V_{OUT} = [(R1+R2)/R2]\,[kT/q]\,2.3 \log_{10}\,(V_X/V_{REF})$$

If we set the resistor ratio $(R1+R2)/R2$ to $1/(2.3kT/q)$ then the equation for V_{OUT} simplifies to

$$V_{OUT} = \log_{10}\,(V_X/V_{REF})$$

$1/(2.3kT/q)$ equals 16.72 at room temperature. If we conveniently set R1 equal to 7.5 kΩ then R2 must be

$$(R1+R2)/R2 = 16.72$$
$$R1+R2 = 16.72\ R2$$
$$R1 = 15.72\ R2$$
$$R2 = R1/15.72 = 7500/15.72 = 477$$

In order to temperature compensate the transfer function, R2 should be a thermistor with a positive temperature coefficient of approximately 0.33 percent per degree Centigrade. Table 12-6 lists available log amps.

MULTIFUNCTION CONVERTERS

Figure 12-10 shows the schematic of the National LH0094 multifunction converter. In this circuit, the basic components of the log-antilog multiplier have been made available in a way that allows the user to create any one of several math functions—multiply, divide, square, square root, exponents, and roots. To facilitate the square and square root functions, the LH0094 includes two 100-ohm precision resistors. The transfer function equation for the LH0094 is

$$E_O = V_Y (V_Z/V_X)^m$$

For the multiply and divide modes, m equals 1. This is achieved by shorting V_A, V_B, and V_C. V_X is set to 10 V (see Fig. 12-11). For the divide mode, V_Y is set to 10 V (see Fig. 12-12).

For the square mode, m equals 2. This is achieved by wiring

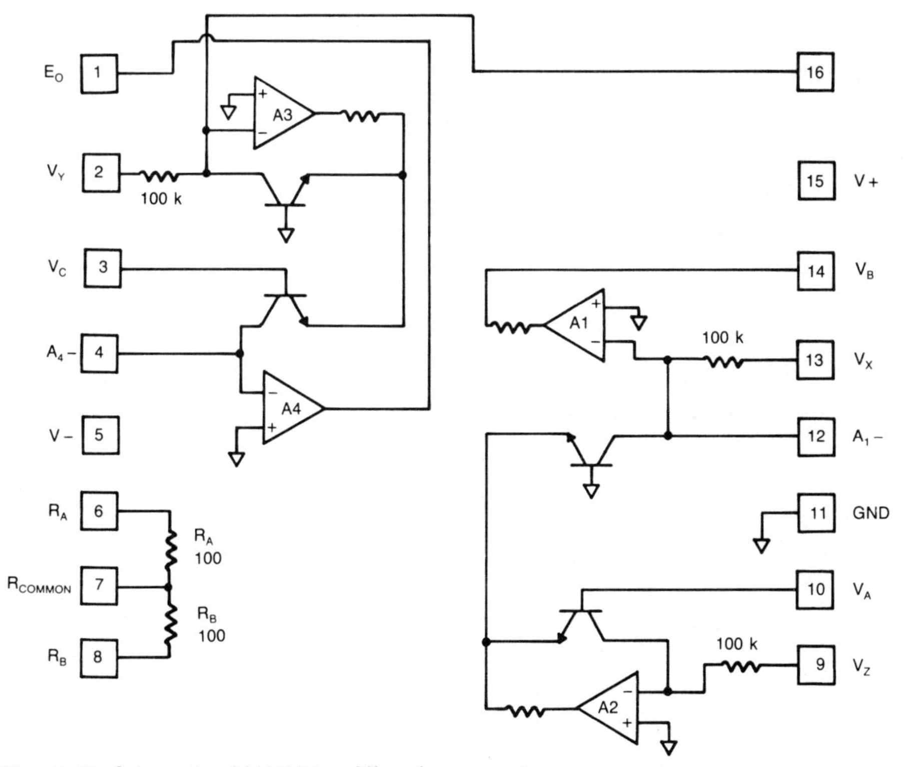

Fig. 12-10. Schematic of LH0094 multifunction converter.

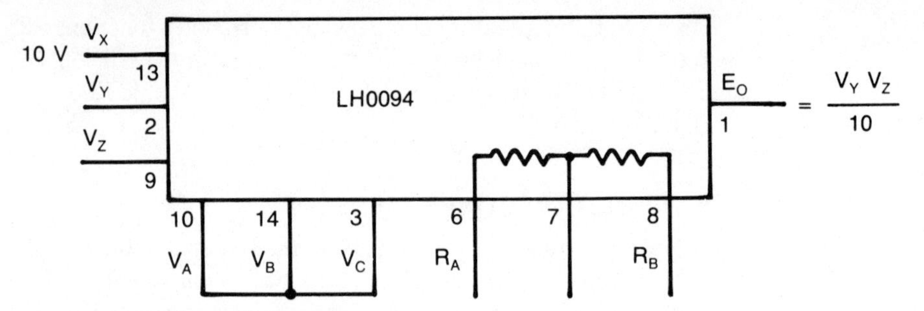

Fig. 12-11. LH0094, multiply mode.

the internal 100-ohm resistors as shown in Fig. 12-13. V_X and V_Y are set to 10 V. Note that you can use external resistors as long as they are the same value. The internal resistors have a 10 percent tolerance but are matched to 0.1 percent.

For the square root mode, m equals 0.5. This is achieved by wiring the internal 100-ohm resistors as shown in Fig. 12-14.

In the exponent mode, m is greater than 1. This is achieved by using external resistors as shown in Fig. 12-15. The value of m is set by the resistor ratio according to the following equation:

$$m = (R1 + R2)/R2$$

For example, for $m = 5$, choose $R1 = 200$ and $R2 = 50$. Note that the square mode is a special case of the exponent mode where $R1 = R2 = 100$.

In the root mode, m is less than 1. See Fig. 12-16. The equation for m is

$$m = R2/(R1 + R2)$$

For example, for $m = 0.2$, choose $R1 = 200$ and $R2 = 50$. Note that

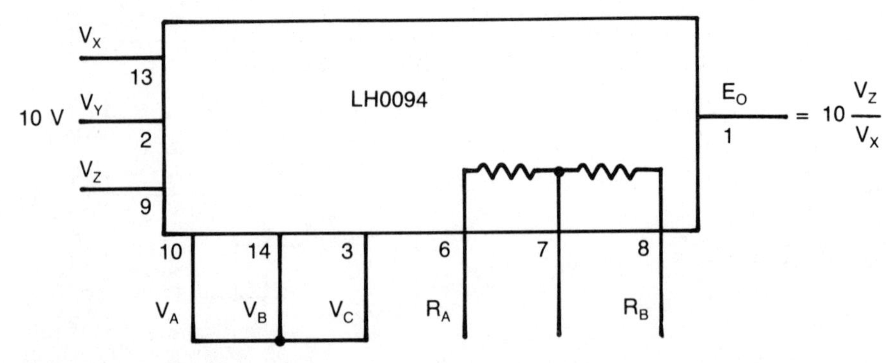

Fig. 12-12. LH0094, divide mode.

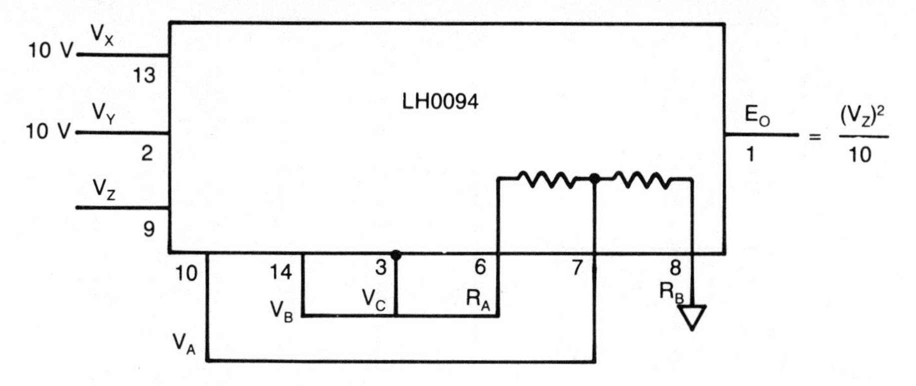

Fig. 12-13. LH0094, square mode.

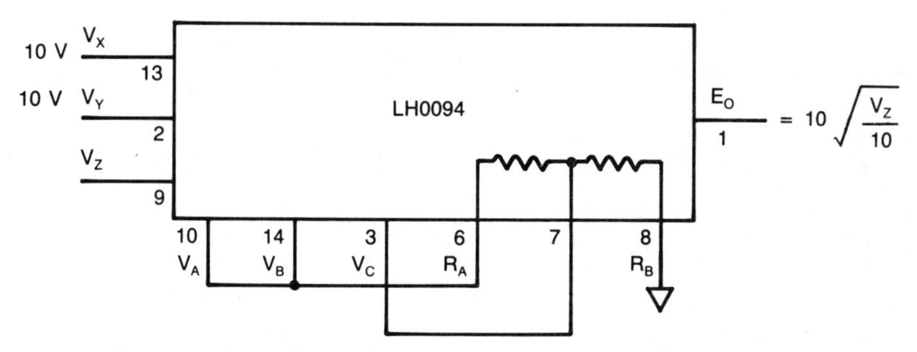

Fig. 12-14. LH0094, square root mode.

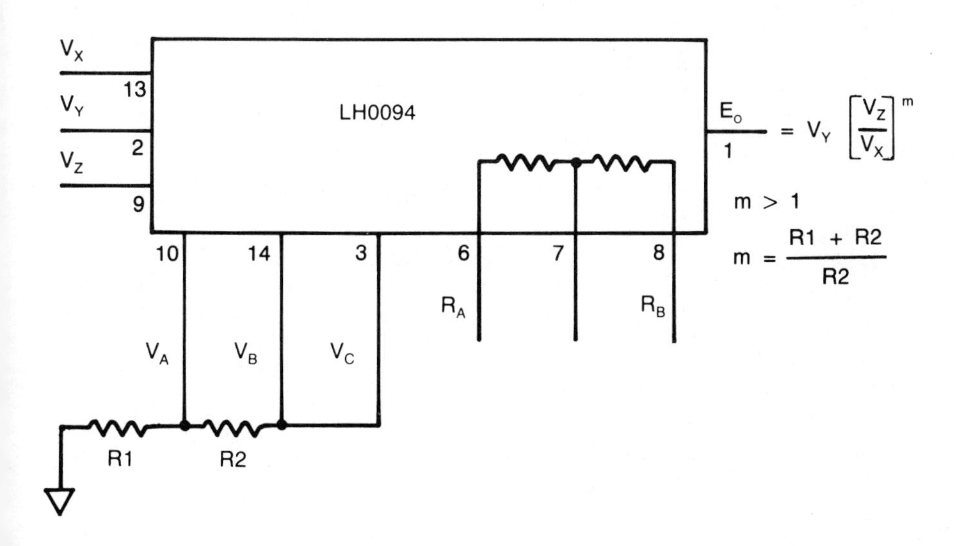

Fig. 12-15. LH0094, exponent mode.

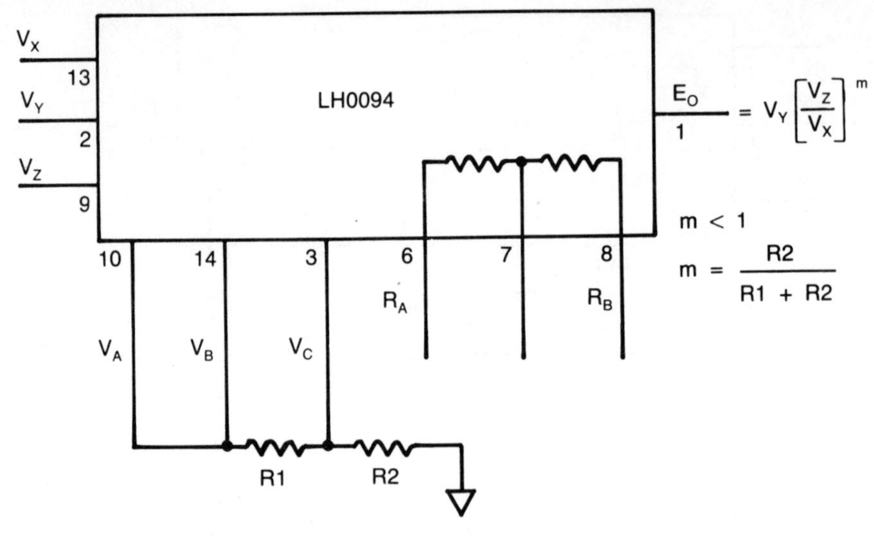

Fig. 12-16. LH0094, root mode.

the square root mode is a special case of the root mode where R1 = R2 = 100.

Table 12-7 lists the two versions (military and commercial temp grades) of the LH0094 and the Burr-Brown 4301 and 4302. The 4301 and 4302 are virtually identical to the LH0094 except that the 4301 and 4302 do not include the 100-ohm matched resistors. The only difference I could find between the 4301 and the 4302 is that the 4301 is in a metal package and the 4302 is in a molded epoxy package.

The maximum error shown in the table is for the multiply mode. Typical accuracies (with external trim) at room temperature for other modes are as follows:

Percent Accuracy with External Trim

Function	4301/4302	LH0094	LH0094C
Multiply	0.25	0.1	0.1
Divide	0.25	0.1	0.1
Square	0.03	0.15	0.15
Square Root	0.07	0.15	0.15
Exponent (m = 5)	0.15	0.05	0.08
Root (m = 0.2)	0.2	0.05	0.08

In addition to these modes, Burr-Brown provides application schematics and accuracy specifications for sine, cosine, arctangent, and vector magnitude:

Function	Accuracy (4301/4302)
Sine	0.5%
Cosine	0.8%
$\tan^{-1}(Y/X)$	0.6%
$\sqrt{X^2 + Y^2}$	0.07%

RMS-TO-DC CONVERTERS

The diode logarithmic characteristic can also be used to perform RMS-to-dc conversion. Recall from basic electronics that the RMS (root-means-square) value of a sine wave is equal to 0.707 of its peak. The general equation for the RMS value of any voltage (dc or ac, sine wave or not, single frequency or multiple frequency over a period of time) is

$$V_{RMS} = \sqrt{\text{Avg.}\ (V_{IN})^2}$$

The circuit in Fig. 12-17 has a dc output voltage proportional to the RMS value of the input, which may be ac or dc:

$$V_{OUT} = \sqrt{\text{Avg.}\ (V_{IN})^2}$$

which is the same as

$$V_{OUT} = \text{Avg.}\ [(V_{IN})^2/V_{OUT}]$$

It is this equation that is implemented by the circuit shown in Fig. 12-17. The absolute value of the input voltage is squared, then divided by the output, and then averaged.

For positive values of V_{IN}, a current I1 flows through R1 and into Q1 and Q2. The output of the op amp A1 is clamped one diode

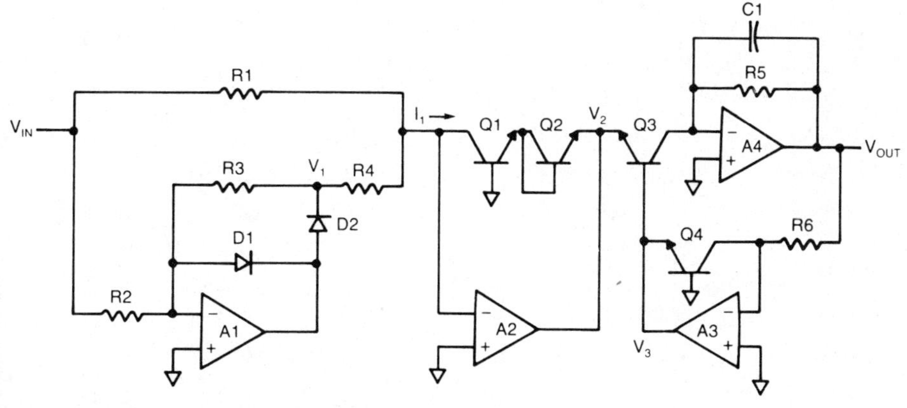

Fig. 12-17. Basic RMS-to-dc converter.

drop negative (about −0.7 V). V1 is approximately equal to zero. No current flows through R3 or R4.

For negative values of V_{IN}, node V1 goes positive and is equal to R3/R2 times V_{IN}. If V1 = $-2\ V_{IN}$, and R1 = R4, then twice as much current flows through R4 than R1, resulting in the current I1 flowing into Q1 and Q2. Thus, for positive and negative values of V_{IN}, the result is that the current I1 always flows into Q1 and Q2 in the direction shown in Fig. 12-17. This means that I1 is proportional to the absolute value of V_{IN}.

Next, the voltage at V2 is proportional to 2 times the log of V_{IN}. Q3 then takes the antilog. This is equivalent to squaring:

$$\text{Antilog}\ (2\ \text{LOG}\ X) = X^2$$

The output voltage is proportional to the square of the input and is converted to a logarithm by Q4 and summed with V2. This is equivalent to division. The external capacitor C averages the result. Table 12-8 lists available RMS-to-dc converters.

Table 12-1. Multipliers, Commercial Temperature Range.

PART NUMBER	MANUFACTURER	MAXIMUM MULTIPLY ERROR (±%)	TYPICAL ERROR DRIFT (±%/°C)	TYPICAL FEED-THROUGH (mV)	UNIT PRICE ($)
AD532JH	ANALOG DEVICES	2	0.04	50	28.55
AD532KH	ANALOG DEVICES	1	0.03	30	39.65
AD533JH	ANALOG DEVICES	2	0.04	200 MAX	9.70
AD533KH	ANALOG DEVICES	1	0.03	150 MAX	16.20
AD533LH	ANALOG DEVICES	0.5	0.01	50 MAX	48.60
AD534JH	ANALOG DEVICES	1	0.022	60	29.95
AD534KH	ANALOG DEVICES	0.5	0.015	30	41.40
AD534LH	ANALOG DEVICES	0.25	0.008	10	66.30
AD539JD	ANALOG DEVICES	2.5	0.03	15	22.45
AD539KD	ANALOG DEVICES	1.5	0.015	15	29.85
4203J	BURR-BROWN	2	0.04	50	36.25
4203K	BURR-BROWN	1	0.04	50	49.65
4205J	BURR-BROWN	2	0.04	50	31.95
4205K	BURR-BROWN	1	0.04	50	46.50
4206J	BURR-BROWN	0.5	0.01	10	48.45
4206K	BURR-BROWN	0.25	0.01	5	68.80

PART NUMBER	MANUFACTURER	MAXIMUM MULTIPLY ERROR (±%)	TYPICAL ERROR DRIFT (±%/°C)	TYPICAL FEED-THROUGH (mV)	UNIT PRICE ($)
ICL8013AC	INTERSIL	0.5	0.02	50 MAX	26.05
ICL8013BC	INTERSIL	1	0.03	100 MAX	11.35
ICL8013CC	INTERSIL	2	0.04	200 MAX	6.30
MC1494L	MOTOROLA	1	0.001	--	6.11
MC1495L	MOTOROLA	4	0.01	--	5.35

Table 12-2. Multipliers, Industrial Temperature Range.

PART NUMBER	MANUFACTURER	MAXIMUM MULTIPLY ERROR (±%)	TYPICAL ERROR DRIFT (±%/°C)	TYPICAL FEED-THROUGH (mV)	UNIT PRICE ($)
AD632AH	ANALOG DEVICES	1	0.02	30	25.00
AD632BH	ANALOG DEVICES	0.5	0.01	10	36.00
4204J	BURR-BROWN	0.5	0.01	10	68.00
4204K	BURR-BROWN	0.25	0.01	5	88.75
4213AM	BURR-BROWN	1	0.008	30	29.35
4213BM	BURR-BROWN	0.5	0.008	30	42.50
4214AP	BURR-BROWN	1	0.008	30	25.25
4214BP	BURR-BROWN	0.5	0.008	30	37.45
MPY100AM	BURR-BROWN	2	0.017	100	10.50
MPY100BM	BURR-BROWN	1	0.008	30	17.00
MPY100CM	BURR-BROWN	0.5	0.008	30	25.50

Table 12-3. Multipliers, Military Temperature Range.

PART NUMBER	MANUFACTURER	MAXIMUM MULTIPLY ERROR (±%)	TYPICAL ERROR DRIFT (±%/°C)	TYPICAL FEED-THROUGH (mV)	UNIT PRICE ($)
AD532SH	ANALOG DEVICES	1	0.01	30	59.30
AD533SH	ANALOG DEVICES	1	0.01	100 MAX	54.70
AD534SH	ANALOG DEVICES	1	0.02	60	84.50
AD534TH	ANALOG DEVICES	0.5	0.01	30	111.80
AD539SD	ANALOG DEVICES	4	0.015	15	45.00
AD632SH	ANALOG DEVICES	1	0.02	30	40.00

PART NUMBER	MANUFACTURER	MAXIMUM MULTIPLY ERROR (±%)	TYPICAL ERROR DRIFT (±%/°C)	TYPICAL FEED-THROUGH (mV)	UNIT PRICE ($)
AD632TH	ANALOG DEVICES	0.5	0.01	10	47.00
4203S	BURR-BROWN	1	0.04	50	77.00
4204S	BURR-BROWN	0.25	0.02	5	101.00
4205S	BURR-BROWN	1	0.04	50	66.45
4213SM	BURR-BROWN	0.5	0.025	30	55.00
4213UM	BURR-BROWN	1	0.04	100 MAX	31.00
4213VM	BURR-BROWN	1	0.02	30	45.00
4213WM	BURR-BROWN	0.5	0.02	30	60.00
4214RM	BURR-BROWN	1	0.025	30	30.50
4214SM	BURR-BROWN	0.5	0.025	30	49.40
MPY100SM	BURR-BROWN	0.5	0.025	30	38.25
ICL8013AM	INTERSIL	0.5	0.008	50 MAX	65.95
ICL8013BM	INTERSIL	1	0.011	100 MAX	58.95
ICL8013CM	INTERSIL	2	0.017	200 MAX	39.70
MC1594L	MOTOROLA	0.5	0.001	--	9.12
MC1595L	MOTOROLA	2	0.005	--	7.61
CA3091D	RCA	4	0.11	9	11.06

Table 12-4. Divider, Squarer, Square Rooter Errors for Various IC Multipliers.

PART NUMBER	MANUFACTURER	DIVIDER ERROR @ Vx=-1V (+/- %)	DIVIDER ERROR @ Vx=-1ØV (+/- %)	SQUARER ERROR (+/- %)	SQUARE ROOTER ERROR (+/- %)
AD532J	ANALOG DEVICES	4	2	Ø.8	1.5
AD532K/S	ANALOG DEVICES	3	1	Ø.4	1
AD533J	ANALOG DEVICES	3	1	Ø.8	Ø.8
AD533K/S	ANALOG DEVICES	2	Ø.5	Ø.4	Ø.4
AD533L	ANALOG DEVICES	1.5	Ø.2	Ø.2	Ø.2
AD534J/S	ANALOG DEVICES	2	Ø.75	Ø.6	1
AD534K	ANALOG DEVICES	1	Ø.35	Ø.3	Ø.5
AD534L	ANALOG DEVICES	Ø.8	Ø.2	Ø.2	Ø.25
AD534T	ANALOG DEVICES	2	Ø.75	Ø.3	Ø.5
4213A	BURR-BROWN	2	Ø.75	Ø.6	1

PART NUMBER	MANUFACTURER	DIVIDER ERROR @ Vx=-1V (+/- %)	DIVIDER ERROR @ Vx=-10V (+/- %)	SQUARER ERROR (+/- %)	SQUARE ROOTER ERROR (+/- %)
4213B/S	BURR-BROWN	1	0.35	0.3	0.5
MPY100A	BURR-BROWN	4	1.5	1.2	2
MPY100B	BURR-BROWN	2	0.75	0.6	1
MPY100C/S	BURR-BROWN	1	0.35	0.3	0.5
ICL8013A/B/C	INTERSIL	1.5	0.3	--	--

Table 12-5. Dividers.

PART NUMBER	MANUFACTURER	TEMP GRADE	MAXIMUM DIVIDER ERROR (±%)	UNIT PRICE ($)
AD535JH	ANALOG DEVICES	COM	1	31.20
AD535KH	ANALOG DEVICES	COM	0.5	43.20
DIV100HP	BURR-BROWN	IND	1	28.75
DIV100JP	BURR-BROWN	IND	0.5	40.25
DIV100KP	BURR-BROWN	IND	0.25	57.50

Table 12-6. Logarithmic Amplifiers.

PART NUMBER	MANUFACTURER	TEMP GRADE	MAXIMUM ERROR (% FS)	UNIT PRICE ($)
4127JG	BURR-BROWN	COM	1	47.40
4127KG	BURR-BROWN	COM	0.5	54.60
LOG100JP	BURR-BROWN	COM	0.25	43.00
ICL8048BC	INTERSIL	COM	0.5	45.30
ICL8048CC	INTERSIL	COM	1	22.75

Table 12-7. Multifunction Converters.

PART NUMBER	MANUFACTURER	TEMP GRADE	MAXIMUM ERROR (% FS)	UNIT PRICE ($)
4301	BURR-BROWN	IND	0.5	90.75
4302	BURR-BROWN	IND	0.5	51.50
LH0094D	NATIONAL	MIL	0.45	111.00
LH0094CD	NATIONAL	IND	0.9	59.15

Table 12-8. RMS-to-Dc Converters.

PART NUMBER	MANUFACTURER	TEMP GRADE	MAXIMUM CONVERSION ERROR	UNIT PRICE ($)
AD536AJH	ANALOG DEVICES	COM	±5mV±0.5% of Reading	11.90
AD536AKH	ANALOG DEVICES	COM	±2mV±0.2% of Reading	20.60
AD536ASH	ANALOG DEVICES	MIL	±5mV±0.5% of Reading	40.40
AD636JH	ANALOG DEVICES	COM	±0.5mV±1% of Reading	8.95
AD636KH	ANALOG DEVICES	COM	±0.2mV±0.5% of Reading	14.95
AD637JD	ANALOG DEVICES	COM	±1mV±0.5% of Reading	22.85
AD637KD	ANALOG DEVICES	COM	±0.5mV±0.2% of Reading	32.85
AD637SD	ANALOG DEVICES	MIL	±1mV±0.2% of Reading	37.85
4340	BURR-BROWN	IND	±0.3mV±0.1% of Reading	90.75
4341	BURR-BROWN	IND	±2mV±0.2% of Reading	28.35
LH0091D	NATIONAL	MIL	±1mV±0.2% of Reading	72.90
LH0091CD	NATIONAL	IND	±1mV±0.2% of Reading	40.00

Chapter 13

Timers and Oscillators

IN THIS CHAPTER, I WILL BRIEFLY DISCUSS AND GIVE SELECTION guides for the following types of linear ICs:

- ☐ Timers
- ☐ Oscillators and function generators
- ☐ Phase-locked loops
- ☐ Voltage-to-frequency converters

TIMERS

Table 13-1 lists available timers. You can see that just about everybody makes a 555, which was originated by Signetics Corporation. You can get one, two, or four 555 type timers in a package. The 555 is the single unit, the 556 is the dual timer, and the 558 is the quad timer. The 555 has a power supply range of 5 to 15 volts (18 V absolute maximum), is guaranteed to operate at frequencies up to 500 kHz (astable mode), and can generate time delays from microseconds to hours (monostable mode).

A block diagram of the 555 is shown in Fig. 13-1. It is composed of two comparators, a flip-flop, an output buffer capable of sourcing or sinking up to 200 mA, and a transistor for discharging an external timing capacitor.

Figure 13-2 shows the 555 operating in the monostable, or one-shot, mode. The timer output is reset to the LOW state at the end

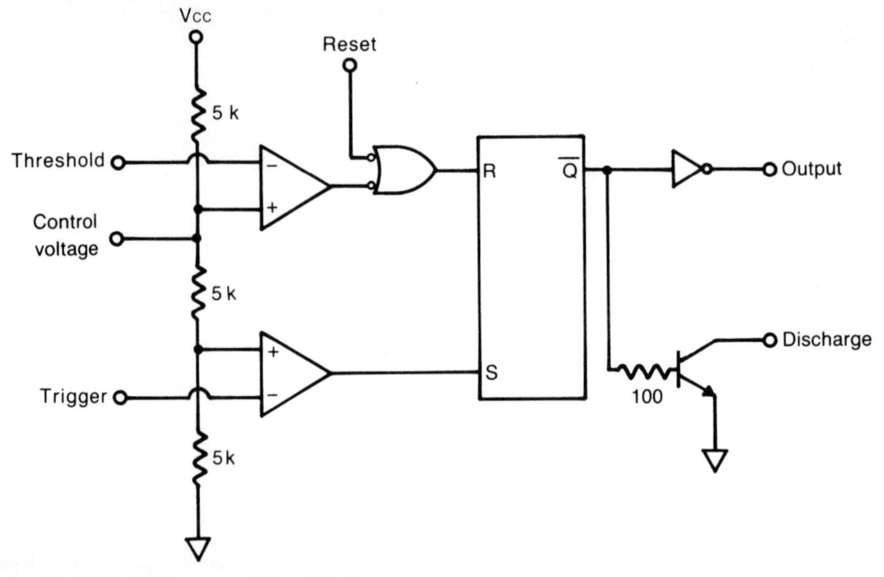

Fig. 13-1. Block diagram of the 555 timer.

of the time delay pulse or may be set LOW by momentarily applying a voltage of less than 1 V at the RESET input. The discharge transistor is ON, and the voltage across the capacitor C is zero. When the TRIGGER input voltage falls below one-third Vcc, (set by the three 5 k Ω resistors) the TRIGGER comparator goes HIGH. This sets the flip-flop and makes the timer output go HIGH. At the same time the discharge transistor turns OFF. The capacitor charges up

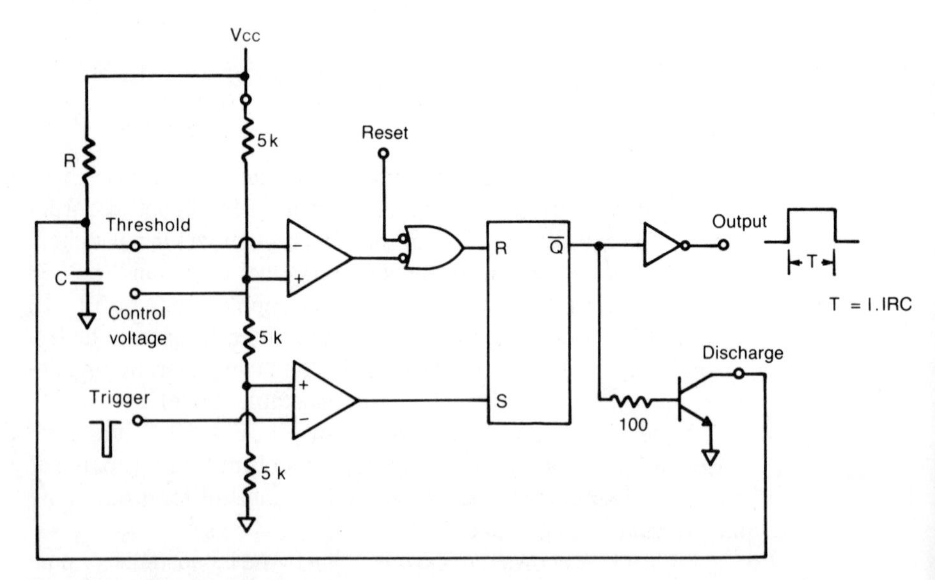

Fig. 13-2. 555, monostable mode.

to just a little more than two-thirds Vcc. The THRESHOLD comparator output goes LOW and resets the flip-flop. The timer output goes LOW and the discharge transistor turns ON and discharges the capacitor. The timer is ready for the next trigger pulse. In the monostable mode the time delay is equal to

$$T = 1.1\ R\ C \text{ (monostable mode)}$$

Timing accuracy for the NE555 is 1% typical, 3% maximum, at room temperature, not counting the errors due to the external R and C. Temperature drift is 50 PPM/°C typical, 150 PPM maximum. Drift with respect to change in power supply voltage is 0.1%/V typical, 0.5%/V maximum. Timing accuracy for the SE555 (military temp version) is 0.05% typical, 0.2% maxmum. Temperature drift is 30 PPM/°C typical, 100 PPM/°C maximum. Power supply drift is 0.15%/V typical, 0.6%/V maximum.

Figure 13-3 shows the 555 in the astable, or oscillator, mode. The capacitor is alternately charged to two-thirds Vcc and discharged to one-third Vcc. The charge time is set by the capacitor C and the sum of R_A and R_B. The discharge time is set by the capacitor C and the single resistor R_B. The sum of the charge and discharge times is equal to the period T:

$$T = 0.693\ (R_A + 2R_B)\ C \text{ (astable mode)}$$

The frequency of oscillation is the reciprocal of this number, or

$$f = 1.443/[(R_A + 2R_B)C] \text{ (astable mode)}$$

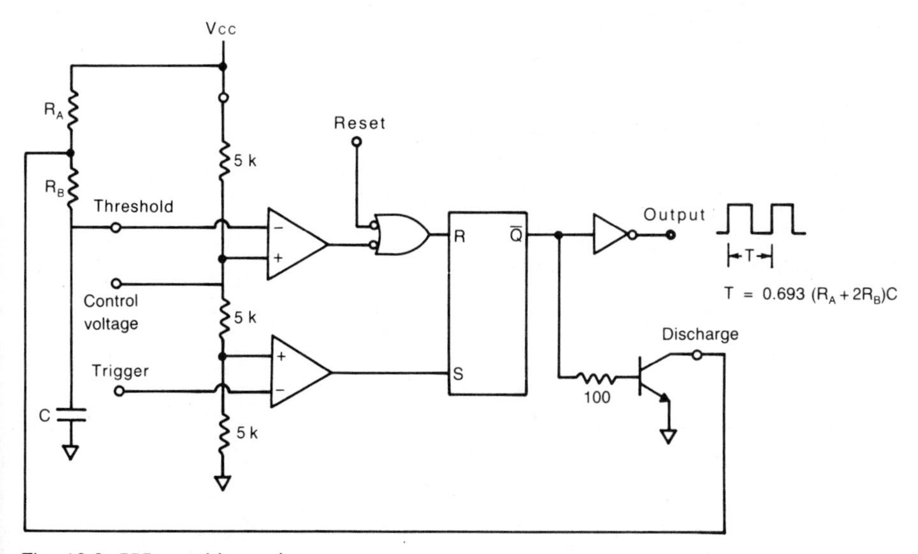

Fig. 13-3. 555, astable mode.

The timing accuracy for the NE555 (commercial) in the astable mode is 5% typical, 13% maximum, not counting errors due to the component tolerances of R_A, R_B, and C. Drift with supply voltage is 0.3%/V typical, 1%/V maximum. Timing accuracy for the SE555 (military) in the astable mode is 4% typical, 6% maximum. Drift with supply voltage is 0.15%/V typical, 0.6%/V maximum. Temperature drift for both the NE555 and SE555 is 500 PPM/°C maximum.

The μA2240 timer (see Table 13-1) combines a timer with a programmable 8-bit counter, allowing time delays of 1RC to 255RC in the monostable mode and up to 255 pulse patterns in the astable mode. The μA2240 will operate at supply voltages from 5 V to 15 V and at frequencies up to 100 kHz. Timing accuracy for either mode for the 2240C (commercial) is 5% maximum. Temperature drift is 200 PPM/°C typical. Supply drift is 0.08%/V typical, 0.3%/V maximum. Timing accuracy for the 2240 (military) is 0.5% typical, 2% maximum. Temperature drift is 150 PPM/°C typical, 300 PPM/°C maximum. Supply drift is 0.05°/V typical, 0.2°/V maximum.

The LM122 timer operates at supply voltages from 4.5 V to 40 V and operates at frequencies up to 300 kHz. It has an internal 3.15 V voltage reference, thus making the timing independent of power supply changes (less than 0.005%/V). Timing accuracy is 2% maximum, not counting errors due to the external timing resistor and capacitor tolerances. Typical temperature drift is 0.003%/°C, or 30 PPM/°C. The LM222 and LM322 are the industrial and commercial versions of the LM122 (military temp. range). The LM2905 (industrial) and LM3905 (commercial) are the same as the LM122 except that two pins (BOOST and V_{ADJ}) are not available. This limits the minimum timing period to 1 millisecond, versus 3 microseconds for the LM122.

OSCILLATORS AND FUNCTION GENERATORS

Table 13-2 lists available oscillators and function generators. Figure 13-4 shows a simplified schematic of the Signetics NE566 voltage-controlled oscillator (VCO) which generates a square wave and a triangle wave with a frequency equal to

$$f = 2\,(V_{CC} - V_C)/(R\ C\ V_{CC})$$

The 566 consists of a voltage-controlled current source, a current-direction switch, a Schmitt trigger, two buffers, and a comparator. The voltage-controlled current source is made up of Q1-Q8, R, R1, R2, D1, and D2. The current is established by the external resistor R, the supply voltage V_{CC}, and the control voltage V_C according to the equation

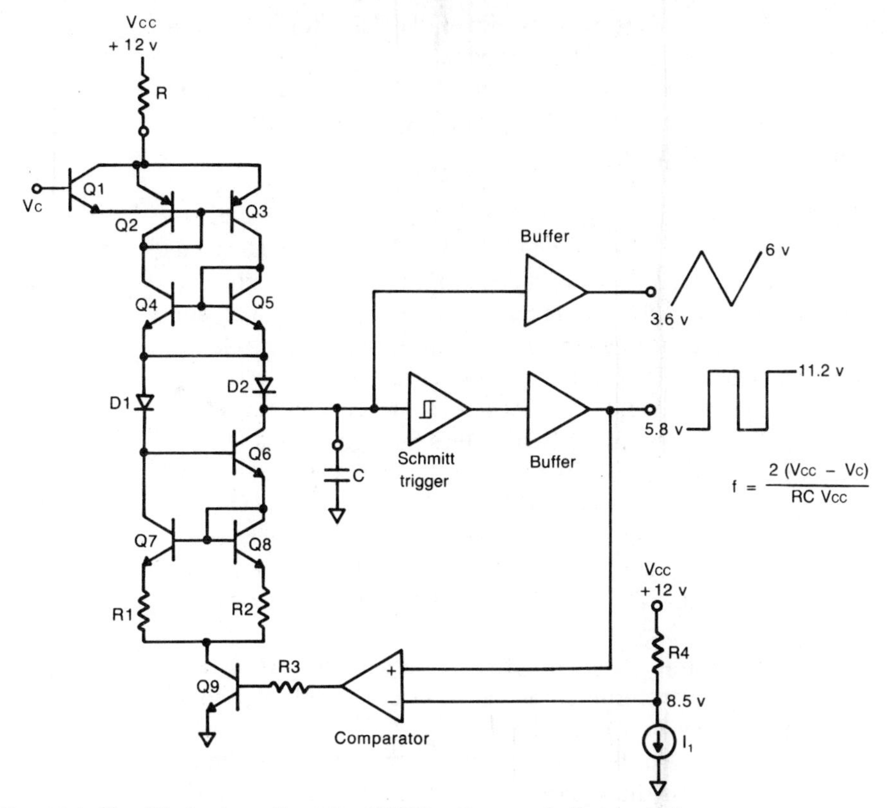

Fig. 13-4. Simplified schematic of the NE566 voltage-controlled oscillator.

$$I = (V_{CC} - V_C)/R$$

This equation assumes that the base-emitter voltages of Q1-Q3 are the same so that the Q1 collector voltage is equal to Vc.

With Q9 OFF, all of the current I will flow through D2 and charge the capacitor C at a linear rate equal to I/C volts/second. The capacitor charges up to about 6 V, the upper threshold of the Schmitt trigger. (Recall from basic electronics that the transfer function of the Schmitt trigger is such that its output goes HIGH at some threshold V_{TH1} when the input is changed from zero to input maximum; and that the Schmitt trigger output goes LOW at some lower threshold V_{TH2} when the input is changed from the input maximum to zero. The difference between V_{TH1} and V_{TH2} is called *hysteresis*.)

The Schmitt trigger goes HIGH. The comparator output goes HIGH. Q9 turns ON. The immediate effect is that D2 is reversed biased. The current I now flows through D1 into Q7 and is mirrored over to Q8. The capacitor begins to discharge through Q6, Q8, and Q9 at a linear rate equal to I/C volts/second (same as the charge rate). When the voltage across the capacitor equals about 3.6 V, the lower threshold of the Schmitt trigger, the Schmitt trigger goes LOW and the cycle starts all over again.

The NE566 triangle wave linearity is 0.5% typical. The SE566 triangle wave linearity is 0.2% typical. Maximum operating frequency is typically 1 MHz.

Figure 13-5 shows a simplified schematic of the Intersil ICL8038 voltage-controlled oscillator which generates three waveforms—square, triangle, and sine. The 8038 uses a dual comparator and flip-flop instead of the Schmitt trigger. Also, the current source is slightly different. In the 8038 the charge and discharge currents are generated independently. R_A controls the charge current and R_B controls the discharge current.

In the charge mode, Q8 is ON and sinks all current generated by R_B. The current generated by R_A flows through Q3 and D2 and charges the capacitor C. In the discharge mode, Q8 is OFF. Current generated by R_B flows into Q5 and is mirrored into Q6 and Q7. With R1-R3 equal to each other, the current in Q4 will be equal to twice the Q5 current. If $R_A = R_B$, then the Q4 current will be twice the Q3 current. Therefore, the capacitor will discharge at the same rate that it charged.

You can see the advantage of the external resistors R_A and R_B: The triangle waveform ramps can be externally adjusted to be equal to each other or a ratio of each other, thus creating a sawtooth waveform. Also, because the voltage waveform on the capacitor also controls the square-wave output, you can adjust the duty cycle of the square wave by adjusting the ratio of the external resistors.

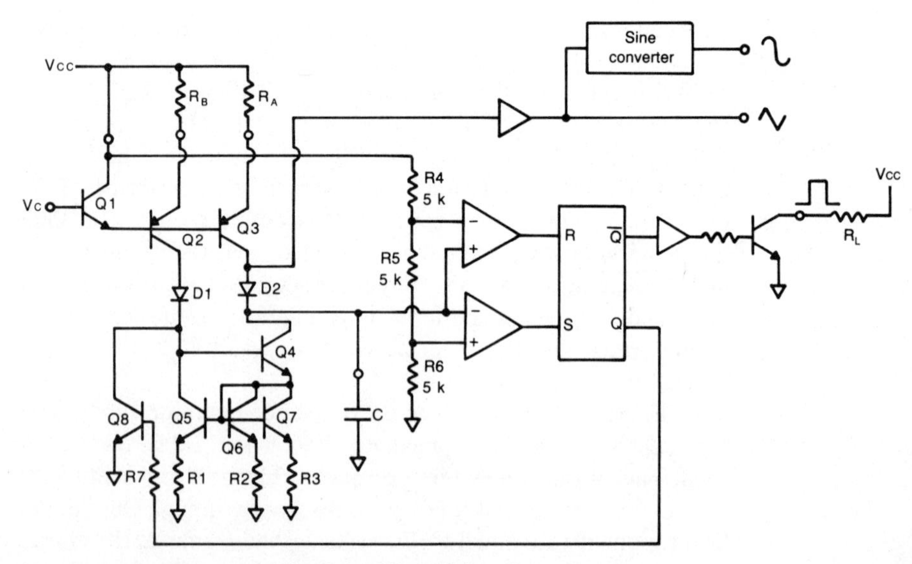

Fig. 13-5. Simplified schematic of the ICL8038 function generator.

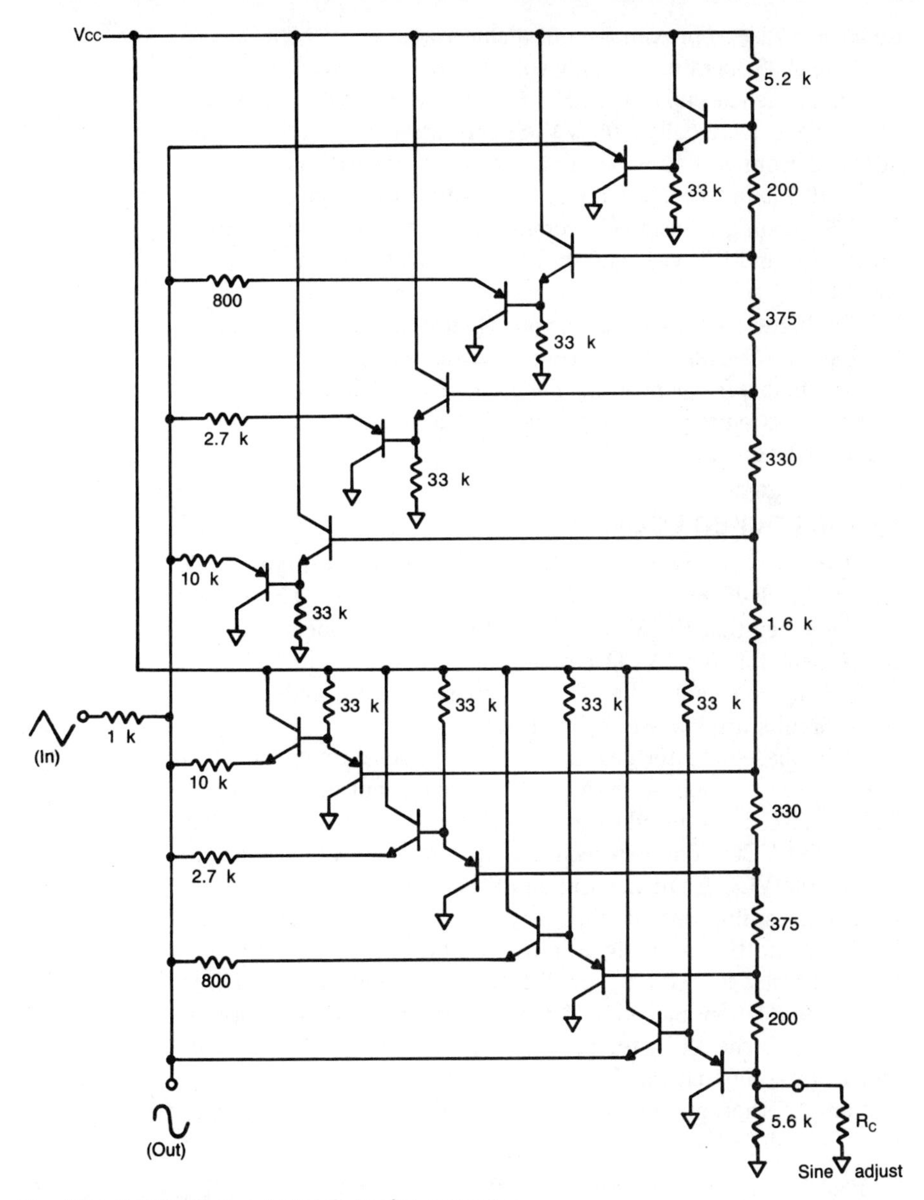

Fig. 13-6. ICL8038 internal triangle-to-sine converter.

For the sinewave output, it is critical to be able to set the ratio of the charge and discharge currents to 1. The sine converter is shown in Fig. 13-6. The circuit acts as a variable attenuator to the triangle wave. As the triangle wave is ramping up, the circuit impedance from the sine wave output node to ground varies from a high impedance around zero volts to a lower impedance as the ramp gets near the peak. The effect is to round the triangle wave into a sine wave. The circuit uses four breakpoints for each polarity to do a piecewise approximation of a sine wave.

The 8038 operates at supply voltages from 10 to 30 V and operates at frequencies up to 100 kHz. Frequency drift with power supply changes is typically 0.05%/V. Square wave duty cycle may be adjusted from 2% to 98%. Triangle wave linearity is typically 0.05% (0.1% for the CC version). Sine wave distortion is 2% typical, 5% maximum for the CC version; 1.5% typical, 3% maximum for the BC and BM versions; and 1% typical, 1.5% maximum for the AC and AM versions.

For better sine wave purity, you may use the Burr-Brown 4423. The 4423 features both sine and cosine outputs; a resistor-programmable frequency range of 0.002 Hz to 20 kHz; and a maximum distortion of 0.2% for frequencies up to 5 kHz, 0.5% maximum for frequencies 5 kHz-20 kHz.

PHASE LOCKED LOOPS

Table 13-3 lists available IC phase-locked loops. Figure 13-7 shows a block diagram of the NE565 phase locked loop. The dashed line indicates that the VCO output is normally connected to the phase detector. The VCO circuitry is virtually identical to the NE566 shown in Fig. 13-4. The 565 phase detector and low-pass filter circuits are shown in Fig. 13-8.

The phase detector generates output pulses corresponding to the difference in phase between the signal input and the VCO. These pulses are filtered to create a dc (or nearly dc) voltage that drives the VCO. The new voltage input to the VCO will act to change the VCO frequency to that of the signal input. Eventually the low-pass filter output will change to that value which "locks" the VCO and the input signal to each other.

The phase-locked loop, or PLL, was designed specifically for FM demodulation and other demodulation applications. With an FM input to the PLL, for example, the output will be the low-frequency audio that was frequency modulated at the transmitting station. Another, more general, application, is frequency multipli-

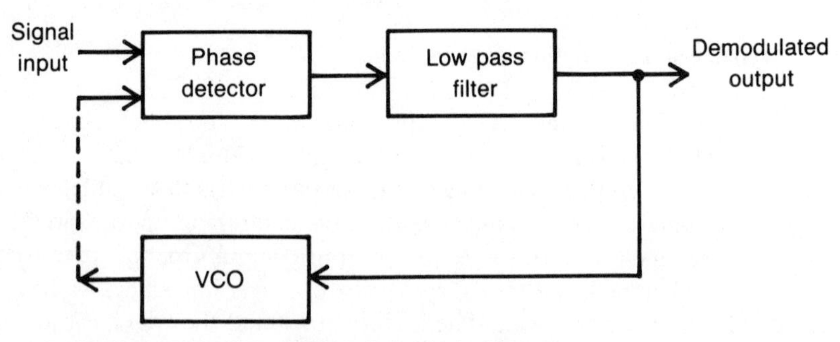

Fig. 13-7. Block diagram of the NE565 phase-locked loop.

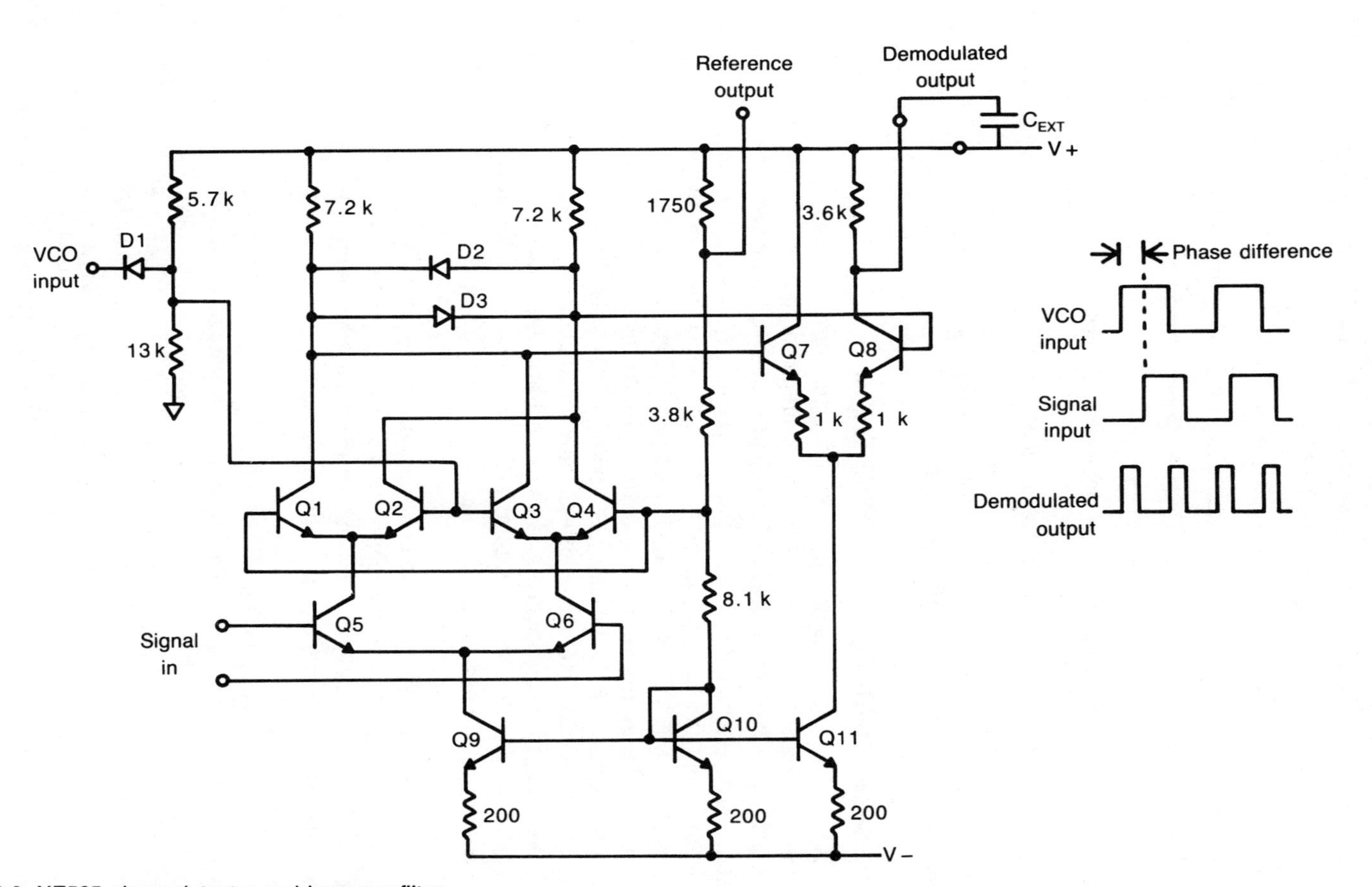

Fig. 13-8. NE565 phase detector and low-pass filter.

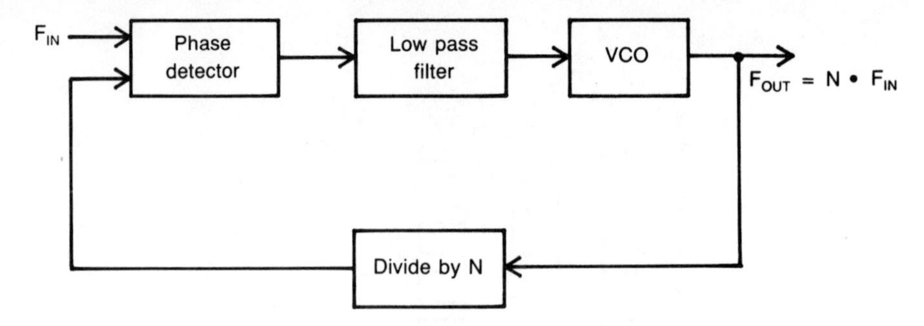

Fig. 13-9. Phase-locked loop used to perform frequency multiplication.

cation, shown in Fig. 13-9. In this application, a divide-by-n counter is inserted between the VCO and the phase detector. The result is an output with a frequency equal to n times the input frequency and synchronized to the input frequency.

VOLTAGE-TO-FREQUENCY CONVERTERS

The voltage-to-frequency, or V/F, converter converts an analog input voltage between zero volts and some maximum voltage to a frequency output that is linearly proportional to the input voltage. A VCO, such as the NE566, may be considered to be a V/F converter, but the control voltage Vc for most VCOs must be in the range of 0.75 Vcc to Vcc. Additional circuitry could be added to convert the control voltage range to a 0-10 V range, for example, but the result would probably be a poor V/F converter. In fact, VCOs and V/F converters have different areas of application: VCOs are designed for such applications as modulators, demodulators, oscillators, and function generators; V/F converters are designed for such applications as high-resolution A/D converters and precision integrators.

Figure 13-10 shows the block diagram of the Burr-Brown VFC32 voltage-to-frequency converter. The VFC32 consists of an op amp (used as an integrator), a comparator, a one-shot (monostable multivibrator, timer), a constant 1 mA current source, and a current-source switch.

Assume that the output of the integrator (the input to the comparator) is a positive voltage. For some positive voltage V_{IN}, a constant current I_{IN} will flow through R1, causing the integrator output to ramp down at a linear rate (see Fig. 13-11). When the integrator output gets to zero volts, the comparator switches and fires the one-shot timer. The one-shot closes the switch S1 (a transistor in actuality) and the capacitor charges at a linear rate for a duration controlled by an external capacitor C1. During this time T_2, the total charge on the capacitor is

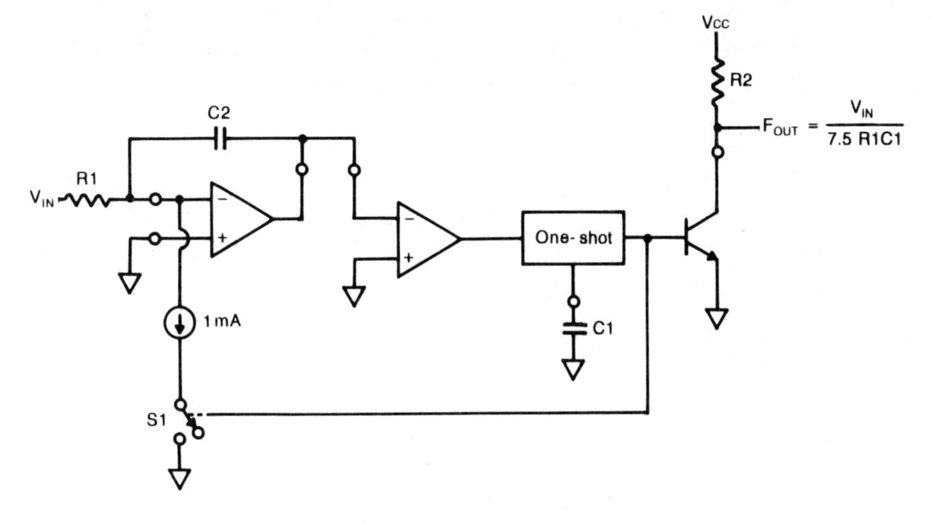

Fig. 13-10. Block diagram of VFC32 voltage-to-frequency converter.

$$Q = \text{current times time} = (1\ \text{mA} - I_{IN})\ T_2$$

At the end of T_2, switch S1 opens and the capacitor begins to be discharged by the integrator. The charge removed during the integration time T_1 is equal to

$$Q = I_{IN}\ T_1 = I_{IN}\ [(1/F_{OUT}) - T_2]$$

For charge balance, the charge put on the capacitor C2 must equal the charge removed, or

$$(1\ \text{mA} - 1_{IN})\ T_2 = I_{IN}\ [(1/F_{OUT}) - T_2]$$

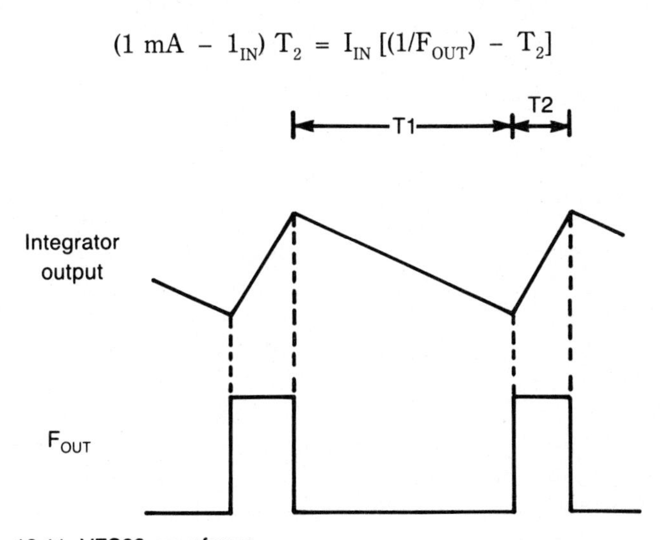

Fig. 13-11. VFC32 waveforms.

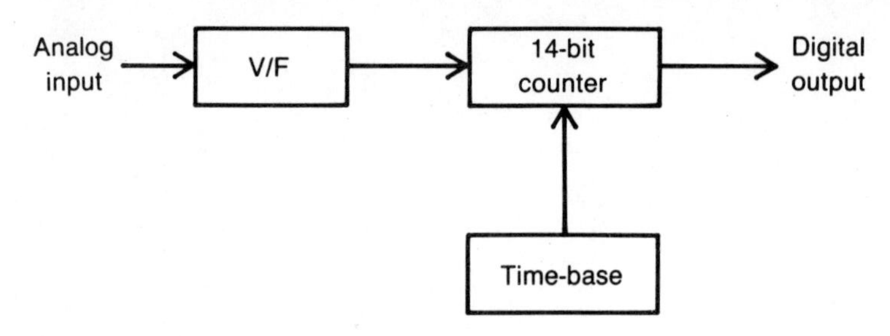

Fig. 13-12. V/F converter used as a 14-bit A/D converter.

Solving for F_{OUT} yields

$$F_{OUT} = I_{IN}/(1 \text{ mA } T_2)$$

According to the VFC32 data sheet T_2 equals

$$T_2 = C1\ (7.5 \text{ V}/1 \text{ mA})$$

Substituting this equation in the F_{OUT} equation, we get

$$F_{OUT} = I_{IN}/(7.5\ C1) = V_{IN}/(7.5\ R1\ C1)$$

Notice that the output frequency does not depend on the value of C2. The value of C2 controls the maximum voltage at the integrator output but does not control frequency. However, C2 is a function of full scale frequency. Burr-Brown recommends a value of 0.001 μF for full-scale frequencies above 100 kHz and a value equal to $10^2/f_{MAX}$ below 100 kHz.

Figure 13-12 shows the V/F converter used as a 14-bit A/D converter. The timebase is an oscillator that resets the counter. The timebase sets the sample rate and scales the output of the counter. For example, let's say you set up the V/F converter to have an output range of 0-10 kHz corresponding to 0-10 V on the input. That's a maximum of 10,000 pulses per second. A 14-bit counter can count up to $2^{14} - 1$, or 16,383. If you set the timebase to put out a reset pulse every second, the counter output in binary would equal the analog input voltage with an accuracy of 1/10,000, or 0.01%.

Table 13-4 lists available V/F converters. Nonlinearity is defined as the amount in percent of full scale that the actual output deviates from the ideal straight-line transfer function for the full input voltage range after the V/F has been adjusted at 10 Hz (not zero) and the full-scale frequency. Note that in the table, I have listed nonlinearity numbers for the largest full-scale frequency given in the data sheets. The nonlinearity of V/F converters is typically a factor of 0.2 to 0.1 times these numbers if you set the full-scale frequency at 0.1 times its maximum full-scale frequency (using 10 kHz at the full-scale frequency instead of 100 kHz, for example).

Table 13-1. Timers.

PART NUMBER	MANUFACTURER	TEMP GRADE	TIMERS PER PKG	UNIT PRICE ($)
UA555TC	FAIRCHILD	COM	1	.57
UA556PC	FAIRCHILD	COM	2	.80
UA2240PC	FAIRCHILD	COM	1	1.22
ICM7555IPA	INTERSIL	IND	1	1.25
ICM7555MTV	INTERSIL	MIL	1	4.25
ICM7556IPD	INTERSIL	IND	2	1.55
ICM7556MJD	INTERSIL	MIL	2	4.20
MC1455P1	MOTOROLA	COM	1	.59
MC1555G	MOTOROLA	MIL	1	3.55
MC3456P	MOTOROLA	COM	2	.97
MC3556L	MOTOROLA	MIL	2	6.19
LM122H	NATIONAL	MIL	1	7.50
LM322N	NATIONAL	COM	1	1.75
LM555J	NATIONAL	MIL	1	3.60
LM555CN	NATIONAL	COM	1	.75
LM556J	NATIONAL	MIL	2	4.28
LM556CN	NATIONAL	COM	2	.93
LM2905N	NATIONAL	IND	1	3.00
LM3905N	NATIONAL	COM	1	1.35
CA555E	RCA	MIL	1	.66
NE555N	SIGNETICS	COM	1	.48
SE555N	SIGNETICS	MIL	1	1.20
SE555CN	SIGNETICS	MIL	1	1.00
NE556N	SIGNETICS	COM	2	.66
NE556-1N	SIGNETICS	COM	2	.70
SA556N	SIGNETICS	IND	2	.96
SA556-1N	SIGNETICS	IND	2	1.02
SE556N	SIGNETICS	MIL	2	3.00
SE556-1N	SIGNETICS	MIL	2	3.20
SE556CN	SIGNETICS	MIL	2	1.30
NE558N	SIGNETICS	COM	4	1.38

PART NUMBER	MANUFACTURER	TEMP GRADE	TIMERS PER PKG	UNIT PRICE ($)
SA558N	SIGNETICS	IND	4	1.68
SE558N	SIGNETICS	MIL	4	3.80
TLC555CP	T I	COM	1	.90

Table 13-2. Oscillators and Function Generators.

PART NUMBER	MANUFACTURER	TEMP GRADE	WAVEFORM OUTPUTS	MAXIMUM DRIFT (PPM/°C)	UNIT PRICE ($)
4023/25	BURR-BROWN	IND	SINE	400	199.50
4423	BURR-BROWN	COM	SINE,COSINE	100	22.00
ICL8038AC	INTERSIL	COM	SINE,SQ,TRI	80	30.00
ICL8038BC	INTERSIL	COM	SINE,SQ,TRI	150	15.00
ICL8038CC	INTERSIL	COM	SINE,SQ,TRI	250 TYP	5.20
ICL8038AM	INTERSIL	MIL	SINE,SQ,TRI	80	32.70
ICL8038BM	INTERSIL	MIL	SINE,SQ,TRI	150	16.35
LM566CN	NATIONAL	COM	SQ,TRI	200 TYP	1.95
NE566N	SIGNETICS	COM	SQ,TRI	300 TYP	.80
SE566N	SIGNETICS	MIL	SQ,TRI	200 TYP	1.20

Table 13-3. Phase-Locked Loops.

PART NUMBER	MANUFACTURER	TEMP GRADE	MAXIMUM VCO FREQ (MHz)	UNIT PRICE ($)
LM565H	NATIONAL	MIL	0.3	15.00
LM565CN	NATIONAL	COM	0.25	1.95
NE564N	SIGNETICS	COM	45	1.96
SE564F	SIGNETICS	MIL	50	6.40
NE565N	SIGNETICS	COM	0.25	.88
SE565N	SIGNETICS	MIL	0.3	1.20

Table 13-4. Voltage-to-Frequency Converters.

PART NUMBER	MANUFACTURER	TEMP GRADE	MAXIMUM NONLINEARITY ERROR	UNIT PRICE ($)
AD537JH	ANALOG DEVICES	COM	0.25% @ 100kHz	8.10
AD537KH	ANALOG DEVICES	COM	0.1% @ 100kHz	12.50

PART NUMBER	MANUFACTURER	TEMP GRADE	MAXIMUM NONLINEARITY ERROR	UNIT PRICE ($)
AD537SH	ANALOG DEVICES	MIL	0.1% @ 100kHz	24.60
AD650AD	ANALOG DEVICES	IND	0.05% @ 500kHz	17.20
AD650BD	ANALOG DEVICES	IND	0.1% @ 1MHz	20.95
AD650JN	ANALOG DEVICES	COM	0.05% @ 500kHz	11.95
AD650KN	ANALOG DEVICES	COM	0.1% @ 1MHz	14.95
AD650SD	ANALOG DEVICES	MIL	0.1% @ 1MHz	28.50
ADVFC32BH	ANALOG DEVICES	IND	0.2% @ 500kHz	11.95
ADVFC32KN	ANALOG DEVICES	COM	0.2% @ 500kHz	8.95
ADVFC32SH	ANALOG DEVICES	MIL	0.2% @ 500kHz	17.25
VFC32BM	BURR-BROWN	IND	0.05% @ 100kHz	14.90
VFC32KP	BURR-BROWN	COM	0.05% @ 100kHz	10.15
VFC32SM	BURR-BROWN	MIL	0.05% @ 100kHz	19.70
VFC32UM	BURR-BROWN	MIL	0.2% @ 500kHz	15.00
VFC32VM	BURR-BROWN	MIL	0.2% @ 500kHz	26.00
VFC32WM	BURR-BROWN	MIL	0.2% @ 500kHz	40.00
VFC42BP	BURR-BROWN	IND	0.01% @ 10kHz	20.35
VFC42SM	BURR-BROWN	MIL	0.01% @ 10kHz	34.40
VFC52BP	BURR-BROWN	IND	0.05% @ 100kHz	20.35
VFC52SM	BURR-BROWN	MIL	0.05% @ 100kHz	34.40
VFC62BM	BURR-BROWN	IND	0.03% @ 100kHz	15.60
VFC62CM	BURR-BROWN	IND	0.03% @ 100kHz	16.60
VFC62SM	BURR-BROWN	MIL	0.03% @ 100kHz	20.65
VFC320BM	BURR-BROWN	IND	0.03% @ 100kHz	15.60
VFC320CM	BURR-BROWN	IND	0.03% @ 100kHz	16.60
VFC320SM	BURR-BROWN	MIL	0.03% @ 100kHz	20.65
LM131H	NATIONAL	MIL	0.01% @ 11kHz	11.55
LM131AH	NATIONAL	MIL	0.01% @ 11kHz	17.30
LM231N	NATIONAL	IND	0.01% @ 11kHz	5.25
LM231AN	NATIONAL	IND	0.01% @ 11kHz	6.50
LM331N	NATIONAL	COM	0.01% @ 11kHz	4.35
LM331AN	NATIONAL	COM	0.01% @ 11kHz	5.65

Chapter 14

Transistor, Amplifier, and Diode Arrays

A TRANSISTOR ARRAY IS AN INTEGRATED CIRCUIT THAT CONtains several independent transistors (usually 4 or more) on a single chip. The collector, base, and emitter connections of each transistor are separately brought out to the IC pins, allowing for maximum circuit design flexibility. In some cases, two of the the transistors may have their emitters connected to a common pin, forming a differential pair. A differential amplifier array contains two or more differential transistor pairs. A diode array contains independent diodes or pairs of diodes with common anodes or common cathodes.

Arrays have two main advantages over their discrete counterparts: (1) the transistor base-emitter voltages and diode voltages are usually matched to within 5 millivolts; and (2) being in a single package, they take up less area on a printed-circuit board. The matching characteristics, in particular, allow you to design your own current mirrors, differential amplifiers, or other special circuits that require matched components. Thus, you can use the same techniques that linear IC designers use and, therefore, design circuits for which there are no readily available off-the-shelf ICs.

TRANSISTOR ARRAYS

Table 14-1 lists available NPN transistor arrays. Most of these are general-purpose arrays—maximum collector current of 50 mA, minimum collector-emitter breakdown voltage of 15 V (called maximum collector voltage in the table because that is the maximum collector-emitter voltage that you the user should put on the device), and a gain-bandwidth product of 500-550 MHz.

The CA1724, CA1725, and CA3138 are high-current NPN transistor arrays, having a maximum collector current of 1 ampere. The 1724 and 1725 are also high-voltage devices, having a minimum collector-emitter breakdown voltage of 40 V and 50 V, respectively. Other high-voltage arrays include: LM3146 (30 V), CA3118 (30 V), CA3118A (40 V), CA3146 (30 V), CA3146A (40 V), CA3183 (30 V), and the CA3183A (40 V). The CA3127, CA3227, and CA3256 are high-frequency NPN transistor arrays, having gain-bandwidth products of 1.15 GHz typical, 3 GHz minimum, and 3 GHz minimum, respectively.

Figure 14-1 shows schematics of NPN transistor arrays. As mentioned above, transistor arrays are independent transistor ar-

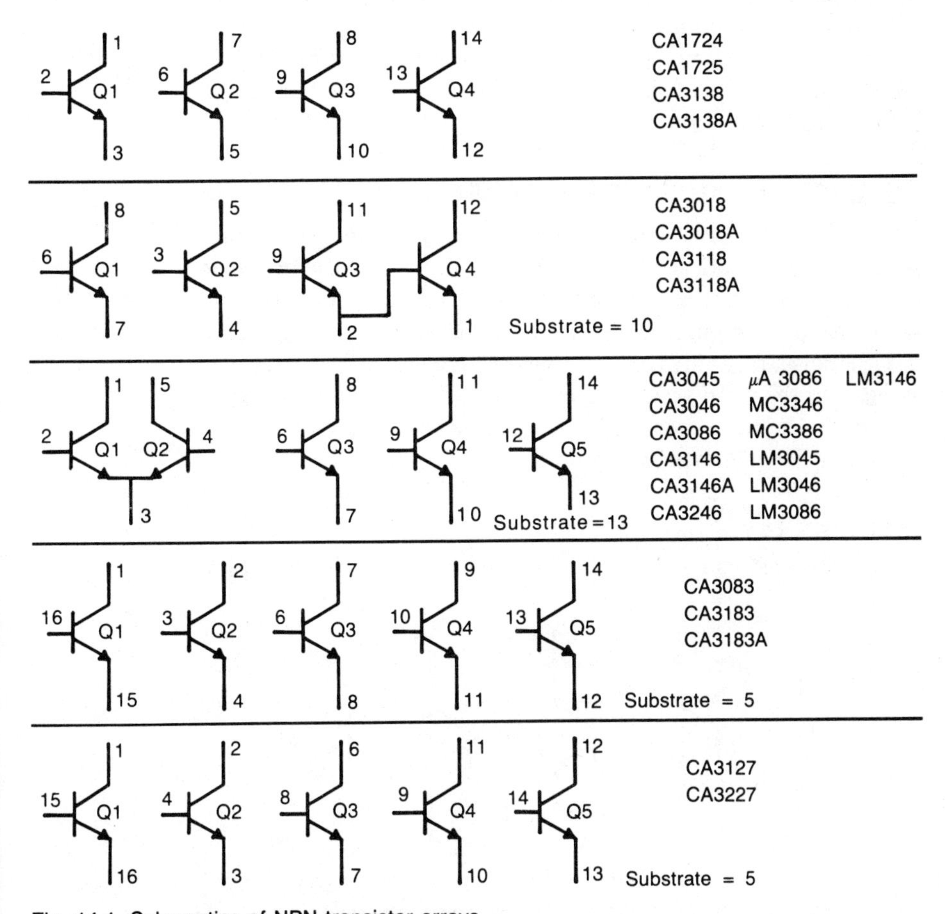

Fig. 14-1. Schematics of NPN transistor arrays.

rays on a single chip. This is not always true, as Fig. 14-1 shows. The 1724, 1725, 3138, and 3183 are actually four separate die in a single package. That is, they are not on a common substrate, but are individual die physically isolated from each other. The schematics for the other arrays show the pin number for the substrate connection. Recall from Chapter 1 that the substrate must be connected to the most negative potential in order to electrically isolate the transistors from each other.

Also note that in the 3018, 3018A, 3118, and 3118A, the emitter of Q3 is tied to the base of Q4, forming a Darlington transistor. In the 3045 and other arrays the emitters of Q1 and Q2 are tied together to form a differential pair.

Table 14-2 lists the only available NPN/PNP transistor array that I could find—the CA3096. Figure 14-2 shows the schematic. The NPNs have a minimum collector-emitter breakdown voltage of 35 V for E and AE versions and 24 V for the CE version. Maximum collector current is 50 mA. Gain-bandwidth product is 280 MHz typical at 1 mA collector current and 335 MHz at 5 mA. The PNPs have a minimum emitter-collector breakdown voltage of 40 V for the E and AE versions and 24 V for the CE version. Maximum collector current is 10 mA. Gain-bandwidth product is 6.8 MHz typical at 100 μA collector current.

DIFFERENTIAL AMPLIFIER ARRAYS

Table 14-3 lists available differential amplifier arrays. Figure 14-3 shows schematics. The CA3049 and CA3102 transistors have typical gain-bandwidth products of 1.35 GHz. The MC3350 transistors have a minimum collector-emitter breakdown voltage of 35 V. Otherwise, these arrays have the same transistor specifications as most of the NPN transistor arrays—15 V minimum collector-emitter breakdown voltage, 50 mA maximum collector current, and 500-550 MHz gain-bandwidth product.

DIODE ARRAYS

Diode arrays are listed in Table 14-4. Schematics are shown in Fig.

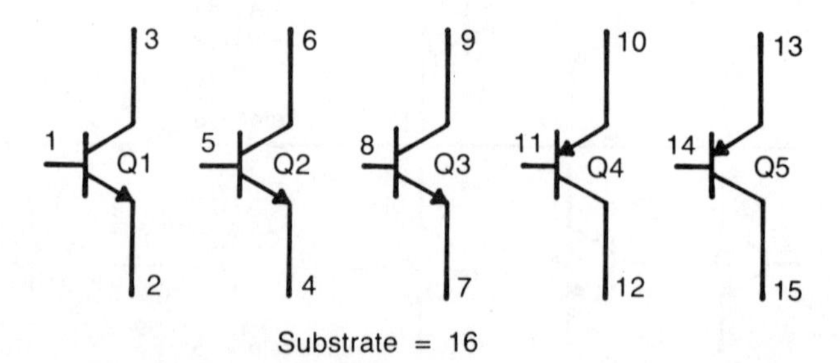

Fig. 14-2. Schematic of the CA3096 NPN/PNP transistor array.

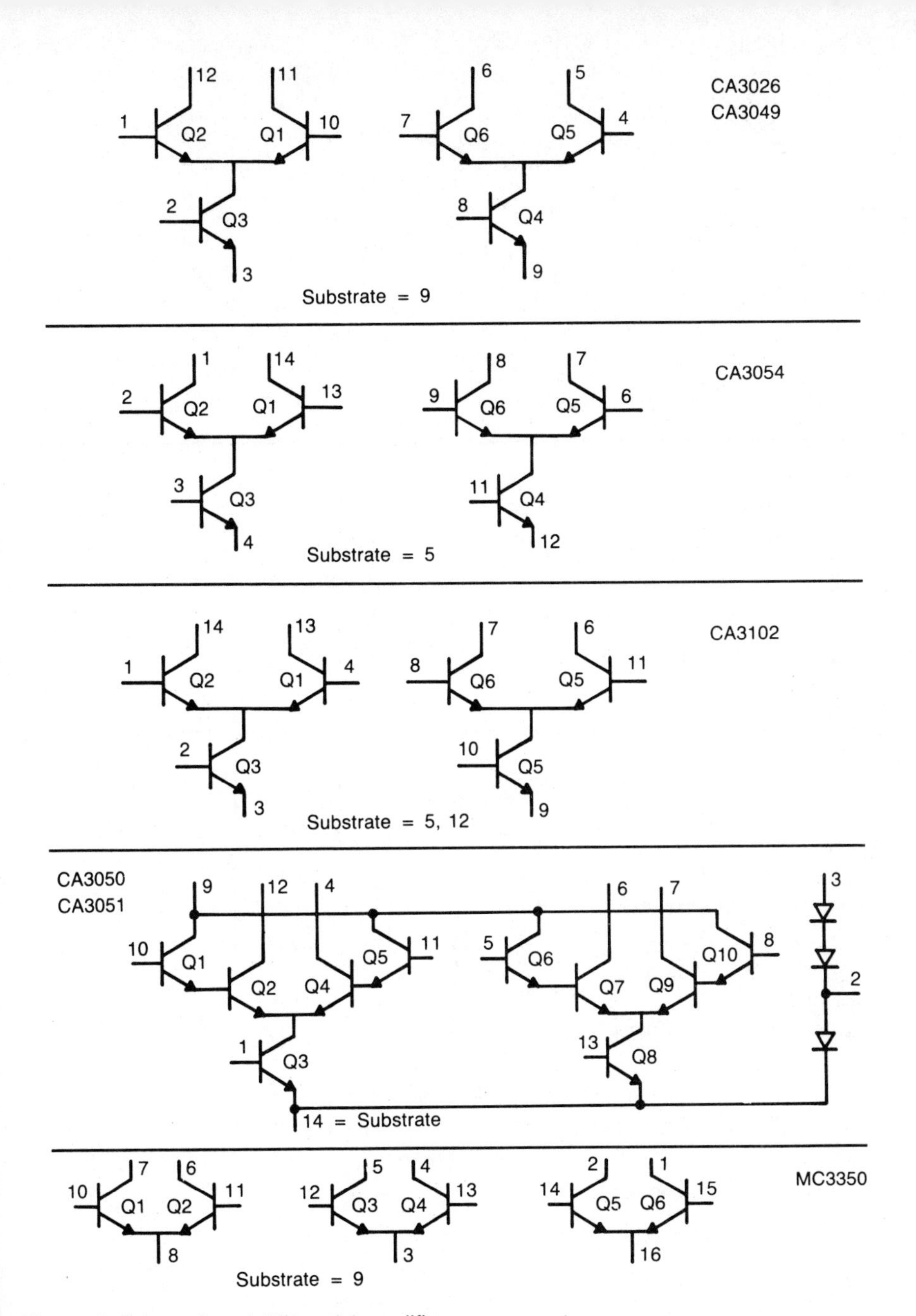

Fig. 14-3. Schematics of differential amplifier arrays.

14-4. Minimum reverse breakdown is 4 V for the CA3019, 5 V for the CA3039, and 30 V for the CA3141. Typical reverse recovery time is not specified for the 3019, 1 ns for the 3039, and 50 ns for the 3141. Maximum dc forward current is 25 mA for all three arrays.

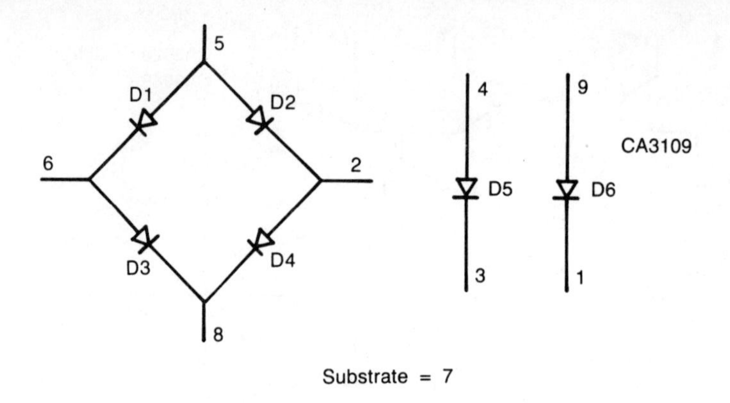

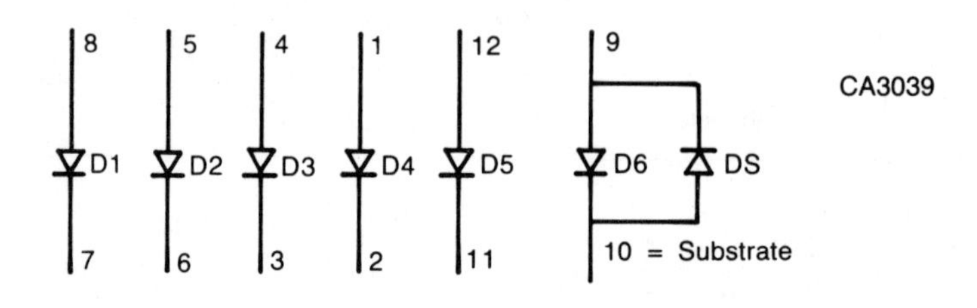

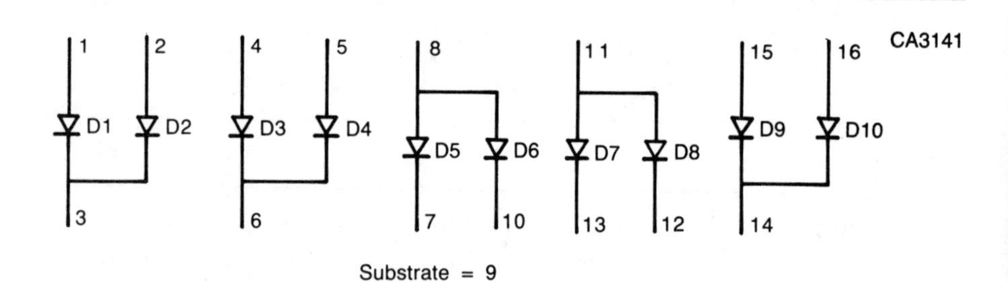

Fig. 14-4. Schematics of diode arrays.

Table 14-1. NPN Transistor Arrays.

PART NUMBER	MANUFACTURER	TEMP GRADE	NPN'S PER PKG	MAXIMUM COLL. CURRENT (mA)	MAXIMUM COLL. VOLTAGE (V)	TYP BAND-WIDTH (MHz)	UNIT PRICE ($)
UA3086PC	FAIRCHILD	COM	5	50	15	550	.55
MC3346P	MOTOROLA	IND	5	50	15	550	1.17
MC3386P	MOTOROLA	IND	5	50	15	550	1.17
LM3045J	NATIONAL	MIL	5	50	15	550	1.60
LM3046N	NATIONAL	IND	5	50	15	550	1.00
LM3086N	NATIONAL	IND	5	50	15	550	.65
LM3146N	NATIONAL	IND	5	50	30	500	1.39

PART NUMBER	MANUFACTURER	TEMP GRADE	NPN'S PER PKG	MAXIMUM COLL. CURRENT (mA)	MAXIMUM COLL. VOLTAGE (V)	TYP BAND-WIDTH (MHz)	UNIT PRICE ($)
CA1724E	RCA	MIL	4	1000	40	--	2.10
CA1725E	RCA	MIL	4	1000	50	--	2.80
CA3018	RCA	MIL	4	50	15	500	1.60
CA3018A	RCA	MIL	4	50	15	500	2.00
CA3045F	RCA	MIL	5	50	15	550	1.40
CA3046	RCA	MIL	5	50	15	550	.76
CA3083	RCA	MIL	5	100	15	450	.88
CA3086	RCA	MIL	5	50	15	550	.64
CA3118AT	RCA	MIL	4	50	40	500	2.80
CA3118T	RCA	MIL	4	50	30	500	2.50
CA3127E	RCA	MIL	5	20	15	1150	2.40
CA3138AE	RCA	MIL	4	1000	15	--	3.76
CA3138E	RCA	MIL	4	1000	15	--	3.40
CA3146AE	RCA	MIL	5	50	40	500	1.60
CA3146E	RCA	MIL	5	50	30	500	1.00
CA3183AE	RCA	MIL	5	75	40	--	1.50
CA3183E	RCA	MIL	5	75	30	--	1.20
CA3227E	RCA	MIL	5	20	8	3000 MIN	3.00
CA3246E	RCA	MIL	5	20	8	3000 MIN	1.50

Table 14-2. NPN/PNP Transistor Arrays.

PART NUMBER	MANUFACTURER	TEMP GRADE	NPN'S PER PKG	PNP'S PER PKG	UNIT PRICE ($)
CA3096AE	RCA	MIL	3	2	1.96
CA3096CE	RCA	MIL	3	2	.62
CA3096E	RCA	MIL	3	2	1.32

Table 14-3. Differential Amplifier Arrays.

PART NUMBER	MANUFACTURER	TEMP GRADE	DIFF AMPS PER PKG	UNIT PRICE ($)
MC3350P	MOTOROLA	IND	3	1.13
CA3026	RCA	MIL	2	1.70

PART NUMBER	MANUFACTURER	TEMP GRADE	DIFF AMPS PER PKG	UNIT PRICE ($)
CA3049T	RCA	MIL	2	2.70
CA3050	RCA	MIL	2	10.00
CA3051	RCA	IND	2	1.20
CA3054	RCA	IND	2	.84
CA3102E	RCA	MIL	2	3.30

Table 14-4. Diode Arrays.

PART NUMBER	MANUFACTURER	TEMP GRADE	DIODES PER PKG	UNIT PRICE ($)
CA3019	RCA	MIL	6	1.50
CA3039	RCA	MIL	6	1.50
CA3141E	RCA	MIL	10	.84

Appendices

Specification Summary Tables

Appendix A
Operational Amplifiers

OPERATIONAL AMPLIFIERS:

PART NUMBER	OP AMPS PER PKG	TEMP GRADE	MAXIMUM OFFSET VOLTAGE (mV)	MAXIMUM OFFSET DRIFT (uV/°C)	MAXIMUM BIAS CURRENT (nA)
AD101A	1	MIL	2	15	75
AD201A	1	IND	2	15	75
AD301A	1	COM	7.5	30	250
AD301AL	1	COM	0.5	5	30
AD3554A	1	IND	2	50	0.05
AD3554B	1	IND	1	15	0.05
AD3554S	1	MIL	1	25	0.05
AD380J	1	COM	2	50	0.1
AD380K	1	COM	1	20	0.1
AD380L	1	COM	1	10	0.1
AD380S	1	MIL	1	50	0.1
AD381J	1	COM	1	15	0.1
AD381K	1	COM	0.5	10	0.1
AD381L	1	COM	0.25	5	0.1
AD381S	1	MIL	1	10	0.1
AD382J	1	COM	1	15	0.1

ANALOG DEVICES

TYP BAND-WIDTH (MHz)	TYP SLEW RATE (V/uS)	TYPICAL INPUT NOISE VOLTAGE ($nV/\sqrt{Hz}$)	MAXIMUM SUPPLY CURRENT (mA)	MAXIMUM SUPPLY VOLTAGE (±V)	OUTPUT CURRENT RATING (A)	UNIT PRICE ($)
1	0.25	15	3	22	0.025	5.20
1	0.25	15	3	22	0.025	3.00
1	0.25	15	3	18	0.025	1.50
1	0.25	15	3	18	0.025	9.80
90	1200	15	45	18	0.1	79.00
90	1200	15	45	18	0.1	90.50
90	1200	15	45	18	0.1	105.50
40	330	15	15	20	0.05	59.00
40	330	15	15	20	0.05	68.00
40	330	15	15	20	0.05	78.00
40	330	15	15	20	0.05	88.00
5	30	18	5	18	0.01	10.25
5	30	18	5	18	0.01	13.10
5	30	18	5	18	0.01	16.75
5	30	18	5	18	0.01	21.00
5	30	18	6	18	0.05	28.00

OPERATIONAL AMPLIFIERS:

PART NUMBER	OP AMPS PER PKG	TEMP GRADE	MAXIMUM OFFSET VOLTAGE (mV)	MAXIMUM OFFSET DRIFT (uV/°C)	MAXIMUM BIAS CURRENT (nA)	TYP BAND-WIDTH (MHz)
AD382K	1	COM	0.5	10	0.05	5
AD382L	1	COM	0.25	5	0.05	5
AD382S	1	MIL	1	10	0.05	5
AD503J	1	COM	50	75	0.015	1
AD503K	1	COM	20	25	0.01	1
AD503S	1	MIL	20	50	0.01	1
AD504J	1	COM	2.5	5	200	0.3
AD504K	1	COM	1.5	3	100	0.3
AD504L	1	COM	0.5	1	80	0.3
AD504M	1	COM	0.5	0.5	80	0.3
AD504S	1	MIL	0.5	1	80	0.3
AD506J	1	COM	3.5	75	0.015	1
AD506K	1	COM	1.5	25	0.01	1
AD506L	1	COM	1	10	0.005	1
AD506S	1	MIL	1.5	50	0.01	1
AD507J	1	COM	5	15	25	35

ANALOG DEVICES

TYP SLEW RATE (V/uS)	TYPICAL INPUT NOISE VOLTAGE (nV/$\sqrt{Hz}$)	MAXIMUM SUPPLY CURRENT (mA)	MAXIMUM SUPPLY VOLTAGE (±V)	OUTPUT CURRENT RATING (A)	UNIT PRICE ($)
30	18	6	18	0.05	33.00
30	18	6	18	0.05	42.00
30	18	6	18	0.05	45.00
6	30	7	18	0.01	17.00
6	30	7	18	0.01	20.20
6	30	7	22	0.01	34.70
0.12	8	4	18	0.01	13.80
0.12	8	3	18	0.01	24.40
0.12	8	3	18	0.01	41.80
0.12	9	3	18	0.01	48.80
0.12	8	3	18	0.01	39.40
6	80	7	18	0.01	17.60
6	30	7	18	0.01	21.00
6	25	7	18	0.01	32.60
6	30	7	18	0.01	36.00
35	12	4	20	0.01	11.90

OPERATIONAL AMPLIFIERS:

PART NUMBER	OP AMPS PER PKG	TEMP GRADE	MAXIMUM OFFSET VOLTAGE (mV)	MAXIMUM OFFSET DRIFT (uV/°C)	MAXIMUM BIAS CURRENT (nA)	TYP BAND-WIDTH (MHz)
AD507K	1	COM	3	15	15	35
AD507S	1	MIL	4	20	15	35
AD509J	1	COM	10		250	20
AD509K	1	COM	8		200	20
AD509S	1	MIL	8		200	20
AD510J	1	COM	0.1	3	25	0.3
AD510K	1	COM	0.05	1	13	0.3
AD510L	1	COM	0.025	0.5	10	0.3
AD510S	1	MIL	0.05	1	13	0.3
AD515J	1	COM	3	50	0.0003	0.35
AD515K	1	COM	1	15	0.00015	0.35
AD515L	1	COM	1	25	0.000075	0.35
AD517J	1	COM	0.15	3	5	0.25
AD517K	1	COM	0.075	1.8	2	0.25
AD517L	1	COM	0.05	1.3	1	0.25
AD517S	1	MIL	0.075	1.8	2	0.25

ANALOG DEVICES

TYP SLEW RATE (V/uS)	TYPICAL INPUT NOISE VOLTAGE (nV/$\sqrt{Hz}$)	MAXIMUM SUPPLY CURRENT (mA)	MAXIMUM SUPPLY VOLTAGE (±V)	OUTPUT CURRENT RATING (A)	UNIT PRICE ($)
35	12	4	20	0.01	18.60
35	12	4	20	0.01	28.20
120	19	6	20	0.005	14.10
120	19	6	20	0.005	22.90
120	19	6	20	0.005	29.15
0.1	10	4	18	0.01	10.70
0.1	10	3	18	0.01	18.00
0.1	10	3	18	0.01	24.80
0.1	10	3	22	0.01	45.50
1	50	1.5	18	0.005	20.10
1	50	1.5	18	0.005	29.00
1	50	1.5	18	0.005	34.70
0.1	20	4	18	0.01	5.80
0.1	20	3	18	0.01	8.30
0.1	20	3	18	0.01	16.90
0.1	20	3	18	0.01	31.10

OPERATIONAL AMPLIFIERS:

PART NUMBER	OP AMPS PER PKG	TEMP GRADE	MAXIMUM OFFSET VOLTAGE (mV)	MAXIMUM OFFSET DRIFT (uV/°C)	MAXIMUM BIAS CURRENT (nA)	TYP BAND-WIDTH (MHz)
AD518J	1	COM	10		500	12
AD518K	1	COM	4		250	12
AD518S	1	MIL	4		250	12
AD542J	1	COM	2	20	0.05	1
AD542K	1	COM	1	10	0.025	1
AD542L	1	COM	0.5	5	0.025	1
AD542S	1	MIL	1	15	0.025	1
AD544J	1	COM	2	20	0.05	2
AD544K	1	COM	1	10	0.025	2
AD544L	1	COM	0.5	5	0.025	2
AD544S	1	MIL	1	15	0.025	2
AD545J	1	COM	1	25	0.002	0.7
AD545K	1	COM	1	15	0.001	0.7
AD545L	1	COM	0.5	5	0.001	0.7
AD545M	1	COM	0.25	3	0.001	0.7
AD547J	1	COM	1	5	0.05	1

ANALOG DEVICES

TYP SLEW RATE (V/uS)	TYPICAL INPUT NOISE VOLTAGE (nV/√Hz)	MAXIMUM SUPPLY CURRENT (mA)	MAXIMUM SUPPLY VOLTAGE (±V)	OUTPUT CURRENT RATING (A)	UNIT PRICE ($)
70		10	20	0.01	2.30
70		7	20	0.01	8.90
70		7	20	0.01	11.40
3	30	1.5	18	0.005	4.60
3	30	1.5	18	0.005	6.60
3	30	1.5	18	0.005	12.90
3	30	1.5	18	0.005	22.50
13	18	2.5	18	0.005	4.30
13	18	2.5	18	0.005	6.30
13	18	2.5	18	0.005	12.40
13	18	2.5	18	0.005	22.20
1	35	1.5	18	0.005	11.00
1	35	1.5	18	0.005	14.00
1	35	1.5	18	0.005	20.00
1	35	1.5	18	0.005	32.00
3	30	1.5	18	0.005	5.10

PART NUMBER	OP AMPS PER PKG	TEMP GRADE	MAXIMUM OFFSET VOLTAGE (mV)	MAXIMUM OFFSET DRIFT (uV/°C)	MAXIMUM BIAS CURRENT (nA)	TYP BAND-WIDTH (MHz)
AD547K	1	COM	0.5	2	0.025	1
AD547K	1	COM	0.5	2	0.025	1
AD547L	1	COM	0.25	1	0.025	1
AD547S	1	MIL	0.5	5	0.05	1
AD642J	2	COM	2		0.075	1
AD642K	2	COM	1		0.035	1
AD642L	2	COM	0.5		0.035	1
AD642S	2	MIL	1		0.035	1
AD644J	2	COM	2		0.075	2
AD644K	2	COM	1		0.035	2
AD644L	2	COM	0.5		0.035	2
AD644S	2	MIL	1		0.035	2
AD647J	2	COM	1	10	0.075	1
AD647K	2	COM	0.5	5	0.035	1
AD647L	2	COM	0.25	2.5	0.035	1
AD647S	2	MIL	0.5	5	0.035	1

ANALOG DEVICES

TYP SLEW RATE (V/uS)	TYPICAL INPUT NOISE VOLTAGE (nV/√Hz)	MAXIMUM SUPPLY CURRENT (mA)	MAXIMUM SUPPLY VOLTAGE (±V)	OUTPUT CURRENT RATING (A)	UNIT PRICE ($)
3	30	1.5	18	0.005	5.10
3	30	1.5	18	0.005	8.70
3	30	1.5	18	0.005	17.00
3	30	1.5	18	0.005	25.50
3	30	2.8	18	0.005	6.90
3	30	2.8	15	0.005	10.40
3	30	2.8	15	0.005	14.56
3	30	2.8	15	0.005	30.00
13	18	4.5	18	0.005	6.90
13	18	4.5	18	0.005	10.40
13	18	4.5	18	0.005	14.56
13	18	4.5	18	0.005	30.00
3	30	2.8	18	0.005	8.25
3	30	2.8	18	0.005	13.50
3	30	2.8	18	0.005	26.25
3	30	2.8	18	0.005	35.95

OPERATIONAL AMPLIFIERS:

PART NUMBER	OP AMPS PER PKG	TEMP GRADE	MAXIMUM OFFSET VOLTAGE (mV)	MAXIMUM OFFSET DRIFT (uV/°C)	MAXIMUM BIAS CURRENT (nA)	TYP BAND-WIDTH (MHz)
AD741	1	MIL	5		500	1
AD741C	1	COM	6		500	1
AD741J	1	COM	3	20	200	1
AD741K	1	COM	2	15	75	1
AD741L	1	COM	0.5	5	50	1
AD741S	1	MIL	2	15	75	1

ANALOG DEVICES

TYP SLEW RATE (V/uS)	TYPICAL INPUT NOISE VOLTAGE (nV/√Hz)	MAXIMUM SUPPLY CURRENT (mA)	MAXIMUM SUPPLY VOLTAGE (±V)	OUTPUT CURRENT RATING (A)	UNIT PRICE ($)
0.5		2.8		0.005	3.00
0.5		2.8		0.005	1.50
0.5		3.3	18	0.01	2.00
0.5		2.8	22	0.005	3.80
0.5		2.8	22	0.005	10.80
0.5		2.8	22	0.005	5.90

PART NUMBER	OP AMPS PER PKG	TEMP GRADE	MAXIMUM OFFSET VOLTAGE (mV)	MAXIMUM OFFSET DRIFT (uV/°C)	MAXIMUM BIAS CURRENT (nA)	TYP BAND-WIDTH (MHz)
3271/25	1	IND	0.05	1	0.08	1 MIN
3354/25	1	IND	0.03	0.1	0.02	3 MIN
3355/25	1	IND	0.05	0.25	0.05	3 MIN
3356/25	1	IND	0.1	1	0.05	3 MIN
3500A	1	IND	5	20	30	1.5
3500B	1	IND	2	5	20	1.5
3500C	1	IND	1	3	15	1.5
3500E	1	IND	0.5	1	50	1.5
3500R	1	MIL	5	20	30	1.5
3500S	1	MIL	2	5	20	1.5
3500T	1	MIL	1	3	15	1.5
3501A	1	IND	5	20	15	0.5
3501B	1	IND	2	10	7	0.5
3501C	1	IND	2	5	3	0.5
3501R	1	MIL	5	20	15	0.5
3501S	1	MIL	2	10	7	0.5

BURR-BROWN

TYP SLEW RATE (V/uS)	TYPICAL INPUT NOISE VOLTAGE (nV/√Hz)	MAXIMUM SUPPLY CURRENT (mA)	MAXIMUM SUPPLY VOLTAGE (±V)	OUTPUT CURRENT RATING (A)	UNIT PRICE ($)
20 MIN			120	0.02	244.00
6 MIN		10	18	0.005	193.00
6 MIN		10	18	0.005	150.00
6 MIN		10	18	0.005	129.50
0.6 MIN		3.5	20	0.01	9.80
0.8 MIN		3.5	20	0.01	17.60
1 MIN		3.5	20	0.01	22.45
0.5 MIN		3.5	20	0.01	36.25
0.6 MIN		3.5	20	0.01	21.25
0.8 MIN		3.5	20	0.01	33.00
1 MIN		3.5	20	0.01	53.25
0.1 MIN		1.5	20	0.005	5.70
0.1 MIN		1.5	20	0.005	12.20
0.1 MIN		1.5	20	0.005	15.95
0.1 MIN		1.5	20	0.005	17.50
0.1 MIN		1.5	20	0.005	33.00

OPERATIONAL AMPLIFIERS:

PART NUMBER	OP AMPS PER PKG	TEMP GRADE	MAXIMUM OFFSET VOLTAGE (mV)	MAXIMUM OFFSET DRIFT (uV/°C)	MAXIMUM BIAS CURRENT (nA)	TYP BAND-WIDTH (MHz)
3507J	1	COM	10	30 TYP	250	10
3508J	1	COM	5	30 TYP	25	20
3510A	1	IND	0.15	2	35	0.4
3510B	1	IND	0.12	1	25	0.4
3510C	1	IND	0.06	0.5	15	0.4
3510S	1	MIL	0.12	1	25	0.4
3521H	1	COM	0.5	10	0.02	1.5
3521J	1	COM	0.25	5	0.02	1.5
3521K	1	COM	0.25	2	0.015	1.5
3521L	1	COM	0.25	1	0.01	1.5
3521R	1	MIL	0.25	5	0.02	1.5
3522J	1	COM	1	50	0.01	1
3522K	1	COM	0.5	10	0.005	1
3522L	1	COM	0.5	10	0.001	1
3522S	1	MIL	0.5	25	0.005	1
3523J	1	COM	1	50	0.0005	1

BURR-BROWN

TYP SLEW RATE (V/uS)	TYPICAL INPUT NOISE VOLTAGE (nV/√Hz)	MAXIMUM SUPPLY CURRENT (mA)	MAXIMUM SUPPLY VOLTAGE (±V)	OUTPUT CURRENT RATING (A)	UNIT PRICE ($)
120		6	20	0.01	12.50
35		4	22	0.01	10.25
0.8	12	3.5	20	0.01	8.90
0.8	12	3.5	20	0.01	11.25
0.8	12	3.5	20	0.01	18.25
0.8	12	3.5	20	0.01	18.25
0.6 MIN		4	20	0.01	22.80
0.6 MIN		4	20	0.01	32.65
0.6 MIN		4	20	0.01	48.95
0.6 MIN		4	20	0.01	72.40
0.6 MIN		4	20	0.01	84.25
0.6 MIN		4 TYP	20	0.01	17.00
0.6 MIN		4 TYP	20	0.01	22.40
0.6 MIN		4 TYP	20	0.01	32.75
0.6 MIN		4 TYP	20	0.01	46.30
0.6 MIN		4 TYP	20	0.01	31.75

PART NUMBER	OP AMPS PER PKG	TEMP GRADE	MAXIMUM OFFSET VOLTAGE (mV)	MAXIMUM OFFSET DRIFT (uV/°C)	MAXIMUM BIAS CURRENT (nA)	TYP BAND-WIDTH (MHz)
3523K	1	COM	0.5	25	0.00025	1
3523L	1	COM	0.5	25	0.0001	1
3527A	1	IND	0.5	10	0.005	1
3527B	1	IND	0.25	5	0.002	1
3527C	1	IND	0.25	2	0.005	1
3528A	1	IND	0.5	15	0.0003	0.7
3528B	1	IND	0.25	5	0.00015	0.7
3528C	1	IND	0.5	10	0.000075	0.7
3542J	1	COM	20	50	0.025	1
3542S	1	MIL	20	50	0.025	1
3550J	1	COM	1	50	0.4	10
3550K	1	COM	1	50	0.4	20
3550S	1	MIL	1	50	0.4	10
3551J	1	COM	1	50 TYP	0.4	50
3551S	1	MIL	1	50 TYP	0.4	50
3554A	1	IND	2	50	0.05	90

BURR-BROWN

TYP SLEW RATE (V/uS)	TYPICAL INPUT NOISE VOLTAGE ($nV/\sqrt{Hz}$)	MAXIMUM SUPPLY CURRENT (mA)	MAXIMUM SUPPLY VOLTAGE (±V)	OUTPUT CURRENT RATING (A)	UNIT PRICE ($)
0.6 MIN		4 TYP	20	0.01	39.70
0.6 MIN		4 TYP	20	0.01	47.60
0.9	30	4	20	0.01	15.50
0.9	30	4	20	0.01	20.75
0.9	30	4	20	0.01	33.15
0.7	40	1.5	20	0.005	19.25
0.7	40	1.5	20	0.005	23.65
0.7	40	1.5	20	0.005	30.25
0.5		4 TYP	20	0.01	9.75
0.5		4 TYP	20	0.01	15.80
65 MIN		11 TYP	20	0.01	31.20
100 MIN		11 TYP	20	0.01	39.75
65 MIN		11 TYP	20	0.01	57.80
250		11 TYP	20	0.01	31.80
250		11 TYP	20	0.01	56.15
1200	15	45	18	0.1	73.20

OPERATIONAL AMPLIFIERS:

PART NUMBER	OP AMPS PER PKG	TEMP GRADE	MAXIMUM OFFSET VOLTAGE (mV)	MAXIMUM OFFSET DRIFT (uV/°C)	MAXIMUM BIAS CURRENT (nA)	TYP BAND-WIDTH (MHz)
3554B	1	IND	1	15	0.05	90
3554S	1	MIL	1	25	0.05	90
3571A	1	COM	2	40	0.1	0.5
3572A	1	COM	2	40	0.1	0.5
3573A	1	COM	10	65	40	1
3580J	1	COM	10	30	0.05	5 MIN
3581J	1	COM	3	25	0.02	5 MIN
3582J	1	COM	3	25	0.02	5 MIN
3583A	1	IND	3	23	0.02	5
3583J	1	COM	3	23	0.02	5
3584J	1	COM	3	25	0.02	7
OPA100A	1	IND	1	15	0.003	1
OPA100B	1	IND	0.5	10	0.002	1
OPA100C	1	IND	0.25	5	0.001	1
OPA100S	1	MIL	0.5	10	0.002	1
OPA101A	1	IND	0.5	10	0.015	10

BURR-BROWN

TYP SLEW RATE (V/uS)	TYPICAL INPUT NOISE VOLTAGE ($nV/\sqrt{Hz}$)	MAXIMUM SUPPLY CURRENT (mA)	MAXIMUM SUPPLY VOLTAGE (±V)	OUTPUT CURRENT RATING (A)	UNIT PRICE ($)
1200	15	45	18	0.1	83.80
1200	15	45	18	0.1	97.60
3		35	40	1	72.45
3		35	40	2	83.00
2.6		5	34	2	36.00
15		10	35	0.06	62.00
20		8	75	0.03	93.45
20		6.5	150	0.015	101.50
30		8.5	150	0.075	100.00
30		8.5	150	0.075	95.00
150		6.5	150	0.015	94.50
2	20	3	18	0.005	11.00
2	20	2	18	0.005	14.00
2	20	1.5	18	0.005	20.00
2	20	2	18	0.005	26.00
6.5	9	8	20	0.012	35.00

PART NUMBER	OP AMPS PER PKG	TEMP GRADE	MAXIMUM OFFSET VOLTAGE (mV)	MAXIMUM OFFSET DRIFT (uV/°C)	MAXIMUM BIAS CURRENT (nA)	TYP BAND-WIDTH (MHz)
OPA101B	1	IND	0.25	5	0.01	10
OPA102A	1	IND	0.5	10	0.015	20
OPA102B	1	IND	0.25	5	0.01	20
OPA103A	1	IND	0.5	25	0.002	1
OPA103B	1	IND	0.5	15	0.001	1
OPA103C	1	IND	0.25	5	0.001	1
OPA103D	1	IND	0.25	2	0.001	1
OPA104A	1	IND	1	25	0.0003	1
OPA104B	1	IND	0.5	15	0.00015	1
OPA104C	1	IND	0.5	10	0.000075	1
OPA11HT	1	MIL	5	5 TYP	25	12
OPA21A	1	MIL	0.1	1	25	0.3
OPA21B	1	MIL	0.2	2	40	0.3
OPA21E	1	IND	0.1	1	25	0.3
OPA21F	1	IND	0.2	2	40	0.3
OPA21G	1	IND	0.5	5	50	0.3

BURR-BROWN

TYP SLEW RATE (V/uS)	TYPICAL INPUT NOISE VOLTAGE (nV/√Hz)	MAXIMUM SUPPLY CURRENT (mA)	MAXIMUM SUPPLY VOLTAGE (±V)	OUTPUT CURRENT RATING (A)	UNIT PRICE ($)
6.5	8	8	20	0.012	43.50
14	9	8	20	0.012	37.00
14	8	8	20	0.012	45.00
1.3	30	1.5	20	0.005	10.20
1.3	30	1.5	20	0.005	13.80
1.3	30	1.5	20	0.005	18.05
1.3	30	1.5	20	0.005	29.85
2.2	35	1.5	20	0.005	17.00
2.2	35	1.5	20	0.005	23.00
2.2	35	1.5	20	0.005	29.50
7		3.7	22	0.015	49.00
0.2	20	0.23	18	0.0014	20.40
0.2	20	0.25	18	0.0014	14.05
0.2	20	0.23	18	0.0014	9.60
0.2	20	0.25	18	0.0014	6.90
0.2	20	0.275	18	0.0014	4.95

PART NUMBER	OP AMPS PER PKG	TEMP GRADE	MAXIMUM OFFSET VOLTAGE (mV)	MAXIMUM OFFSET DRIFT (uV/°C)	MAXIMUM BIAS CURRENT (nA)	TYP BAND-WIDTH (MHz)
OPA27A	1	MIL	0.025	0.6	40	8
OPA27B	1	MIL	0.06	1.3	55	8
OPA27C	1	MIL	0.1	1.8	80	8
OPA27E	1	IND	0.025	0.6	40	8
OPA27F	1	IND	0.06	1.3	55	8
OPA27G	1	IND	0.1	1.8	80	8
OPA37A	1	MIL	0.025	0.6	40	40
OPA37B	1	MIL	0.06	1.3	55	40
OPA37C	1	MIL	0.1	1.8	80	40
OPA37E	1	IND	0.025	0.6	40	40
OPA37F	1	IND	0.06	1.3	55	40
OPA37G	1	IND	0.1	1.8	80	40
OPA501A	1	IND	10	65	40	1
OPA501B	1	IND	5	40	3	1
OPA501R	1	MIL	10	65	40	1
OPA501S	1	MIL	5	40	3	1

BURR-BROWN

TYP SLEW RATE (V/uS)	TYPICAL INPUT NOISE VOLTAGE (nV/√Hz)	MAXIMUM SUPPLY CURRENT (mA)	MAXIMUM SUPPLY VOLTAGE (±V)	OUTPUT CURRENT RATING (A)	UNIT PRICE ($)
2.8	3	4.7	22	0.017	63.75
2.8	3	4.7	22	0.017	28.75
2.8	3.2	5.7	22	0.017	16.60
2.8	3	4.7	22	0.017	19.80
2.8	3	4.7	22	0.017	12.00
2.8	3.2	5.7	22	0.017	8.25
17	3	4.7	22	0.017	63.75
17	3	4.7	22	0.017	28.75
17	3.2	5.7	22	0.017	16.60
17	3	4.7	22	0.017	19.80
17	3	4.7	22	0.017	12.00
17	3.2	5.7	22	0.017	8.25
1.5 MIN		10	36	10	53.40
1.5 MIN		10	40	10	63.00
1.5 MIN		10	36	10	68.50
1.5 MIN		10	40	10	82.00

PART NUMBER	OP AMPS PER PKG	TEMP GRADE	MAXIMUM OFFSET VOLTAGE (mV)	MAXIMUM OFFSET DRIFT (uV/°C)	MAXIMUM BIAS CURRENT (nA)	TYP BAND-WIDTH (MHz)
OPA605A	1	IND	1	25	0.035	20
OPA605C	1	IND	0.5	5	0.035	20
OPA605H	1	COM	1	25	0.035	20
OPA605K	1	COM	0.5	5	0.035	20

BURR-BROWN

TYP SLEW RATE (V/uS)	TYPICAL INPUT NOISE VOLTAGE ($nV/\sqrt{Hz}$)	MAXIMUM SUPPLY CURRENT (mA)	MAXIMUM SUPPLY VOLTAGE (±V)	OUTPUT CURRENT RATING (A)	UNIT PRICE ($)
94	20	9	18	0.03	64.00
94	20	9	18	0.03	89.00
94	20	9	18	0.03	51.50
94	20	9	18	0.03	76.65

OPERATIONAL AMPLIFIERS:

PART NUMBER	OP AMPS PER PKG	TEMP GRADE	MAXIMUM OFFSET VOLTAGE (mV)	MAXIMUM OFFSET DRIFT (uV/°C)	MAXIMUM BIAS CURRENT (nA)	TYP BAND-WIDTH (MHz)
UA101	1	MIL	5	3 TYP	500	1
UA101A	1	MIL	2	15	75	1
UA107	1	MIL	2	15	75	1
UA108	1	MIL	2	15	2	1
UA108A	1	MIL	0.5	5	2	1
UA124	4	MIL	5	7 TYP	150	1
UA1458	2	COM	6	15 TYP	500	1.1
UA1458C	2	COM	10	15 TYP	700	1.1
UA148	4	MIL	5		100	1
UA1558	2	MIL	5	15 TYP	500	1.1
UA201	1	COM	7.5	6 TYP	1500	1
UA201A	1	IND	2	15	75	1
UA208	1	IND	2	15	2	1
UA208A	1	IND	0.5	5	2	1
UA224	4	IND	5	7 TYP	150	1
UA248	4	IND	6		200	1

FAIRCHILD

TYP SLEW RATE (V/uS)	TYPICAL INPUT NOISE VOLTAGE (nV/√Hz)	MAXIMUM SUPPLY CURRENT (mA)	MAXIMUM SUPPLY VOLTAGE (±V)	OUTPUT CURRENT RATING (A)	UNIT PRICE ($)
0.5		3	22	0.005	1.62
0.5	15	3	22	0.005	2.09
0.5	15	3	22	0.005	3.71
0.35	30	0.6	20	0.0013	2.28
0.35	40	0.6	20	0.0013	4.75
0.25		1.2	16	0.01	3.80
0.8	45	5.6	18	0.0012	.76
0.8	45	8	18	0.0011	.76
0.5	60	3.6	22	0.005	3.99
0.8	45	5	22	0.0012	3.33
0.5		3	22	0.005	.38
0.5	15	3	22	0.005	1.81
0.35	40	0.6	20	0.0013	1.71
0.35	40	0.6	20	0.0013	3.80
0.25		1.2	16	0.01	5.32
0.5	60	4.5	18	0.005	6.04

OPERATIONAL AMPLIFIERS:

PART NUMBER	OP AMPS PER PKG	TEMP GRADE	MAXIMUM OFFSET VOLTAGE (mV)	MAXIMUM OFFSET DRIFT (uV/°C)	MAXIMUM BIAS CURRENT (nA)	TYP BAND-WIDTH (MHz)
UA2902	4	IND	7	7 TYP	250	1
UA301A	1	COM	7.5	30	250	1
UA308	1	COM	7.5	30	7	1
UA308A	1	COM	0.5	5	7	1
UA324	4	COM	7	7 TYP	250	1
UA3303	4	IND	8	10 TYP	500	1
UA3403	4	COM	8	10 TYP	500	1
UA348	4	COM	6		200	1
UA4136	4	MIL	5		500	3
UA4136C	4	COM	6		500	3
UA709	1	MIL	5	6 TYP	500	1
UA709A	1	MIL	2	25	200	1
UA709C	1	COM	7.5	6 TYP	1500	1
UA714	1	MIL	0.075	1.3	3	0.6
UA714C	1	COM	0.15	1.8	7	0.6
UA714E	1	COM	0.075	1.3	4	0.6

FAIRCHILD

TYP SLEW RATE (V/uS)	TYPICAL INPUT NOISE VOLTAGE (nV/√Hz)	MAXIMUM SUPPLY CURRENT (mA)	MAXIMUM SUPPLY VOLTAGE (±V)	OUTPUT CURRENT RATING (A)	UNIT PRICE ($)
0.25		1.2	16	0.01	1.24
0.5	15	3	18	0.005	.63
0.35	40	0.8	18	0.0013	.57
0.35	40	0.8	18	0.0013	1.81
0.25		1.2	16	0.01	.68
0.6		7	18	0.005	1.33
0.6		7	18	0.005	1.10
0.5	60	4.5	18	0.005	1.29
1	10	12	22	0.006	6.65
1	10	12	18	0.005	1.31
0.3		3.6	18	0.005	1.33
0.3		3.6	18	0.005	1.71
0.3		3.6	18	0.005	.76
0.17	9.6	4	18	0.01	6.25
0.17	9.8	5	18	0.012	2.47
0.17	9.6	4	18	0.01	5.70

PART NUMBER	OP AMPS PER PKG	TEMP GRADE	MAXIMUM OFFSET VOLTAGE (mV)	MAXIMUM OFFSET DRIFT (uV/°C)	MAXIMUM BIAS CURRENT (nA)	TYP BAND-WIDTH (MHz)
UA714L	1	COM	0.25	3	30	0.6
UA715	1	MIL	5		750	40
UA715C	1	COM	7.5		1500	40
UA725	1	MIL	1	5	100	0.05
UA725A	1	MIL	0.5	2	75	0.05
UA725C	1	COM	2.5	2 TYP	125	0.05
UA725E	1	COM	0.5	2	75	0.05
UA741	1	MIL	5		500	1
UA741A	1	MIL	3	15	80	1.5
UA741C	1	COM	6		500	1
UA741E	1	COM	3	15	80	1.5
UA747	2	MIL	5		500	1
UA747A	2	MIL	3	15	30	1.5
UA747C	2	COM	6		500	1
UA747E	2	COM	3	15	30	1.5
UA748	1	MIL	5		500	1

FAIRCHILD

TYP SLEW RATE (V/uS)	TYPICAL INPUT NOISE VOLTAGE ($nV/\sqrt{Hz}$)	MAXIMUM SUPPLY CURRENT (mA)	MAXIMUM SUPPLY VOLTAGE (±V)	OUTPUT CURRENT RATING (A)	UNIT PRICE ($)
0.17	9.8	6	18	0.012	2.09
18		7	18	0.005	4.28
18		10	18	0.005	2.66
0.005	8	3.5	20	0.005	5.70
0.005	12 MAX	4	20	0.005	7.41
0.005	8	5	20	0.005	3.80
0.005	12 MAX	4	20	0.005	5.32
0.5	70	2.8	20		1.33
0.7	70	5	20	0.0075	3.33
0.5	70	2.8	20	0.005	.57
0.7	70	5	20	0.0075	1.33
0.5	70	5.6	20	0.005	2.00
.7	70	7.5	20	0.0075	3.52
0.5	70	5.6	20	0.005	.76
.7	70	7.5	20	0.0075	1.71
0.5	70	2.8	20	0.005	1.24

PART NUMBER	OP AMPS PER PKG	TEMP GRADE	MAXIMUM OFFSET VOLTAGE (mV)	MAXIMUM OFFSET DRIFT (uV/°C)	MAXIMUM BIAS CURRENT (nA)	TYP BAND-WIDTH (MHz)
UA748C	1	COM	6		500	1
UA759	1	MIL	3		150	1
UA759C	1	COM	6		250	1
UA771	1	COM	10	10 TYP	0.2	3
UA771A	1	COM	2	10 TYP	0.1	3
UA771AM	1	MIL	2	10 TYP	0.1	3
UA771B	1	COM	5	10 TYP	0.1	3
UA771BM	1	MIL	5	10 TYP	0.1	3
UA771L	1	COM	15	10 TYP	0.2	3
UA772	2	COM	10	10 TYP	0.2	3
UA772A	2	COM	2	10 TYP	0.1	3
UA772AM	2	MIL	2	10 TYP	0.1	3
UA772B	2	COM	5	10 TYP	0.1	3
UA772BM	2	MIL	5	10 TYP	0.1	3
UA772L	2	COM	15	10 TYP	0.2	3
UA774	4	COM	10	10 TYP	0.2	3

FAIRCHILD

TYP SLEW RATE (V/uS)	TYPICAL INPUT NOISE VOLTAGE (nV/√Hz)	MAXIMUM SUPPLY CURRENT (mA)	MAXIMUM SUPPLY VOLTAGE (±V)	OUTPUT CURRENT RATING (A)	UNIT PRICE ($)
0.5	70	2.8	20	0.005	.91
0.6	30	18	18	0.325	3.33
0.5	30	18	18	0.325	2.09
13	16	3	18	0.005	.76
13	16	3	18	0.005	2.28
13	16	3.4	18	0.005	14.25
13	16	3	18	0.005	2.53
13	16	3.4	18	0.005	2.53
13	16	3	18	0.005	.57
13	16	6	18	0.005	.99
13	16	6	18	0.005	3.99
13	16	6.8	18	0.005	8.46
13	16	6	18	0.005	1.86
13	16	6.8	18	0.005	6.18
13	16	6	18	0.005	.89
13	16	12	18	0.005	2.15

OPERATIONAL AMPLIFIERS:

PART NUMBER	OP AMPS PER PKG	TEMP GRADE	MAXIMUM OFFSET VOLTAGE (mV)	MAXIMUM OFFSET DRIFT (uV/°C)	MAXIMUM BIAS CURRENT (nA)	TYP BAND-WIDTH (MHz)
UA774B	4	COM	5	10 TYP	0.1	3
UA774L	4	COM	15	10 TYP	0.2	3
UA776	1	MIL	5		15	1
UA776C	1	COM	6		50	1

FAIRCHILD

TYP SLEW RATE (V/uS)	TYPICAL INPUT NOISE VOLTAGE (nV/ Hz)	MAXIMUM SUPPLY CURRENT (mA)	MAXIMUM SUPPLY VOLTAGE (±V)	OUTPUT CURRENT RATING (A)	UNIT PRICE ($)
13	16	12	18	0.005	3.80
13	16	12	18	0.005	1.96
0.8	20	0.18	18	0.002	4.75
0.8	20	0.19	18	0.002	1.24

PART NUMBER	OP AMPS PER PKG	TEMP GRADE	MAXIMUM OFFSET VOLTAGE (mV)	MAXIMUM OFFSET DRIFT (uV/°C)	MAXIMUM BIAS CURRENT (nA)	TYP BAND-WIDTH (MHz)
HA-2500	1	MIL	5	20 TYP	200	12
HA-2502	1	MIL	8	20 TYP	250	12
HA-2505	1	COM	8	20 TYP	250	12
HA-2510	1	MIL	8	20 TYP	200	12
HA-2512	1	MIL	10	25 TYP	250	12
HA-2515	1	COM	10	30 TYP	250	12
HA-2520	1	MIL	8	20 TYP	200	20
HA-2522	1	MIL	10	25 TYP	250	20
HA-2525	1	COM	10	30 TYP	250	20
HA-2539-2	1	MIL	5	20 TYP	20000	600
HA-2539-5	1	COM	15	20 TYP	20000	600
HA-2540-2	1	MIL	5	20 TYP	20000	400
HA-2540-5	1	COM	15	20 TYP	20000	400
HA-2600	1	MIL	4	5 TYP	10	12
HA-2602	1	MIL	5		25	12
HA-2605	1	COM	5		25	12

HARRIS

TYP SLEW RATE (V/uS)	TYPICAL INPUT NOISE VOLTAGE (nV/$\sqrt{Hz}$)	MAXIMUM SUPPLY CURRENT (mA)	MAXIMUM SUPPLY VOLTAGE (±V)	OUTPUT CURRENT RATING (A)	UNIT PRICE ($)
30		6	20	0.01	18.00
30		6	20	0.01	10.30
30		6	20	0.01	1.87
60		6	20	0.01	22.13
60		6	20	0.01	15.59
60		6	20	0.01	2.30
120		6	20	0.01	19.54
120		6	20	0.01	15.20
120		6	20	0.01	3.59
600	15	25	20	0.01	33.78
600	15	25	20	0.01	8.44
400	15	25	17.5	0.01	18.79
400	15	25	17.5	0.01	5.64
7		3.7	22.5	0.015	20.12
7		4	22.5	0.01	10.09
7		4	22.5	0.01	1.97

OPERATIONAL AMPLIFIERS:

PART NUMBER	OP AMPS PER PKG	TEMP GRADE	MAXIMUM OFFSET VOLTAGE (mV)	MAXIMUM OFFSET DRIFT (uV/°C)	MAXIMUM BIAS CURRENT (nA)	TYP BAND-WIDTH (MHz)
HA-2620	1	MIL	4		15	100
HA-2622	1	MIL	5		25	100
HA-2625	1	COM	5		25	100
HA-2640	1	MIL	4	15 TYP	25	4
HA-2645	1	COM	6	15 TYP	30	4
HA-2650	2	MIL	3	8 TYP	100	8
HA-2655	1	COM	5	8 TYP	200	8
HA-2720	1	MIL	3		20	1.5
HA-2725	1	COM	5		30	1.5
HA-2730	2	MIL	3		20	1.5
HA-2735	2	COM	5		30	1.5
HA-2740-2	4	MIL	3		20	1
HA-2740-5	4	COM	5		30	1
HA-4156-5	4	COM	5	5 TYP	300	3.5
HA-4600-2	4	MIL	2.5	2 TYP	200	8
HA-4600-5	4	COM	2.5	2 TYP	200	8

HARRIS

TYP SLEW RATE (V/uS)	TYPICAL INPUT NOISE VOLTAGE (nV/$\sqrt{Hz}$)	MAXIMUM SUPPLY CURRENT (mA)	MAXIMUM SUPPLY VOLTAGE (±V)	OUTPUT CURRENT RATING (A)	UNIT PRICE ($)
35		3.7	22.5	0.015	18.13
35		4	22.5	0.01	10.95
35		4	22.5	0.01	3.33
5		3.8	50	0.012	15.40
5		4.5	50	0.01	5.67
5	25	4	20	0.02	16.33
5	25	5	20	0.018	5.50
0.8	35	0.2 TYP	22.5	0.005	17.69
0.8	35	0.2 TYP	22.5	0.005	6.30
0.8	35	0.4 TYP	22.5	0.005	42.10
0.8	35	0.4 TYP	22.5	0.005	18.22
0.8		1 TYP	22.5	0.005	36.04
0.8		1 TYP	22.5	0.005	12.02
1.6	9	7	20	0.005	1.59
4	8	5.5	20	0.01	32.33
4	8	5.5	20	0.01	14.71

OPERATIONAL AMPLIFIERS:

PART NUMBER	OP AMPS PER PKG	TEMP GRADE	MAXIMUM OFFSET VOLTAGE (mV)	MAXIMUM OFFSET DRIFT (uV/°C)	MAXIMUM BIAS CURRENT (nA)	TYP BAND-WIDTH (MHz)
HA-4602-2	4	MIL	9	5 TYP	400	8
HA-4605-5	4	COM	9	5 TYP	400	8
HA-4620-2	4	MIL	2.5	2 TYP	200	70
HA-4620-5	4	COM	2.5	2 TYP	200	70
HA-4622-2	4	MIL	9	5 TYP	400	70
HA-4625-5	4	COM	9	5 TYP	400	70
HA-4741-2	4	MIL	3	5 TYP	200	3.5
HA-4741-5	4	COM	5	5 TYP	300	3.5
HA-5062-2	2	MIL	6	10 TYP	0.2	1
HA-5062-5	2	COM	15	10 TYP	0.4	1
HA-5062A-5	2	COM	6	10 TYP	0.2	1
HA-5062B-5	2	COM	3	10 TYP	0.2	1
HA-5064-2	4	MIL	6	10 TYP	0.2	1
HA-5064-5	4	COM	15	20 TYP	0.4	1
HA-5064A-5	4	COM	6	10 TYP	0.2	1
HA-5064B-5	4	COM	3	10 TYP	0.2	1

HARRIS

TYP SLEW RATE (V/uS)	TYPICAL INPUT NOISE VOLTAGE (nV/√Hz)	MAXIMUM SUPPLY CURRENT (mA)	MAXIMUM SUPPLY VOLTAGE (±V)	OUTPUT CURRENT RATING (A)	UNIT PRICE ($)
4	8	7.5	20	0.008	17.09
4	8	7.5	20	0.008	4.86
20	8	5.5	20	0.01	37.57
20	8	5.5	20	0.01	17.43
20	8	7.5	20	0.008	15.50
20	8	7.5	20	0.008	4.38
1.6	9	5	20	0.005	6.98
1.6	9	7	20	0.005	1.62
4		0.4	20	0.001	8.02
4		0.5	20	0.001	1.19
4		0.4	20	0.001	2.31
4		0.4	20	0.001	5.84
4		0.8	20	0.005	13.43
4		1	20	0.005	1.85
4		0.8	20	0.005	3.68
4		0.8	20	0.005	9.24

OPERATIONAL AMPLIFIERS:

PART NUMBER	OP AMPS PER PKG	TEMP GRADE	MAXIMUM OFFSET VOLTAGE (mV)	MAXIMUM OFFSET DRIFT (uV/°C)	MAXIMUM BIAS CURRENT (nA)	TYP BAND-WIDTH (MHz)
HA-5082-2	2	MIL	5	10 TYP	0.2	4
HA-5082-5	2	COM	15	10 TYP	0.4	4
HA-5082A-5	2	COM	5	10 TYP	0.2	4
HA-5082B-5	2	COM	2	10 TYP	0.2	4
HA-5084-2	4	MIL	5	8.3 TYP	0.2	4
HA-5084-5	4	COM	15	8.3 TYP	0.4	4
HA-5084A-5	4	COM	5	8.3 TYP	0.2	4
HA-5084B-5	4	COM	2	8.3 TYP	0.2	4
HA-5100-2	1	MIL	1	5 TYP	0.05	18
HA-5100-5	1	COM	1	10 TYP	0.05	18
HA-5105-5	1	COM	1.5	15 TYP	0.1	18
HA-5110-2	1	MIL	1	5 TYP	0.05	60
HA-5110-5	1	COM	1	5 TYP	0.05	60
HA-5115-5	1	COM	1.5	15 TYP	0.1	60
HA-5130-2	1	MIL	0.025	0.6	2	2.5
HA-5130-5	1	COM	0.025	0.6	2	2.5

HARRIS

TYP SLEW RATE (V/uS)	TYPICAL INPUT NOISE VOLTAGE (nV/$\sqrt{Hz}$)	MAXIMUM SUPPLY CURRENT (mA)	MAXIMUM SUPPLY VOLTAGE (±V)	OUTPUT CURRENT RATING (A)	UNIT PRICE ($)
15		5.6	20	0.005	6.81
15		5.6	20	0.005	1.51
15		5.6	20	0.005	1.89
15		5.6	20	0.005	4.69
15		11	20	0.005	11.79
15		12	20	0.005	1.69
15		11	20	0.005	3.36
15		11	20	0.005	8.40
8	20	7	20	0.0012	15.35
8	20	7	20	0.0012	9.66
8	20	8	20	0.0011	4.93
50	20	7	20	0.0012	15.26
50	20	7	20	0.0012	9.66
50	20	8	20	0.0011	4.93
0.8	9	1.3	20	0.025	22.74
0.8	9	1.3	20	0.025	7.52

PART NUMBER	OP AMPS PER PKG	TEMP GRADE	MAXIMUM OFFSET VOLTAGE (mV)	MAXIMUM OFFSET DRIFT (uV/°C)	MAXIMUM BIAS CURRENT (nA)	TYP BAND-WIDTH (MHz)
HA-5135-2	1	MIL	0.075	1.3	4	2.5
HA-5135-5	1	COM	0.075	1.3	4	2.5
HA-5160-2	1	MIL	3	10 TYP	0.05	100
HA-5160-5	1	COM	3	20 TYP	0.05	100
HA-5162-5	1	COM	15	35	0.065	100
HA-5170-2	1	MIL	0.5	5	0.03	5
HA-5170-5	1	COM	0.5	5	0.06	5
HA-5190	1	MIL	5	20 TYP	15000	150
HA-5195	1	COM	6	20 TYP	15000	150

HARRIS

TYP SLEW RATE (V/uS)	TYPICAL INPUT NOISE VOLTAGE (nV/√Hz)	MAXIMUM SUPPLY CURRENT (mA)	MAXIMUM SUPPLY VOLTAGE (±V)	OUTPUT CURRENT RATING (A)	UNIT PRICE ($)
0.8	9	1.7	20	0.025	13.27
0.8	9	1.7	20	0.025	3.99
120	35	10	20	0.005	27.68
120	35	10	20	0.005	11.25
70	35	12	20	0.005	6.26
8	12	2.1	22	0.01	32.63
8	12	2.1	22	0.01	9.79
200	15	28	17.5	0.025	30.01
200	15	28	17.5	0.025	12.15

OPERATIONAL AMPLIFIERS:

PART NUMBER	OP AMPS PER PKG	TEMP GRADE	MAXIMUM OFFSET VOLTAGE (mV)	MAXIMUM OFFSET DRIFT (uV/°C)	MAXIMUM BIAS CURRENT (nA)	TYP BAND- WIDTH (MHz)
ICL7611AC	1	COM	2	10 TYP	0.05	1.4
ICL7611AM	1	MIL	2	10 TYP	0.05	1.4
ICL7611BC	1	COM	5	15 TYP	0.05	1.4
ICL7611BM	1	MIL	5	15 TYP	0.05	1.4
ICL7611DC	1	COM	15	25 TYP	0.05	1.4
ICL7611DM	1	MIL	15	25 TYP	0.05	1.4
ICL7612AC	1	COM	2	10 TYP	0.05	1.4
ICL7612AM	1	MIL	2	10 TYP	0.05	1.4
ICL7612BC	1	COM	5	15 TYP	0.05	1.4
ICL7612BM	1	MIL	5	15 TYP	0.05	1.4
ICL7612DC	1	COM	15	25 TYP	0.05	1.4
ICL7612DM	1	MIL	15	25 TYP	0.05	1.4
ICL7613AC	1	COM	2	10 TYP	0.05	1.4
ICL7613AM	1	MIL	2	10 TYP	0.05	1.4
ICL7613BC	1	COM	5	15 TYP	0.05	1.4
ICL7613BM	1	MIL	5	15 TYP	0.05	1.4

INTERSIL

TYP SLEW RATE (V/uS)	TYPICAL INPUT NOISE VOLTAGE (nV/√Hz)	MAXIMUM SUPPLY CURRENT (mA)	MAXIMUM SUPPLY VOLTAGE (±V)	OUTPUT CURRENT RATING (A)	UNIT PRICE ($)
1.6	100	2.5	9		7.45
1.6	100	2.5	9		18.45
1.6	100	2.5	9		2.25
1.6	100	2.5	9		5.55
1.6	100	2.5	9		1.15
1.6	100	2.5	9		3.40
1.6	100	2.5	9		9.90
1.6	100	2.5	9		24.65
1.6	100	2.5	9		3.05
1.6	100	2.5	9		7.45
1.6	100	2.5	9		1.50
1.6	100	2.5	9		4.30
1.6	100	2.5	9		8.90
1.6	100	2.5	9		22.15
1.6	100	2.5	9		2.70
1.6	100	2.5	9		6.65

OPERATIONAL AMPLIFIERS:

PART NUMBER	OP AMPS PER PKG	TEMP GRADE	MAXIMUM OFFSET VOLTAGE (mV)	MAXIMUM OFFSET DRIFT (uV/°C)	MAXIMUM BIAS CURRENT (nA)	TYP BAND-WIDTH (MHz)
ICL7613DC	1	COM	15	25 TYP	0.05	1.4
ICL7613DM	1	MIL	15	25 TYP	0.05	1.4
ICL7614AC	1	COM	2	10 TYP	0.05	0.48
ICL7614AM	1	MIL	2	10 TYP	0.05	0.48
ICL7614BC	1	COM	5	15 TYP	0.05	0.48
ICL7614BM	1	MIL	5	15 TYP	0.05	0.48
ICL7614DC	1	COM	15	25 TYP	0.05	0.48
ICL7614DM	1	MIL	15	25 TYP	0.05	0.48
ICL7615AC	1	COM	2	10 TYP	0.05	0.48
ICL7615AM	1	MIL	2	10 TYP	0.05	0.48
ICL7615BC	1	COM	5	15 TYP	0.05	0.48
ICL7615BM	1	MIL	5	15 TYP	0.05	0.48
ICL7615DC	1	COM	15	25 TYP	0.05	0.48
ICL7615DM	1	COM	15	25 TYP	0.05	0.48
ICL7621AC	2	COM	2	10 TYP	0.05	0.48
ICL7621AM	2	MIL	2	10 TYP	0.05	0.48

INTERSIL

TYP SLEW RATE (V/uS)	TYPICAL INPUT NOISE VOLTAGE (nV/√Hz)	MAXIMUM SUPPLY CURRENT (mA)	MAXIMUM SUPPLY VOLTAGE (±V)	OUTPUT CURRENT RATING (A)	UNIT PRICE ($)
1.6	100	2.5	9		1.35
1.6	100	2.5	9		4.10
0.16	100	0.25	9		7.45
0.16	100	0.25	9		18.45
0.16	100	0.25	9		2.25
0.16	100	0.25	9		5.55
0.16	100	0.25	9		1.15
0.16	100	0.25	9		3.20
0.16	100	0.25	9		8.90
0.16	100	0.25	9		22.15
0.16	100	0.25	9		2.70
0.16	100	0.25	9		6.65
0.16	100	0.25	9		1.35
0.16	100	0.25	9		4.10
0.16	100	0.5	9		13.80
0.16	100	0.5	9		34.45

OPERATIONAL AMPLIFIERS:

PART NUMBER	OP AMPS PER PKG	TEMP GRADE	MAXIMUM OFFSET VOLTAGE (mV)	MAXIMUM OFFSET DRIFT (uV/°C)	MAXIMUM BIAS CURRENT (nA)	TYP BAND-WIDTH (MHz)
ICL7621BC	2	COM	5	15 TYP	0.05	0.48
ICL7621BM	2	MIL	5	15 TYP	0.05	0.48
ICL7621DC	2	COM	15	25 TYP	0.05	0.48
ICL7621DM	2	MIL	15	25 TYP	0.05	0.48
ICL7622AC	2	COM	2	10 TYP	0.05	0.48
ICL7622AM	2	MIL	2	10 TYP	0.05	0.48
ICL7622BC	2	COM	5	15 TYP	0.05	0.48
ICL7622BM	2	MIL	5	15 TYP	0.05	0.48
ICL7622DC	2	COM	15	25 TYP	0.05	0.48
ICL7622DM	2	MIL	15	25 TYP	0.05	0.48
ICL7631BC	3	COM	5	15 TYP	0.05	1.4
ICL7631BM	3	MIL	5	15 TYP	0.05	1.4
ICL7631CC	3	COM	10	20 TYP	0.05	1.4
ICL7631CM	3	MIL	10	20 TYP	0.05	1.4
ICL7631EC	3	COM	20	30 TYP	0.05	1.4
ICL7631EM	3	MIL	20	30 TYP	0.05	1.4

INTERSIL

TYP SLEW RATE (V/uS)	TYPICAL INPUT NOISE VOLTAGE (nV/√Hz)	MAXIMUM SUPPLY CURRENT (mA)	MAXIMUM SUPPLY VOLTAGE (±V)	OUTPUT CURRENT RATING (A)	UNIT PRICE ($)
0.16	100	0.5	9		4.15
0.16	100	0.5	9		10.40
0.16	100	0.5	9		2.05
0.16	100	0.5	9		6.20
0.16	100	0.5	9		19.20
0.16	100	0.5	9		38.40
0.16	100	0.5	9		7.00
0.16	100	0.5	9		11.65
0.16	100	0.5	9		2.10
0.16	100	0.5	9		5.25
1.6	100	7.5	9		7.45
1.6	100	7.5	9		14.80
1.6	100	7.5	9		5.70
1.6	100	7.5	9		11.30
1.6	100	7.5	9		3.10
1.6	100	7.5	9		5.95

OPERATIONAL AMPLIFIERS:

PART NUMBER	OP AMPS PER PKG	TEMP GRADE	MAXIMUM OFFSET VOLTAGE (mV)	MAXIMUM OFFSET DRIFT (uV/°C)	MAXIMUM BIAS CURRENT (nA)	TYP BAND-WIDTH (MHz)
ICL7632BC	3	COM	5	15 TYP	0.05	1.4
ICL7632BM	3	MIL	5	15 TYP	0.05	1.4
ICL7632CC	3	COM	10	20 TYP	0.05	1.4
ICL7632CM	3	MIL	10	20 TYP	0.05	1.4
ICL7632EC	3	COM	20	30 TYP	0.05	1.4
ICL7632EM	3	MIL	20	30 TYP	0.05	1.4
ICL7641BC	4	COM	5	15 TYP	0.05	1.4
ICL7641BM	4	MIL	5	15 TYP	0.05	1.4
ICL7641CC	4	COM	10	20 TYP	0.05	1.4
ICL7641CM	4	MIL	10	20 TYP	0.05	1.4
ICL7641EC	4	COM	20	30 TYP	0.05	1.4
ICL7641EM	4	MIL	20	30 TYP	0.05	1.4
ICL7642BC	4	COM	5	15 TYP	0.05	0.044
ICL7642BM	4	MIL	5	15 TYP	0.05	0.044
ICL7642CC	4	COM	10	20 TYP	0.05	0.044
ICL7642CM	4	MIL	10	20 TYP	0.05	0.044

INTERSIL

TYP SLEW RATE (V/uS)	TYPICAL INPUT NOISE VOLTAGE $(nV/\sqrt{Hz})$	MAXIMUM SUPPLY CURRENT (mA)	MAXIMUM SUPPLY VOLTAGE (±V)	OUTPUT CURRENT RATING (A)	UNIT PRICE ($)
1.6	100	7.5	9		7.45
1.6	100	7.5	9		14.80
1.6	100	7.5	9		5.70
1.6	100	7.5	9		11.30
1.6	100	7.5	9		3.10
1.6	100	7.5	9		5.95
1.6	100	10	9		10.50
1.6	100	10	9		20.95
1.6	100	10	9		7.95
1.6	100	10	9		15.85
1.6	100	10	9		3.95
1.6	100	10	9		7.60
0.016	100	0.088	9		10.50
0.016	100	0.088	9		20.95
0.016	100	0.088	9		7.95
0.016	100	0.088	9		14.95

OPERATIONAL AMPLIFIERS:

PART NUMBER	OP AMPS PER PKG	TEMP GRADE	MAXIMUM OFFSET VOLTAGE (mV)	MAXIMUM OFFSET DRIFT (uV/°C)	MAXIMUM BIAS CURRENT (nA)	TYP BAND-WIDTH (MHz)
ICL7642EC	4	COM	20	30 TYP	0.05	0.044
ICL7642EM	4	MIL	20	30 TYP	0.05	0.044
ICL7650C	1	COM	0.005	0.05	0.01	2
ICL7650I	1	IND	0.005	0.05	0.01	2
ICL7650M	1	MIL	0.005	0.05	0.01	2
ICL7652C	1	COM	0.005	0.05	0.03	0.45
ICL7652I	1	IND	0.005	0.05	0.03	0.45
ICL8007AC	1	COM	30	50	0.004	1
ICL8007AM	1	MIL	30	50	0.004	1
ICL8007C	1	COM	50	75	0.05	1
ICL8007M	1	MIL	20	75	0.02	1
ICL8008C	1	COM	6	15 TYP	25	1
ICL8008M	1	MIL	5	7 TYP	10	1
ICL8021C	1	COM	6	5 TYP	30	0.27
ICL8021M	1	MIL	3	5 TYP	20	0.27
ICL8022C	2	COM	6	5 TYP	30	0.27

INTERSIL

TYP SLEW RATE (V/uS)	TYPICAL INPUT NOISE VOLTAGE (nV/√Hz)	MAXIMUM SUPPLY CURRENT (mA)	MAXIMUM SUPPLY VOLTAGE (±V)	OUTPUT CURRENT RATING (A)	UNIT PRICE ($)
0.016	100	0.088	9		3.95
0.016	100	0.088	9		7.60
2.5		3.5	9	0.00047	4.15
2.5		3.5	9	0.00047	9.75
2.5		3.5	9	0.00047	15.75
0.5		3.5	9	0.00047	6.75
0.5		3.5	9	0.00047	11.25
6	60	6	18	0.005	41.65
6	60	6	18	0.005	83.25
6	60	6	18	0.005	13.65
6	60	5.2	18	0.005	24.70
0.5		2.8	18	0.005	4.20
0.5		2.8	18	0.005	16.60
0.16	40	0.05	18	0.001	1.90
0.16	40	0.04	18	0.001	6.55
0.16	40	0.1	18	0.001	7.75

OPERATIONAL AMPLIFIERS:

PART NUMBER	OP AMPS PER PKG	TEMP GRADE	MAXIMUM OFFSET VOLTAGE (mV)	MAXIMUM OFFSET DRIFT (uV/°C)	MAXIMUM BIAS CURRENT (nA)	TYP BAND-WIDTH (MHz)
ICL8022M	2	MIL	3	5 TYP	20	0.27
ICL8023C	3	COM	6	5 TYP	30	0.27
ICL8023M	3	MIL	3	5 TYP	20	0.27
ICL8043C	2	COM	50	75	0.05	1
ICL8043M	2	MIL	20	75	0.02	1

INTERSIL

TYP SLEW RATE (V/uS)	TYPICAL INPUT NOISE VOLTAGE (nV/√Hz)	MAXIMUM SUPPLY CURRENT (mA)	MAXIMUM SUPPLY VOLTAGE (±V)	OUTPUT CURRENT RATING (A)	UNIT PRICE ($)
0.16	40	0.08	18	0.001	15.55
0.16	40	0.15	18	0.001	8.80
0.16	40	0.12	18	0.001	17.55
6	60	6.8	18	0.005	11.85
6	60	6	18	0.005	32.70

OPERATIONAL AMPLIFIERS:

PART NUMBER	OP AMPS PER PKG	TEMP GRADE	MAXIMUM OFFSET VOLTAGE (mV)	MAXIMUM OFFSET DRIFT (uV/°C)	MAXIMUM BIAS CURRENT (nA)	TYP BAND-WIDTH (MHz)
MC1436	1	COM	10		40	1
MC1436C	1	COM	12		90	1
MC1437	2	COM	7.5	1.5 TYP	1500	1
MC1439	1	COM	7.5	3 TYP	1000	1
MC1456	1	COM	10		30	1
MC1456C	1	COM	12		90	1
MC1458	2	COM	6		500	1
MC1458C	2	COM	10		700	1
MC1458N	2	COM	6		500	1
MC1458S	2	COM	6		500	1
MC1536	1	MIL	5		20	1
MC1537	2	MIL	5	1.5 TYP	500	1
MC1539	1	MIL	3	3 TYP	500	1
MC1556	1	MIL	4		15	1
MC1558	2	MIL	5		500	1
MC1558N	2	MIL	5		500	1

MOTOROLA

TYP SLEW RATE (V/uS)	TYPICAL INPUT NOISE VOLTAGE (nV/√Hz)	MAXIMUM SUPPLY CURRENT (mA)	MAXIMUM SUPPLY VOLTAGE (±V)	OUTPUT CURRENT RATING (A)	UNIT PRICE ($)
2	50	5	34	0.004	5.72
2	50	5	30	0.004	3.55
0.25		7.5	18	0.0012	2.25
34	30	6.7	18	0.005	2.84
2.5	45	3	18	0.0055	1.40
2.5	45	4	18	0.005	1.20
0.5	45	5.6	18	0.005	.60
0.5	45	8	18	0.0045	.58
0.5	45	5.6	18	0.005	2.40
12		5.4	18	0.005	2.40
2	50	4	40	0.0044	16.13
0.25		7.5	18	0.005	3.18
34	30	5	18	0.01	5.14
2.5	45	1.5	22	0.0012	2.04
0.5	45	5	22	0.005	1.56
0.5	45	5	22	0.005	5.33

OPERATIONAL AMPLIFIERS:

PART NUMBER	OP AMPS PER PKG	TEMP GRADE	MAXIMUM OFFSET VOLTAGE (mV)	MAXIMUM OFFSET DRIFT (uV/°C)	MAXIMUM BIAS CURRENT (nA)	TYP BAND-WIDTH (MHz)
MC1558S	2	MIL	5		500	1
MC1709	1	MIL	5	3 TYP	500	1
MC1709C	1	COM	7.5		1500	1
MC1741	1	MIL	5		500	1
MC1741C	1	COM	6		500	1
MC1741N	1	MIL	5		500	1
MC1741NC	1	COM	6		500	1
MC1741S	1	MIL	5		500	1
MC1741SC	1	COM	6		500	1
MC1747	2	MIL	5		500	1
MC1747C	2	COM	6		500	1
MC1748	1	MIL	5		500	1
MC1748C	1	COM	6		500	1
MC1776	1	MIL	5		50	1
MC1776C	1	COM	6		50	1
MC3303	4	IND	8	10 TYP	500	1

MOTOROLA

TYP SLEW RATE (V/uS)	TYPICAL INPUT NOISE VOLTAGE (nV/√Hz)	MAXIMUM SUPPLY CURRENT (mA)	MAXIMUM SUPPLY VOLTAGE (±V)	OUTPUT CURRENT RATING (A)	UNIT PRICE ($)
12		5	22	0.005	4.60
0.4		5.5	18	0.005	1.50
0.4		6.7	18	0.005	.73
0.5	45	2.8	22	0.005	1.35
0.5	45	2.8	18	0.005	.47
0.5	45	2.8	22	0.005	1.61
0.5	45	2.8	18	0.005	.90
12	45	2.8	22	0.005	2.13
12	45	2.8	18	0.005	1.35
0.5		5.6	22	0.005	2.45
0.5		5.6	18	0.005	1.04
0.8		2.8	22	0.005	1.40
0.8		2.8	18	0.005	.47
0.8	30	0.18	18	0.005	5.14
0.8	30	0.19	18	0.005	1.51
0.6		7	18	0.005	4.15

OPERATIONAL AMPLIFIERS:

PART NUMBER	OP AMPS PER PKG	TEMP GRADE	MAXIMUM OFFSET VOLTAGE (mV)	MAXIMUM OFFSET DRIFT (uV/°C)	MAXIMUM BIAS CURRENT (nA)	TYP BAND-WIDTH (MHz)
MC33074	4	IND	4.5	10 TYP	500	4.5
MC33074A	4	IND	2	10 TYP	500	4.5
MC3358	2	IND	8	10 TYP	500	1
MC34001	1	COM	10	10 TYP	0.2	4
MC34001A	1	COM	2	10 TYP	0.1	4
MC34001B	1	COM	5	10 TYP	0.2	4
MC34002	2	COM	10	10 TYP	0.2	4
MC34002A	2	COM	2	10 TYP	0.1	4
MC34002B	2	COM	5	10 TYP	0.2	4
MC34004	4	COM	10	10 TYP	0.2	4
MC34004B	4	COM	5	10 TYP	0.2	4
MC3403	4	COM	10	10 TYP	500	1
MC34074	4	COM	4.5	10 TYP	500	4.5
MC34074A	4	COM	2	10 TYP	500	4.5
MC3458	2	COM	10	10 TYP	500	1
MC3476	1	COM	6		50	1

MOTOROLA

TYP SLEW RATE (V/uS)	TYPICAL INPUT NOISE VOLTAGE (nV/√Hz)	MAXIMUM SUPPLY CURRENT (mA)	MAXIMUM SUPPLY VOLTAGE (±V)	OUTPUT CURRENT RATING (A)	UNIT PRICE ($)
10	32	10	22	0.0014	2.54
10	32	10	22	0.0014	5.22
0.6		3.7	18	0.005	1.55
13	25	2.7	18	0.005	.60
13	25	2.5	18	0.005	2.55
13	25	2.5	18	0.005	1.05
13	25	5.4	18	0.005	.97
13	25	5	18	0.005	4.85
13	25	5	18	0.005	1.93
13	25	10.8	18	0.005	1.35
13	25	10	18	0.005	3.48
0.6		7	18	0.005	.84
10	32	10	22	0.0014	2.00
10	32	10	22	0.0014	3.01
0.6		3.7	18	0.005	.82
0.8		0.2	18	0.0012	1.48

PART NUMBER	OP AMPS PER PKG	TEMP GRADE	MAXIMUM OFFSET VOLTAGE (mV)	MAXIMUM OFFSET DRIFT (uV/°C)	MAXIMUM BIAS CURRENT (nA)	TYP BAND-WIDTH (MHz)
MC35001	1	MIL	10	10 TYP	0.2	4
MC35001A	1	MIL	2	10 TYP	0.075	4
MC35001B	1	MIL	5	10 TYP	0.1	4
MC35002	2	MIL	10	10 TYP	0.2	4
MC35002A	2	MIL	2	10 TYP	0.075	4
MC35002B	2	MIL	5	10 TYP	0.1	4
MC35004	4	MIL	10	10 TYP	0.2	4
MC35004B	4	MIL	5	10 TYP	0.1	4
MC3503	4	MIL	5	10 TYP	500	1
MC35074	4	MIL	4.5	10 TYP	500	4.5
MC35074A	4	MIL	2	10 TYP	500	4.5
MC3558	2	MIL	5	10 TYP	500	1
MC4558	2	MIL	5		500	2.8
MC4558AC	2	COM	5		500	2.8
MC4558C	2	COM	6		500	2.8
MC4558NC	2	COM	6		500	2.8

MOTOROLA

TYP SLEW RATE (V/uS)	TYPICAL INPUT NOISE VOLTAGE (nV/√Hz)	MAXIMUM SUPPLY CURRENT (mA)	MAXIMUM SUPPLY VOLTAGE (±V)	OUTPUT CURRENT RATING (A)	UNIT PRICE ($)
13	25	2.7	22	0.005	3.85
13	25	2.5	22	0.005	8.55
13	25	2.5	22	0.005	4.70
13	25	5.4	22	0.005	5.75
13	25	5	22	0.005	12.90
13	25	5	22	0.005	7.10
13	25	10.8	22	0.005	9.90
13	25	10	22	0.005	16.10
0.6		4	18	0.005	13.33
10	32	10	22	0.0014	9.28
10	32	10	22	0.0014	13.12
0.6		2.2	18	0.005	4.06
1.6	45	5	22	0.005	2.77
1.6	45	5	22	0.005	2.40
1.6	45	5.6	18	0.005	.80
1.6	45	5.6	18	0.005	2.28

OPERATIONAL AMPLIFIERS:

PART NUMBER	OP AMPS PER PKG	TEMP GRADE	MAXIMUM OFFSET VOLTAGE (mV)	MAXIMUM OFFSET DRIFT (uV/°C)	MAXIMUM BIAS CURRENT (nA)	TYP BAND-WIDTH (MHz)
MC4558N	2	MIL	5		500	2.8
MC4741	4	MIL	5		500	1
MC4741C	4	COM	6		500	1

MOTOROLA

TYP SLEW RATE (V/uS)	TYPICAL INPUT NOISE VOLTAGE ($nV/\sqrt{Hz}$)	MAXIMUM SUPPLY CURRENT (mA)	MAXIMUM SUPPLY VOLTAGE (±V)	OUTPUT CURRENT RATING (A)	UNIT PRICE ($)
1.6	45	5	22	0.005	4.75
0.5		4	22	0.005	6.08
0.5		7	18	0.005	1.40

PART NUMBER	OP AMPS PER PKG	TEMP GRADE	MAXIMUM OFFSET VOLTAGE (mV)	MAXIMUM OFFSET DRIFT (uV/°C)	MAXIMUM BIAS CURRENT (nA)	TYP BAND-WIDTH (MHz)
LF13741	1	COM	15	10 TYP	0.2	1
LF147	4	MIL	5	10 TYP	0.2	4
LF155	1	MIL	5	5 TYP	0.1	2.5
LF155A	1	MIL	2	5	0.05	2.5
LF156	1	MIL	5	5 TYP	0.1	5
LF156A	1	MIL	2	5	0.05	4.5
LF157	1	MIL	5	5 TYP	0.1	20
LF157A	1	MIL	2	5	0.05	20
LF255	1	IND	5	5 TYP	0.1	2.5
LF256	1	IND	5	5 TYP	0.1	5
LF257	1	IND	5	5 TYP	0.1	20
LF347	4	COM	10	10 TYP	0.2	4
LF347B	4	COM	5	10 TYP	0.2	4
LF351	1	COM	10	10 TYP	0.2	4
LF353	2	COM	10	10 TYP	0.2	4
LF355	1	COM	10	5 TYP	0.2	2.5

NATIONAL

TYP SLEW RATE (V/uS)	TYPICAL INPUT NOISE VOLTAGE ($nV/\sqrt{Hz}$)	MAXIMUM SUPPLY CURRENT (mA)	MAXIMUM SUPPLY VOLTAGE (±V)	OUTPUT CURRENT RATING (A)	UNIT PRICE ($)
0.5	37	4	18	0.0012	.56
13	20	11	22	0.0012	16.70
5	20	4	22	0.005	5.05
5	25	4	22	0.005	39.10
12	12	7	22	0.005	5.05
12	12	7	22	0.005	39.10
50	12	7	22	0.005	5.05
50	12	7	22	0.005	41.20
5	20	4	22	0.005	3.60
12	12	7	22	0.005	3.60
50	12	7	22	0.005	3.60
13	20	11	18	0.0012	1.80
13	20	11	18	0.0012	3.60
13	16	3.4	18	0.0012	.60
13	16	6.5	18	0.0012	.98
5	20	4	18	0.005	1.10

PART NUMBER	OP AMPS PER PKG	TEMP GRADE	MAXIMUM OFFSET VOLTAGE (mV)	MAXIMUM OFFSET DRIFT (uV/°C)	MAXIMUM BIAS CURRENT (nA)	TYP BAND-WIDTH (MHz)
LF355A	1	COM	2	5	0.05	2.5
LF355B	1	COM	5	5 TYP	0.1	2.5
LF356	1	COM	10	5 TYP	0.2	5
LF356A	1	COM	2	5	0.05	4.5
LF356B	1	COM	5	5 TYP	0.1	5
LF357	1	COM	10	5 TYP	0.2	20
LF357A	1	COM	2	5	0.05	20
LF357B	1	COM	5	5 TYP	0.1	20
LF411AC	1	COM	0.5	10	0.2	4
LF411AM	1	MIL	0.5	10	0.2	4
LF411C	1	COM	2	20	0.2	4
LF411M	1	MIL	2	20	0.2	4
LF412AC	2	COM	1	10	0.2	4
LF412AM	2	MIL	1	10	0.2	4
LF412C	2	COM	3	20	0.2	4
LF412M	2	MIL	3	20	0.2	4

NATIONAL

TYP SLEW RATE (V/uS)	TYPICAL INPUT NOISE VOLTAGE (nV/√Hz)	MAXIMUM SUPPLY CURRENT (mA)	MAXIMUM SUPPLY VOLTAGE (±V)	OUTPUT CURRENT RATING (A)	UNIT PRICE ($)
5	25	4	18	0.005	21.60
5	20	4	22	0.005	1.60
12	12	10	18	0.005	1.10
12	12	10	18	0.005	21.60
12	12	7	22	0.005	1.60
50	12	10	18	0.005	1.10
50	12	10	18	0.005	24.48
50	12	7	22	0.005	1.60
15	25	2.8	22	0.0012	4.35
15	25	2.8	22	0.0012	16.15
15	25	3.4	18	0.0012	.95
15	25	3.4	18	0.0012	5.30
15	25	5.6	22	0.0012	6.60
15	25	5.6	22	0.0012	22.50
15	25	6.8	18	0.0012	1.85
15	25	6.8	18	0.0012	7.65

OPERATIONAL AMPLIFIERS:

PART NUMBER	OP AMPS PER PKG	TEMP GRADE	MAXIMUM OFFSET VOLTAGE (mV)	MAXIMUM OFFSET DRIFT (uV/°C)	MAXIMUM BIAS CURRENT (nA)	TYP BAND-WIDTH (MHz)
LF441AC	1	COM	0.5	10	0.05	1
LF441AM	1	MIL	0.5	10	0.05	1
LF441C	1	COM	5	20	0.1	1
LF442AC	2	COM	1	10	0.05	1
LF442AM	2	MIL	1	10	0.05	1
LF442C	2	COM	5	7 TYP	0.1	1
LF444AC	4	COM	5	10 TYP	0.05	1
LF444AM	4	MIL	5	10 TYP	0.05	1
LF444C	4	COM	10	10 TYP	0.1	1
LH0021	1	MIL	3	25	300	1
LH0021C	1	IND	6	30	500	1
LH0022	1	MIL	4	10	0.01	1
LH0022C	1	IND	6	15	0.025	1
LH0024	1	MIL	4	20 TYP	30000	70
LH0024C	1	IND	8	25 TYP	40000	70
LH0032	1	MIL	5	25 TYP	0.1	70

NATIONAL

TYP SLEW RATE (V/uS)	TYPICAL INPUT NOISE VOLTAGE (nV/√Hz)	MAXIMUM SUPPLY CURRENT (mA)	MAXIMUM SUPPLY VOLTAGE (±V)	OUTPUT CURRENT RATING (A)	UNIT PRICE ($)
1	35	0.2	22	0.0012	4.00
1	35	0.2	22	0.0012	16.15
1	35	0.25	18	0.0012	.75
1	35	0.4	22	0.0012	6.10
1	35	0.4	22	0.0012	21.35
1	35	0.5	18	0.0012	1.40
1	35	0.8	22	0.0012	3.80
1	35	0.8	22	0.0012	16.60
1	35	1	18	0.0012	2.25
3	15	3.5	18	1.1	52.50
3	15	4	18	1	29.70
3	35	2.5	22	0.01	22.25
3	35	2.8	22	0.01	12.00
500		15	18	0.012	37.00
400		15	18	0.01	22.50
500	15	20	18	0.01	40.00

PART NUMBER	OP AMPS PER PKG	TEMP GRADE	MAXIMUM OFFSET VOLTAGE (mV)	MAXIMUM OFFSET DRIFT (uV/°C)	MAXIMUM BIAS CURRENT (nA)	TYP BAND-WIDTH (MHz)
LH0032C	1	IND	15	25 TYP	0.5	70
LH0041	1	MIL	3	3 TYP	300	1
LH0041C	1	IND	6	3 TYP	500	1
LH0042	1	MIL	20	5 TYP	0.025	1
LH0042C	1	IND	20	10 TYP	0.05	1
LH0044	1	MIL	0.05	1.3	30	0.4
LH0044A	1	MIL	0.025	0.5	15	0.4
LH0044AC	1	IND	0.025	0.5	15	0.4
LH0044B	1	IND	0.05	0.5	30	0.4
LH0044C	1	IND	0.1	1.3	30	0.4
LH0052	1	MIL	0.5	5	0.0025	1
LH0052C	1	IND	1	10	0.005	1
LH0061	1	MIL	4	5 TYP	300	1
LH0061C	1	IND	10	5 TYP	500	1
LH0062	1	MIL	5	25	0.01	2
LH0062C	1	IND	15	35	0.065	2

NATIONAL

TYP SLEW RATE (V/uS)	TYPICAL INPUT NOISE VOLTAGE ($nV/\sqrt{Hz}$)	MAXIMUM SUPPLY CURRENT (mA)	MAXIMUM SUPPLY VOLTAGE (±V)	OUTPUT CURRENT RATING (A)	UNIT PRICE ($)
500	15	22	18	0.01	23.80
3	15	3.5	18	0.13	27.60
3	15	4	18	0.13	20.60
3	35	3.5	22	0.01	11.50
3	35	4	22	0.01	6.60
0.06	9	4	20	0.0012	24.60
0.06	9	3	20	0.0013	68.90
0.06	9	3	20	0.0013	36.10
0.06	9	4	20	0.0012	20.50
0.06	9	4	20	0.0012	14.40
3	35	3.5	22	0.01	22.50
3	35	3.8	22	0.01	17.00
70		10	18	0.5	32.40
70		15	18	0.5	24.00
70	35	8	20	0.01	30.00
70	35	12	20	0.01	20.40

PART NUMBER	OP AMPS PER PKG	TEMP GRADE	MAXIMUM OFFSET VOLTAGE (mV)	MAXIMUM OFFSET DRIFT (uV/°C)	MAXIMUM BIAS CURRENT (nA)	TYP BAND-WIDTH (MHz)
LH0101	1	MIL	10	10 TYP	1	5
LH0101A	1	MIL	3	10 TYP	0.3	5
LH0101AC	1	IND	3	10 TYP	0.3	5
LH0101C	1	IND	10	10 TYP	1	5
LH2011	2	MIL	0.3	3	0.05	0.8
LH2011B	2	MIL	0.6	5	0.1	0.8
LH2011C	2	IND	1	3 TYP	0.18	0.8
LH2101A	2	MIL	2	15	75	1
LH2108	2	MIL	2	15	2	1
LH2108A	2	MIL	0.5	5	2	1
LH2201A	2	IND	2	15	75	1
LH2208	2	IND	2	15	2	1
LH2208A	2	IND	0.5	5	2	1
LH2301A	2	COM	7.5	30	250	1
LH2308	2	COM	7.5	30	7	1
LH2308A	2	COM	0.5	5	7	1

NATIONAL

TYP SLEW RATE (V/uS)	TYPICAL INPUT NOISE VOLTAGE ($nV/\sqrt{Hz}$)	MAXIMUM SUPPLY CURRENT (mA)	MAXIMUM SUPPLY VOLTAGE (±V)	OUTPUT CURRENT RATING (A)	UNIT PRICE ($)
10	25	35	22	2.1	32.40
10	25	35	22	2.1	40.50
10	25	35	22	2.1	32.00
10	25	35	22	2.1	27.00
0.3	150	1.2	20	0.002	34.50
0.3	150	1.6	20	0.002	33.45
0.3	150	1.6	20	0.002	19.10
0.5	15	5	22	0.005	14.00
0.3	35	0.6	20	0.0013	27.90
0.3	35	0.6	20	0.0013	32.10
0.5	15	5	22	0.005	11.40
0.3	35	0.6	20	0.0013	19.10
0.3	35	0.6	20	0.0013	26.50
0.5	15	6	22	0.005	7.80
0.3	35	0.8	20	0.0013	11.60
0.3	35	0.8	20	0.0013	20.75

PART NUMBER	OP AMPS PER PKG	TEMP GRADE	MAXIMUM OFFSET VOLTAGE (mV)	MAXIMUM OFFSET DRIFT (uV/°C)	MAXIMUM BIAS CURRENT (nA)	TYP BAND-WIDTH (MHz)
LH24250	2	MIL	5		50	0.25
LH24250C	2	COM	6		75	0.25
LH740A	1	MIL	15	5 TYP	0.2	1
LH740AC	1	COM	20	5 TYP	0.5	1
LM101A	1	MIL	2	15	75	1
LM107	1	MIL	2	15	75	1
LM108	1	MIL	2	15	2	1
LM108A	1	MIL	0.5	5	2	1
LM11	1	MIL	0.3	3	0.05	0.8
LM112	1	MIL	2	15	2	1
LM118	1	MIL	4		250	15
LM11C	1	COM	0.6	5	0.1	0.8
LM11CL	1	COM	5	3 TYP	0.2	0.8
LM124	4	MIL	5	7 TYP	150	1
LM124A	4	MIL	2	20	50	1
LM13080	1	COM	7	5 TYP	400	1

NATIONAL

TYP SLEW RATE (V/uS)	TYPICAL INPUT NOISE VOLTAGE (nV/$\sqrt{Hz}$)	MAXIMUM SUPPLY CURRENT (mA)	MAXIMUM SUPPLY VOLTAGE (±V)	OUTPUT CURRENT RATING (A)	UNIT PRICE ($)
0.16	35	0.2	18	0.0012	25.50
0.16	35	0.2	18	0.0012	15.20
6		4	22	0.005	18.80
6		5	22	0.005	13.25
0.5	15	3	22	0.005	2.40
0.5	15	3	22	0.005	2.40
0.3	35	0.6	20	0.0013	5.75
0.3	35	0.6	20	0.0013	11.20
0.3	150	0.6	20	0.002	18.00
0.2	35	0.6	20	0.0013	15.90
70	15	8	20	0.006	11.40
0.3	150	0.6	20	0.002	2.00
0.3	150	0.8	20	0.002	1.90
0.25		1.2	16	0.01	3.19
0.25		1.2	16	0.01	12.40
0.4		6	7.5	0.25	1.29

OPERATIONAL AMPLIFIERS:

PART NUMBER	OP AMPS PER PKG	TEMP GRADE	MAXIMUM OFFSET VOLTAGE (mV)	MAXIMUM OFFSET DRIFT (uV/°C)	MAXIMUM BIAS CURRENT (nA)	TYP BAND- WIDTH (MHz)
LM143	1	MIL	5		20	1
LM144	1	MIL	5		20	1
LM1458	2	COM	6		500	1
LM146	4	MIL	5		100	1.2
LM148	4	MIL	5		100	1
LM149	4	MIL	5		100	4
LM1558	2	MIL	5		500	1
LM158	2	MIL	5	7 TYP	150	1
LM158A	2	MIL	2	15	50	1
LM201A	1	IND	2	15	75	1
LM207	1	IND	2	15	75	1
LM208	1	IND	2	15	2	1
LM208A	1	IND	0.5	5	2	1
LM212	1	IND	2	15	2	1
LM216	1	IND	10		0.15	1
LM216A	1	IND	3		0.05	1

NATIONAL

TYP SLEW RATE (V/uS)	TYPICAL INPUT NOISE VOLTAGE (nV/√Hz)	MAXIMUM SUPPLY CURRENT (mA)	MAXIMUM SUPPLY VOLTAGE (±V)	OUTPUT CURRENT RATING (A)	UNIT PRICE ($)
2.5	35	4	40	0.0044	23.95
30	35	4	40	0.0044	23.95
0.2		5.6	18	0.005	.60
0.4	28	2	22	0.0012	7.20
0.5	60	3.6	22	0.005	5.30
2	60	3.6	22	0.005	5.80
0.2		5	22	0.005	3.10
0.25		1.2	16	0.01	3.45
0.25		1.2	16	0.01	7.25
0.5	15	3	22	0.005	.80
0.5	15	3	22	0.005	2.15
0.3	35	0.6	20	0.0013	3.87
0.3	35	0.6	20	0.0013	11.30
0.2	35	0.6	20	0.0013	12.75
0.3	300	0.8	20	0.0013	10.45
0.3	300	0.6	20	0.0013	14.95

OPERATIONAL AMPLIFIERS:

PART NUMBER	OP AMPS PER PKG	TEMP GRADE	MAXIMUM OFFSET VOLTAGE (mV)	MAXIMUM OFFSET DRIFT (uV/°C)	MAXIMUM BIAS CURRENT (nA)	TYP BAND-WIDTH (MHz)
LM218	1	IND	4		250	15
LM224	4	IND	5	7 TYP	150	1
LM224A	4	IND	3	20	80	1
LM246	4	IND	6		250	1.2
LM248	4	IND	6		200	1
LM249	4	IND	6		200	4
LM258	2	IND	5	7 TYP	150	1
LM258A	2	IND	3	15	80	1
LM2902	4	IND	7	7 TYP	250	1
LM2904	2	IND	7	7 TYP	250	1
LM301A	1	COM	7.5	30	250	1
LM307	1	COM	7.5	30	250	1
LM308	1	COM	7.5	30	7	1
LM308A	1	COM	0.5	5	7	1
LM308A-1	1	COM	0.5	1	7	1
LM308A-2	1	COM	0.5	2	7	1

NATIONAL

TYP SLEW RATE (V/uS)	TYPICAL INPUT NOISE VOLTAGE ($nV/\sqrt{Hz}$)	MAXIMUM SUPPLY CURRENT (mA)	MAXIMUM SUPPLY VOLTAGE (±V)	OUTPUT CURRENT RATING (A)	UNIT PRICE ($)
70	15	8	20	0.006	10.60
0.25		1.2	16	0.01	1.90
0.25		1.2	16	0.01	11.60
0.4	28	2.5	18	0.0012	4.35
0.5	60	4.5	18	0.005	3.60
2	60	4.5	18	0.005	3.60
0.25		1.2	16	0.01	1.82
0.25		1.2	16	0.01	5.30
0.25		1.2	13	0.01	1.10
0.25		1.2	13	0.01	1.10
0.5	15	3	18	0.005	.55
0.5	15	3	18	0.005	.56
0.3	35	0.8	18	0.0013	.71
0.3	35	0.8	18	0.0013	2.90
0.3	35	0.8	18	0.0013	34.50
0.3	35	0.8	18	0.0013	22.40

OPERATIONAL AMPLIFIERS:

PART NUMBER	OP AMPS PER PKG	TEMP GRADE	MAXIMUM OFFSET VOLTAGE (mV)	MAXIMUM OFFSET DRIFT (uV/°C)	MAXIMUM BIAS CURRENT (nA)	TYP BAND-WIDTH (MHz)
LM312	1	COM	7.5	30	7	1
LM316	1	COM	10		0.15	1
LM316A	1	COM	3		0.05	1
LM318	1	COM	10		500	15
LM324	4	COM	7	7 TYP	250	1
LM324A	4	COM	3	30	100	1
LM343	1	COM	8		40	1
LM344	1	COM	8		40	1
LM346	4	COM	6		250	1.2
LM348	4	COM	6		200	1
LM349	4	COM	6		200	4
LM358	2	COM	7	7 TYP	250	1
LM358A	2	COM	3	20	100	1
LM4250	1	MIL	5		50	0.1
LM4250C	1	COM	6		75	0.1
LM709	1	MIL	5	6 TYP	500	1

NATIONAL

TYP SLEW RATE (V/uS)	TYPICAL INPUT NOISE VOLTAGE ($nV/\sqrt{Hz}$)	MAXIMUM SUPPLY CURRENT (mA)	MAXIMUM SUPPLY VOLTAGE (±V)	OUTPUT CURRENT RATING (A)	UNIT PRICE ($)
0.2	35	0.8	18	0.0013	3.40
0.3	300	0.8	20	0.0013	6.45
0.3	300	0.6	20	0.0013	12.00
70	15	10	20	0.006	2.50
0.25		1.2	16	0.01	.75
0.25		1.2	16	0.01	2.68
2.5	35	5	34	0.0044	6.38
30	35	5	34	0.0044	6.38
0.4	28	2.5	18	0.0012	2.70
0.5	60	4.5	18	0.005	1.10
2	60	4.5	18	0.005	1.25
0.25		1.2	16	0.01	.65
0.25		1.2	16	0.01	1.65
0.03	35	0.1	18	0.0012	6.70
0.03	35	0.1	18	0.0012	1.75
0.3		5.5	18	0.005	2.50

OPERATIONAL AMPLIFIERS:

PART NUMBER	OP AMPS PER PKG	TEMP GRADE	MAXIMUM OFFSET VOLTAGE (mV)	MAXIMUM OFFSET DRIFT (uV/°C)	MAXIMUM BIAS CURRENT (nA)	TYP BAND-WIDTH (MHz)
LM709A	1	MIL	2	10	200	1
LM709C	1	COM	7.5	12 TYP	1500	1
LM725	1	MIL	1	5	100	0.5
LM725A	1	MIL	0.5	2	80	0.5
LM725C	1	COM	2.5	2 TYP	125	0.5
LM741	1	MIL	5		500	1
LM741A	1	MIL	3	15	80	1.5
LM741C	1	COM	6		500	1
LM741E	1	COM	3	15	80	1.5
LM747	2	MIL	5		500	1
LM747A	2	MIL	3	15	80	1.5
LM747C	2	COM	6		500	1
LM747E	2	COM	3	15	80	1.5
LM748	1	MIL	5	3 TYP	500	1
LM748C	1	COM	5	3 TYP	500	1

NATIONAL

TYP SLEW RATE (V/uS)	TYPICAL INPUT NOISE VOLTAGE (nV/√Hz)	MAXIMUM SUPPLY CURRENT (mA)	MAXIMUM SUPPLY VOLTAGE (±V)	OUTPUT CURRENT RATING (A)	UNIT PRICE ($)
0.3		3.6	18	0.005	3.40
0.3		6.6	18	0.005	.70
0.005	8	3.5	22	0.005	17.30
0.005	8	3.5	22	0.006	30.00
0.005	8	5	22	0.005	3.40
0.5		2.8	22	0.005	2.15
0.7		3.8	22	0.0075	5.75
0.5		2.8	18	0.005	.50
0.7		3.8	22	0.0075	1.15
0.5	25	5.6	22	0.005	2.10
0.7	25	5	22	0.0075	11.55
0.5	25	5.6	18	0.005	.86
0.7	25	5	18	0.0075	6.90
0.5		2.8	22	0.005	2.00
0.5		2.8	22	0.005	.60

OPERATIONAL AMPLIFIERS:

PART NUMBER	OP AMPS PER PKG	TEMP GRADE	MAXIMUM OFFSET VOLTAGE (mV)	MAXIMUM OFFSET DRIFT (uV/°C)	MAXIMUM BIAS CURRENT (nA)	TYP BAND-WIDTH (MHz)
OP-01	1	MIL	0.7	8	30	2.5
OP-01C	1	COM	5	20	100	2.5
OP-01G	1	MIL	5	20	100	2.5
OP-01H	1	COM	0.7	8	30	2.5
OP-02	1	MIL	2	10	50	1.3
OP-02A	1	MIL	0.5	8	30	1.3
OP-02B	1	MIL	5	20	100	1.3
OP-02C	1	COM	2	10	50	1.3
OP-02D	1	COM	5	20	100	1.3
OP-02E	1	COM	0.5	8	30	1.3
OP-04	2	MIL	2	10	75	1.3
OP-04A	2	MIL	0.75	8	50	1.3
OP-04B	2	MIL	5	20	100	1.3
OP-04C	2	COM	2	10	75	1.3
OP-04D	2	COM	5	20	100	1.3
OP-04E	2	COM	0.75	8	50	1.3

P M I

TYP SLEW RATE (V/uS)	TYPICAL INPUT NOISE VOLTAGE (nV/√Hz)	MAXIMUM SUPPLY CURRENT (mA)	MAXIMUM SUPPLY VOLTAGE (±V)	OUTPUT CURRENT RATING (A)	UNIT PRICE ($)
18		3	22	0.006	14.85
18		3	20	0.006	2.78
18		3	20	0.006	8.93
18		3	22	0.006	7.88
0.5	21	3	22	0.006	10.13
0.5	21	2.4	22	0.006	20.40
0.5	21	3	22	0.006	7.65
0.5	21	3	22	0.006	3.98
0.5	21	3	22	0.006	2.63
0.5	21	2.4	22	0.006	5.48
0.5	21	6	22	0.006	16.50
0.5	21	6	22	0.006	36.00
0.5	21	6	22	0.006	12.00
0.5	21	6	22	0.006	6.38
0.5	21	6	22	0.006	4.43
0.5	21	6	22	0.006	8.40

PART NUMBER	OP AMPS PER PKG	TEMP GRADE	MAXIMUM OFFSET VOLTAGE (mV)	MAXIMUM OFFSET DRIFT (uV/°C)	MAXIMUM BIAS CURRENT (nA)	TYP BAND-WIDTH (MHz)
OP-05	1	MIL	0.5	2	3	0.6
OP-05A	1	MIL	0.15	0.9	2	0.6
OP-05C	1	COM	1.3	4.5	7	0.6
OP-05E	1	COM	0.5	2	4	0.6
OP-06A	1	MIL	0.2	0.8	70	0.5
OP-06B	1	MIL	0.5	2	80	0.5
OP-06C	1	MIL	1.3	4.5	110	0.5
OP-06E	1	COM	0.2	0.8	70	0.5
OP-06F	1	COM	0.5	2	80	0.5
OP-06G	1	COM	1.3	4.5	110	0.5
OP-07	1	MIL	0.075	1.3	3	0.6
OP-07A	1	MIL	0.025	0.6	2	0.6
OP-07C	1	COM	0.15	1.8	7	0.6
OP-07D	1	COM	0.15	2.5	12	0.6
OP-07E	1	COM	0.075	1.3	4	0.6
OP-08A	1	MIL	0.15	2.5	2	0.8

P M I

TYP SLEW RATE (V/uS)	TYPICAL INPUT NOISE VOLTAGE (nV/√Hz)	MAXIMUM SUPPLY CURRENT (mA)	MAXIMUM SUPPLY VOLTAGE (±V)	OUTPUT CURRENT RATING (A)	UNIT PRICE ($)
0.3	9.6	4	22	0.0105	21.68
0.3	9.6	4	22	0.0105	51.00
0.3	9.8	5	22	0.0057	4.88
0.3	9.6	4	22	0.0105	7.80
0.005	7	4	22	0.011	53.55
0.005	7	4	22	0.011	20.40
0.005	7	5	22	0.0057	15.30
0.005	7	4	22	0.011	21.60
0.005	7	4	22	0.011	18.00
0.005	7	5	22	0.0057	5.10
0.3	9.6	4	22	0.0105	12.75
0.3	9.6	4	22	0.0105	28.05
0.3	9.8	5	22	0.0057	4.28
0.3	9.8	5	22	0.0057	3.23
0.3	9.6	5	22	0.0105	5.85
0.12	20	0.6	20	0.005	38.25

OPERATIONAL AMPLIFIERS:

PART NUMBER	OP AMPS PER PKG	TEMP GRADE	MAXIMUM OFFSET VOLTAGE (mV)	MAXIMUM OFFSET DRIFT (uV/°C)	MAXIMUM BIAS CURRENT (nA)	TYP BAND-WIDTH (MHz)
OP-08C	1	MIL	1	10	5	0.8
OP-08E	1	COM	0.15	2.5	2	0.8
OP-08G	1	COM	1	10	5	0.8
OP-09A	4	MIL	0.5	10	300	2
OP-09B	4	MIL	2.5	15	500	2
OP-09E	4	COM	0.5	10	300	2
OP-09F	4	COM	2.5	15	500	2
OP-10	2	MIL	0.5	2	3	0.6
OP-10A	2	MIL	0.5	2	3	0.6
OP-10C	2	COM	0.5	4.5	7	0.6
OP-10E	2	COM	0.5	2	4	0.6
OP-11A	4	MIL	0.5	10	300	2
OP-11B	4	MIL	2.5	15	500	2
OP-11C	4	MIL	5	4 TYP	500	2
OP-11E	4	COM	0.5	10	300	2
OP-11F	4	COM	2.5	15	500	2

P M I

TYP SLEW RATE (V/uS)	TYPICAL INPUT NOISE VOLTAGE (nV/√Hz)	MAXIMUM SUPPLY CURRENT (mA)	MAXIMUM SUPPLY VOLTAGE (±V)	OUTPUT CURRENT RATING (A)	UNIT PRICE ($)
0.12	20	0.8	18	0.005	12.75
0.12	20	0.6	20	0.005	11.70
0.12	20	0.8	18	0.005	4.50
1	12	6	22	0.0055	30.00
1	12	6	22	0.0055	15.00
1	12	6	22	0.0055	12.75
1	12	6	22	0.0055	8.25
0.17	9.6	8	22	0.0105	33.75
0.17	9.6	8	22	0.0105	60.00
0.17	9.8	10	22	0.0057	15.00
0.17	9.6	8	22	0.0105	26.25
1	12	6	22	0.0055	30.00
1	12	6	22	0.0055	15.00
1	12	6	22	0.0055	8.10
1	12	6	22	0.0055	12.75
1	12	6	22	0.0055	8.25

OPERATIONAL AMPLIFIERS:

PART NUMBER	OP AMPS PER PKG	TEMP GRADE	MAXIMUM OFFSET VOLTAGE (mV)	MAXIMUM OFFSET DRIFT (uV/°C)	MAXIMUM BIAS CURRENT (nA)	TYP BAND-WIDTH (MHz)
OP-11G	4	COM	5	4 TYP	500	2
OP-12A	1	MIL	0.15	2.5	2	0.8
OP-12B	1	MIL	0.3	3.5	2	0.8
OP-12C	1	MIL	1	10	5	0.8
OP-12E	1	COM	0.15	2.5	2	0.8
OP-12F	1	COM	0.3	3.5	2	0.8
OP-12G	1	COM	1	10	5	0.8
OP-14	2	MIL	2	10	75	1.3
OP-14A	2	MIL	0.75	8	50	1.3
OP-14B	2	MIL	5	20	100	1.3
OP-14C	2	COM	2	10	75	1.3
OP-14D	2	COM	5	20	100	1.3
OP-14E	2	COM	0.75	8	50	1.3
OP-15A	1	MIL	0.5	5	0.05	6
OP-15B	1	MIL	1	10	0.1	5.7
OP-15C	1	MIL	2	15	0.2	5.4

P M I

TYP SLEW RATE (V/uS)	TYPICAL INPUT NOISE VOLTAGE (nV/$\sqrt{Hz}$)	MAXIMUM SUPPLY CURRENT (mA)	MAXIMUM SUPPLY VOLTAGE (±V)	OUTPUT CURRENT RATING (A)	UNIT PRICE ($)
1	12	6	22	0.0055	4.50
0.12	20	0.6	20	0.005	32.55
0.12	20	0.6	20	0.005	25.50
0.12	20	0.8	18	0.005	12.75
0.12	20	0.6	20	0.005	14.40
0.12	20	0.6	20	0.005	7.20
0.12	20	0.8	18	0.005	6.00
0.5	21	6	22	0.006	12.75
0.5	21	6	22	0.006	33.15
0.5	21	6	22	0.006	10.20
0.5	21	6	22	0.006	4.88
0.5	21	6	22	0.006	3.75
0.5	21	6	22	0.006	7.80
13	15	4	22	0.0055	32.40
11	15	4	22	0.0055	16.20
9	15	5	18	0.0055	8.10

OPERATIONAL AMPLIFIERS:

PART NUMBER	OP AMPS PER PKG	TEMP GRADE	MAXIMUM OFFSET VOLTAGE (mV)	MAXIMUM OFFSET DRIFT (uV/°C)	MAXIMUM BIAS CURRENT (nA)	TYP BAND-WIDTH (MHz)
OP-15E	1	COM	0.5	5	0.05	6
OP-15F	1	COM	1	10	0.1	5.7
OP-15G	1	COM	2	15	0.2	5.4
OP-16A	1	MIL	0.5	5	0.05	8
OP-16B	1	MIL	1	10	0.1	7.6
OP-16C	1	MIL	2	15	0.2	7.2
OP-16E	1	COM	0.5	5	0.05	8
OP-16F	1	COM	1	10	0.1	7.6
OP-16G	1	COM	2	15	0.2	7.2
OP-17A	1	MIL	0.5	5	0.05	30
OP-17B	1	MIL	1	10	0.1	28
OP-17C	1	MIL	2	15	0.2	26
OP-17E	1	COM	0.5	5	0.05	30
OP-17F	1	COM	1	10	0.1	28
OP-17G	1	COM	2	15	0.2	26
OP-207A	2	MIL	0.1	1.3	3	0.6

P M I

TYP SLEW RATE (V/uS)	TYPICAL INPUT NOISE VOLTAGE (nV/√Hz)	MAXIMUM SUPPLY CURRENT (mA)	MAXIMUM SUPPLY VOLTAGE (±V)	OUTPUT CURRENT RATING (A)	UNIT PRICE ($)
13	15	4	22	0.0055	10.13
11	15	4	22	0.0055	6.00
9	15	5	18	0.0055	3.60
25	15	7	22	0.0055	32.40
21	15	7	22	0.0055	16.20
17	15	8	18	0.0055	8.10
25	15	7	22	0.0055	10.13
21	15	7	22	0.0055	6.00
17	15	8	18	0.0055	3.60
60	15	7	22	0.0055	32.40
50	15	7	22	0.0055	16.20
40	15	8	18	0.0055	8.10
60	15	7	22	0.0055	10.13
50	15	7	22	0.0055	6.00
40	15	8	18	0.0055	3.60
0.2	9.6	4	22	0.01	63.00

OPERATIONAL AMPLIFIERS:

PART NUMBER	OP AMPS PER PKG	TEMP GRADE	MAXIMUM OFFSET VOLTAGE (mV)	MAXIMUM OFFSET DRIFT (uV/°C)	MAXIMUM BIAS CURRENT (nA)	TYP BAND-WIDTH (MHz)
OP-207B	2	MIL	0.2	1.8	7	0.6
OP-207E	2	COM	0.1	1.3	3	0.6
OP-207F	2	COM	0.2	1.8	7	0.6
OP-20B	1	MIL	0.25	1.5	25	0.1
OP-20C	1	MIL	0.5	3	30	0.1
OP-20F	1	COM	0.25	1.5	25	0.1
OP-20G	1	COM	0.5	3	30	0.1
OP-20H	1	COM	1	7	40	0.1
OP-215A	2	MIL	1	10	0.1	5.7
OP-215B	2	MIL	2	10	0.2	5.7
OP-215C	2	MIL	4	6 TYP	0.3	5.4
OP-215E	2	COM	1	10	0.1	5.7
OP-215F	2	COM	2	10	0.2	5.7
OP-215G	2	COM	4	6 TYP	0.3	5.4
OP-21A	1	MIL	0.1	1	100	0.6
OP-21B	1	MIL	0.2	2	120	0.6

P M I

TYP SLEW RATE (V/uS)	TYPICAL INPUT NOISE VOLTAGE (nV/√Hz)	MAXIMUM SUPPLY CURRENT (mA)	MAXIMUM SUPPLY VOLTAGE (±V)	OUTPUT CURRENT RATING (A)	UNIT PRICE ($)
0.2	9.6	5	22	0.01	49.50
0.2	9.6	4	22	0.01	33.00
0.2	9.6	5	22	0.01	25.50
0.05	60	0.08	18	0.0008	20.40
0.05	60	0.085	18	0.0008	14.03
0.05	60	0.08	18	0.0008	7.80
0.05	60	0.085	18	0.0008	5.63
0.05	60	0.095	18	0.0008	4.05
18	15	8.5	22	0.0055	38.25
18	15	8.5	22	0.0055	26.78
15	15	10	18	0.0055	12.75
18	15	8.5	22	0.0055	18.00
18	15	8.5	22	0.0055	12.60
15	15	10	18	0.0055	6.60
0.25	20	0.3	18	0.0013	20.40
0.25	20	0.36	18	0.0013	14.03

PART NUMBER	OP AMPS PER PKG	TEMP GRADE	MAXIMUM OFFSET VOLTAGE (mV)	MAXIMUM OFFSET DRIFT (uV/°C)	MAXIMUM BIAS CURRENT (nA)	TYP BAND-WIDTH (MHz)
OP-21E	1	IND	0.1	1	100	0.6
OP-21F	1	IND	0.2	2	120	0.6
OP-21G	1	IND	0.5	5	150	0.6
OP-220A	2	MIL	0.15	1.5	20	0.1
OP-220B	2	MIL	0.3	2	25	0.1
OP-220C	2	MIL	0.75	3	30	0.1
OP-220E	2	IND	0.15	1.5	20	0.1
OP-220F	2	IND	0.3	2	25	0.1
OP-220G	2	IND	0.75	3	30	0.1
OP-221A	2	MIL	0.15	1.5	80	0.6
OP-221B	2	MIL	0.3	2	100	0.6
OP-221C	2	MIL	0.5	3	120	0.6
OP-221E	2	IND	0.15	1.5	80	0.6
OP-221F	2	IND	0.3	2	100	0.6
OP-221G	2	IND	0.5	3	120	0.6
OP-227A	2	MIL	0.08	1	40	8

P M I

TYP SLEW RATE (V/uS)	TYPICAL INPUT NOISE VOLTAGE (nV/√Hz)	MAXIMUM SUPPLY CURRENT (mA)	MAXIMUM SUPPLY VOLTAGE (±V)	OUTPUT CURRENT RATING (A)	UNIT PRICE ($)
0.25	20	0.3	18	0.0013	7.80
0.25	20	0.36	18	0.0013	5.63
0.25	20	0.42	18	0.0013	4.05
0.05	40	0.17	18	0.002	33.75
0.05	40	0.19	18	0.002	24.08
0.05	40	0.22	18	0.002	17.25
0.05	40	0.17	18	0.002	19.88
0.05	40	0.19	18	0.002	14.18
0.05	40	0.22	18	0.002	8.10
0.3	18	0.8	18	0.0025	27.00
0.3	18	0.85	18	0.0025	18.90
0.3	18	0.9	18	0.0025	13.50
0.3	18	0.8	18	0.0025	14.85
0.3	18	0.85	18	0.0025	10.80
0.3	18	0.9	18	0.0025	6.75
2.8	3	9.4	22	0.017	75.00

PART NUMBER	OP AMPS PER PKG	TEMP GRADE	MAXIMUM OFFSET VOLTAGE (mV)	MAXIMUM OFFSET DRIFT (uV/°C)	MAXIMUM BIAS CURRENT (nA)	TYP BAND-WIDTH (MHz)
OP-227B	2	MIL	0.12	1.5	55	8
OP-227C	2	MIL	0.18	1.8	80	8
OP-227E	2	IND	0.08	1	40	8
OP-227F	2	IND	0.12	1.5	55	8
OP-227G	2	IND	0.18	1.8	80	8
OP-22A	1	MIL	0.3	1.5	30	0.25
OP-22B	1	MIL	0.5	2	35	0.25
OP-22E	1	IND	0.3	1.5	30	0.25
OP-22F	1	IND	0.5	2	35	0.25
OP-22H	1	COM	1	3	50	0.25
OP-27A	1	MIL	0.025	0.6	40	8
OP-27B	1	MIL	0.06	1.3	50	8
OP-27C	1	MIL	0.1	1.8	75	8
OP-27EP	1	COM	0.025	0.6	40	8
OP-27EZ	1	IND	0.025	0.6	40	8
OP-27FP	1	COM	0.06	1.3	50	8

P M I

TYP SLEW RATE (V/uS)	TYPICAL INPUT NOISE VOLTAGE (nV/√Hz)	MAXIMUM SUPPLY CURRENT (mA)	MAXIMUM SUPPLY VOLTAGE (±V)	OUTPUT CURRENT RATING (A)	UNIT PRICE ($)
2.8	3	9.4	22	0.017	37.50
2.8	3.2	10.8	22	0.017	21.00
2.8	3	9.4	22	0.017	22.50
2.8	3	9.4	22	0.017	13.50
2.8	3.2	10.8	22	0.017	10.50
0.08	50	0.17	18	0.0014	13.58
0.08	50	0.19	18	0.0014	9.68
0.08	50	0.17	18	0.0014	7.95
0.08	50	0.19	18	0.0014	5.78
0.08	50	0.21	18	0.0013	4.13
2.8	3	4.7	22	0.017	63.75
2.8	3	4.7	22	0.017	28.73
2.8	3.2	5.4	22	0.017	16.58
2.8	3	4.7	22	0.017	15.84
2.8	3	4.7	22	0.017	19.80
2.8	3	4.7	22	0.017	9.60

PART NUMBER	OP AMPS PER PKG	TEMP GRADE	MAXIMUM OFFSET VOLTAGE (mV)	MAXIMUM OFFSET DRIFT (uV/°C)	MAXIMUM BIAS CURRENT (nA)	TYP BAND-WIDTH (MHz)
OP-27FZ	1	IND	0.06	1.3	50	8
OP-27GP	1	COM	0.1	1.8	75	8
OP-27GZ	1	IND	0.1	1.8	75	8
OP-37A	1	MIL	0.025	0.6	40	63
OP-37B	1	MIL	0.06	1.3	50	63
OP-37C	1	MIL	0.1	1.8	75	63
OP-37EP	1	COM	0.025	0.6	40	63
OP-37EZ	1	IND	0.025	0.6	40	63
OP-37FP	1	COM	0.06	1.3	50	63
OP-37FZ	1	IND	0.06	1.3	50	63
OP-37GP	1	COM	0.1	1.8	75	63
OP-37GZ	1	IND	0.1	1.8	75	63
OP-420B	4	MIL	2.5	10	20	0.15
OP-420C	4	MIL	4	15	30	0.15
OP-420F	4	IND	2.5	10	20	0.15
OP-420G	4	IND	4	15	30	0.15

P M I

TYP SLEW RATE (V/uS)	TYPICAL INPUT NOISE VOLTAGE (nV/$\sqrt{Hz}$)	MAXIMUM SUPPLY CURRENT (mA)	MAXIMUM SUPPLY VOLTAGE (±V)	OUTPUT CURRENT RATING (A)	UNIT PRICE ($)
2.8	3	4.7	22	0.017	12.00
2.8	3.2	5.4	22	0.017	6.60
2.8	3.2	5.4	22	0.017	8.25
17	3	4.7	22	0.017	63.75
17	3	4.7	22	0.017	28.73
17	3.2	5.4	22	0.017	16.58
17	3	4.7	22	0.017	15.84
17	3	4.7	22	0.017	19.80
17	3	4.7	22	0.017	9.60
17	3	4.7	22	0.017	12.00
17	3.2	5.4	22	0.017	6.60
17	3.2	5.4	22	0.017	8.25
0.05	50	0.36	18	0.002	27.75
0.05	50	0.46	18	0.002	16.50
0.05	50	0.36	18	0.002	18.00
0.05	50	0.46	18	0.002	11.70

OPERATIONAL AMPLIFIERS:

PART NUMBER	OP AMPS PER PKG	TEMP GRADE	MAXIMUM OFFSET VOLTAGE (mV)	MAXIMUM OFFSET DRIFT (uV/°C)	MAXIMUM BIAS CURRENT (nA)	TYP BAND-WIDTH (MHz)
OP-420H	4	COM	6	25	40	0.15
OP-421B	4	MIL	2.5	10	50	1
OP-421C	4	MIL	4	15	80	1
OP-421F	4	IND	2.5	10	50	1
OP-421G	4	IND	4	15	80	1
OP-421H	4	COM	6	15	150	1

P M I

TYP SLEW RATE (V/uS)	TYPICAL INPUT NOISE VOLTAGE (nV/√Hz)	MAXIMUM SUPPLY CURRENT (mA)	MAXIMUM SUPPLY VOLTAGE (±V)	OUTPUT CURRENT RATING (A)	UNIT PRICE ($)
0.05	50	0.6	18	0.002	7.80
0.5	15	1.8	18	0.003	13.50
0.5	15	2.3	18	0.003	9.00
0.5	15	1.8	18	0.003	9.00
0.5	15	2.3	18	0.003	6.45
0.5	15	3	18	0.003	3.60

OPERATIONAL AMPLIFIERS:

PART NUMBER	OP AMPS PER PKG	TEMP GRADE	MAXIMUM OFFSET VOLTAGE (mV)	MAXIMUM OFFSET DRIFT (uV/°C)	MAXIMUM BIAS CURRENT (nA)	TYP BAND-WIDTH (MHz)
CA081AE	1	COM	6	10 TYP	0.04	5
CA081BE	1	COM	3	10 TYP	0.03	5
CA081E	1	COM	15	10 TYP	0.05	5
CA082AE	2	COM	6	10 TYP	0.04	5
CA082BE	2	COM	3	10 TYP	0.03	5
CA082E	2	COM	15	10 TYP	0.05	5
CA084E	4	COM	15	10 TYP	0.05	5
CA101	1	MIL	5	6 TYP	500	1
CA124	4	MIL	5	7 TYP	150	1
CA1458	2	COM	6		500	1
CA1558	2	MIL	5		500	1
CA158	2	MIL	5	7 TYP	150	1
CA158A	2	MIL	2	15	50	1
CA201	1	COM	7.5	10 TYP	1500	1
CA224	4	IND	7	7 TYP	250	1
CA258	2	IND	5	7 TYP	150	1

RCA

TYP SLEW RATE (V/uS)	TYPICAL INPUT NOISE VOLTAGE (nV/√Hz)	MAXIMUM SUPPLY CURRENT (mA)	MAXIMUM SUPPLY VOLTAGE (±V)	OUTPUT CURRENT RATING (A)	UNIT PRICE ($)
13	40	2.8	18	0.005	1.00
13	40	2.8	18	0.005	2.72
13	40	2.8	18	0.005	.52
13	40	5.6	18	0.005	2.00
13	40	5.6	18	0.005	5.00
13	40	5.6	18	0.005	.90
13	40	11.2	18	0.005	2.02
0.5	15	3	22	0.005	.80
0.5		4	16	0.01	2.30
0.5		5.6	18	0.005	.62
0.5		5.6	22	0.005	.96
0.5		2.4	16	0.01	1.20
0.5		2.4	16	0.01	1.70
0.5	15	3	22	0.005	1.20
0.5		4	16	0.01	.90
0.5		2.4	16	0.01	.90

PART NUMBER	OP AMPS PER PKG	TEMP GRADE	MAXIMUM OFFSET VOLTAGE (mV)	MAXIMUM OFFSET DRIFT (uV/°C)	MAXIMUM BIAS CURRENT (nA)	TYP BAND-WIDTH (MHz)
CA258A	2	IND	3	15	80	1
CA2904	2	IND	7	7 TYP	250	1
CA3010	1	MIL	5		12000	20
CA3010A	1	MIL	2		4000	20
CA3015	1	MIL	5		24000	50
CA3015A	1	MIL	2		6000	50
CA301A	1	COM	7.5	30	250	1
CA3029	1	COM	5		12000	20
CA3029A	1	COM	2		4000	20
CA3030	1	COM	5		24000	50
CA3030A	1	COM	2		6000	50
CA3037	1	MIL	5		12000	20
CA3037A	1	MIL	2		4000	20
CA3038	1	MIL	5		24000	50
CA3038A	1	MIL	2		6000	50
CA307	1	COM	7.5	30	250	1

RCA

TYP SLEW RATE (V/uS)	TYPICAL INPUT NOISE VOLTAGE (nV/√Hz)	MAXIMUM SUPPLY CURRENT (mA)	MAXIMUM SUPPLY VOLTAGE (±V)	OUTPUT CURRENT RATING (A)	UNIT PRICE ($)
0.5		2.4	16	0.01	1.50
0.5		2.4	13	0.01	.60
3		9	8	0.001	2.50
3		9	8	0.001	2.80
7		21	16	0.003	3.60
7		21	16	0.003	4.00
0.5	15	3	18	0.005	.50
3		9	8	0.001	.78
3		9	8	0.001	.96
7		21	16	0.003	1.56
7		21	16	0.003	1.80
3		9	8	0.001	7.00
3		9	8	0.001	8.00
7		21	16	0.003	9.00
7		21	16	0.003	10.00
0.5	15	3	18	0.005	.50

OPERATIONAL AMPLIFIERS:

PART NUMBER	OP AMPS PER PKG	TEMP GRADE	MAXIMUM OFFSET VOLTAGE (mV)	MAXIMUM OFFSET DRIFT (uV/°C)	MAXIMUM BIAS CURRENT (nA)	TYP BAND-WIDTH (MHz)
CA3078	1	COM	4.5	6 TYP	170	0.5
CA3078A	1	MIL	3.5	5 TYP	12	0.04
CA3100E	1	IND	5		2000	38
CA3100S	1	MIL	5		2000	38
CA3100T	1	MIL	5		2000	38
CA3130	1	MIL	15	10 TYP	0.05	15
CA3130A	1	MIL	5	10 TYP	0.03	15
CA3130B	1	MIL	2	10 TYP	0.02	15
CA3140	1	MIL	15	8 TYP	0.05	4.5
CA3140A	1	MIL	5	6 TYP	0.04	4.5
CA3140B	1	MIL	2	5 TYP	0.03	4.5
CA3160	1	MIL	15	8 TYP	0.05	4
CA3160A	1	MIL	5	6 TYP	0.03	4
CA3160B	1	MIL	2	5 TYP	0.02	4
CA3193	1	COM	0.5	5	40	1.2
CA3193A	1	IND	0.2	3	20	1.2

RCA

TYP SLEW RATE (V/uS)	TYPICAL INPUT NOISE VOLTAGE (nV/√Hz)	MAXIMUM SUPPLY CURRENT (mA)	MAXIMUM SUPPLY VOLTAGE (±V)	OUTPUT CURRENT RATING (A)	UNIT PRICE ($)
1.5	18	0.13	7	0.012	.92
0.5	35	0.025	18	0.012	1.56
70		10.5	18	0.0045	1.42
70		10.5	18	0.0045	2.82
70		10.5	18	0.0045	2.64
30		15	8	0.006	1.10
30		15	8	0.006	1.50
30		15	8	0.006	13.54
9	40	6	18	0.006	.64
9	40	6	18	0.006	1.00
9	40	6	22	0.006	13.54
10	72	15	8	0.006	1.00
10	72	15	8	0.006	1.58
10	72	15	8	0.006	13.54
0.25	24	3.5	18	0.0065	1.52
0.25	24	3.5	18	0.0065	3.08

OPERATIONAL AMPLIFIERS:

PART NUMBER	OP AMPS PER PKG	TEMP GRADE	MAXIMUM OFFSET VOLTAGE (mV)	MAXIMUM OFFSET DRIFT (uV/°C)	MAXIMUM BIAS CURRENT (nA)	TYP BAND-WIDTH (MHz)
CA3193B	1	MIL	0.075	2	15	1.2
CA324	4	COM	7	7 TYP	250	1
CA3240	2	IND	15	15 TYP	0.05	4.5
CA3240A	2	IND	5	15 TYP	0.04	4.5
CA3260	2	MIL	15	8 TYP	0.05	4
CA3260A	2	MIL	5	6 TYP	0.03	4
CA3260B	2	MIL	2	5 TYP	0.02	4
CA3420	1	MIL	10	4 TYP	0.005	0.5
CA3420A	1	MIL	5	4 TYP	0.005	0.5
CA3420B	1	MIL	2	4 TYP	0.005	0.5
CA3440	1	MIL	10	4 TYP	0.05	0.063
CA3440A	1	MIL	5	4 TYP	0.04	0.063
CA3440B	1	MIL	2	4 TYP	0.03	0.063
CA3493	1	COM	0.5	5	40	1.2
CA358	2	COM	7	7 TYP	250	1
CA358A	2	COM	3	20	100	1

RCA

TYP SLEW RATE (V/uS)	TYPICAL INPUT NOISE VOLTAGE ($nV/\sqrt{Hz}$)	MAXIMUM SUPPLY CURRENT (mA)	MAXIMUM SUPPLY VOLTAGE (±V)	OUTPUT CURRENT RATING (A)	UNIT PRICE ($)
0.25	24	3.5	22	0.0065	5.44
0.5		4	16	0.01	.72
9	40	12	18	0.006	1.48
9	40	12	18	0.006	2.36
10		31	8	0.0011	2.00
10		31	8	0.0011	3.12
10		31	8	0.0011	5.40
0.5	62	1	11		1.80
0.5	62	1	11		3.30
0.5	62	1	11		5.20
0.03	110	0.17	12.5	0.003	1.06
0.03	110	0.17	12.5	0.003	2.12
0.03	110	0.17	12.5	0.003	3.40
0.25	24	3.5	18	0.0065	1.40
0.5		2.4	16	0.01	.60
0.5		2.4	16	0.01	1.30

PART NUMBER	OP AMPS PER PKG	TEMP GRADE	MAXIMUM OFFSET VOLTAGE (mV)	MAXIMUM OFFSET DRIFT (uV/°C)	MAXIMUM BIAS CURRENT (nA)	TYP BAND-WIDTH (MHz)
CA741	1	MIL	5		500	1
CA741C	1	COM	6		500	1
CA747	2	MIL	5		500	1
CA747C	2	COM	6		500	1
CA748	1	MIL	5		500	1
CA748C	1	COM	6		500	1

RCA

TYP SLEW RATE (V/uS)	TYPICAL INPUT NOISE VOLTAGE (nV/√Hz)	MAXIMUM SUPPLY CURRENT (mA)	MAXIMUM SUPPLY VOLTAGE (±V)	OUTPUT CURRENT RATING (A)	UNIT PRICE ($)
0.5		2.8	22	0.005	.60
0.5		2.8	18	0.005	.50
0.5		5.6	22	0.005	.74
0.5		5.6	18	0.005	.66
0.5		2.8	22	0.005	.64
0.5		2.8	18	0.005	.54

OPERATIONAL AMPLIFIERS:

PART NUMBER	OP AMPS PER PKG	TEMP GRADE	MAXIMUM OFFSET VOLTAGE (mV)	MAXIMUM OFFSET DRIFT (uV/°C)	MAXIMUM BIAS CURRENT (nA)	TYP BAND-WIDTH (MHz)
NE4558	2	COM	6	4 TYP	500	3
NE530	1	COM	6	6 TYP	150	3
NE531	1	COM	6	10 TYP	1500	1
NE532	2	COM	7	7 TYP	250	1
NE538	1	COM	6	6 TYP	150	6
NE5512	2	COM	5	5 TYP	20	3
NE5514	4	COM	5	5 TYP	20	3
NE5532	2	COM	4	5 TYP	800	10
NE5532A	2	COM	4	5 TYP	800	10
NE5533	2	COM	4	5 TYP	1500	10
NE5533A	2	COM	4	5 TYP	1500	10
NE5534	1	COM	4	5 TYP	1500	10
NE5534A	1	COM	4	5 TYP	1500	10
NE5535	2	COM	6	6 TYP	150	1
SA1458	2	IND	6	12 TYP	500	1
SA532	2	IND	7	7 TYP	250	1

SIGNETICS

TYP SLEW RATE (V/uS)	TYPICAL INPUT NOISE VOLTAGE ($nV/\sqrt{Hz}$)	MAXIMUM SUPPLY CURRENT (mA)	MAXIMUM SUPPLY VOLTAGE (±V)	OUTPUT CURRENT RATING (A)	UNIT PRICE ($)
1	25	5.6	18	0.005	.58
35	30	3	18	0.005	1.16
35	20	10	22	0.001	1.40
0.3	40	4	16	0.01	.60
60	30	3	18	0.005	1.26
1	30	5	16	0.0167	1.00
1	30	10	16	0.0167	1.30
9	5	16	22	0.02	1.60
9	5	16	22	0.02	3.00
13	4	16	22	0.02	2.60
13	3.5	16	22	0.02	3.62
13	4	8	22	0.02	1.10
13	3.5	8	22	0.02	2.15
15	30	5.6	18	0.005	1.78
0.8	30	5.6	18	0.005	.88
0.3	40	4	16	0.01	.94

OPERATIONAL AMPLIFIERS:

PART NUMBER	OP AMPS PER PKG	TEMP GRADE	MAXIMUM OFFSET VOLTAGE (mV)	MAXIMUM OFFSET DRIFT (uV/°C)	MAXIMUM BIAS CURRENT (nA)	TYP BAND-WIDTH (MHz)
SA534	4	IND	7	7 TYP	250	1
SA741C	1	IND	6	10 TYP	500	1
SA747C	2	IND	6	10 TYP	500	1
SE530	1	MIL	4	15	80	3
SE531	1	MIL	5	10 TYP	1500	1
SE532	2	MIL	5	7 TYP	150	1
SE538	1	MIL	4	4 TYP	80	6
SE5512	2	MIL	2	4 TYP	10	3
SE5514	4	MIL	2	4 TYP	10	3
SE5532	2	MIL	2	5 TYP	400	10
SE5532A	2	MIL	2	5 TYP	400	10
SE5534	1	MIL	2	5 TYP	800	10
SE5534A	1	MIL	2	5 TYP	800	10
SE5535	2	MIL	4	4 TYP	80	1

SIGNETICS

TYP SLEW RATE (V/uS)	TYPICAL INPUT NOISE VOLTAGE ($nV/\sqrt{Hz}$)	MAXIMUM SUPPLY CURRENT (mA)	MAXIMUM SUPPLY VOLTAGE (±V)	OUTPUT CURRENT RATING (A)	UNIT PRICE ($)
0.3	40	4.8	16	0.01	.96
0.5	20	2.8	18	0.005	.76
0.5	20	5.6	18	0.005	.92
35	30	3	22	0.005	2.90
35	20	7	22	0.001	5.70
0.3	40	4	16	0.01	2.20
60	30	3	22	0.005	6.00
1	30	5	16	0.0167	2.70
1	30	10	16	0.0167	7.58
9	5	13	22	0.02	8.60
9	5	13	22	0.02	10.50
13	4	6.5	22	0.02	4.73
13	3.5	6.5	22	0.02	6.00
15	30	5.6	22	0.005	4.50

OPERATIONAL AMPLIFIERS:

PART NUMBER	OP AMPS PER PKG	TEMP GRADE	MAXIMUM OFFSET VOLTAGE (mV)	MAXIMUM OFFSET DRIFT (uV/°C)	MAXIMUM BIAS CURRENT (nA)	TYP BAND-WIDTH (MHz)
TL022C	2	COM	5		250	0.5
TL044C	4	COM	5		250	0.5
TL060AC	1	COM	6	10 TYP	0.2	1
TL060C	1	COM	15	10 TYP	0.2	1
TL060I	1	IND	6	10 TYP	0.2	1
TL061AC	1	COM	6	10 TYP	0.2	1
TL061BC	1	COM	3	10 TYP	0.2	1
TL061C	1	COM	15	10 TYP	0.2	1
TL061I	1	IND	6	10 TYP	0.2	1
TL061M	1	MIL	6	10 TYP	0.2	1
TL062AC	2	COM	6	10 TYP	0.2	1
TL062BC	2	COM	3	10 TYP	0.2	1
TL062C	2	COM	15	10 TYP	0.2	1
TL062I	2	IND	6	10 TYP	0.2	1
TL062M	2	MIL	6	10 TYP	0.2	1
TL064AC	4	COM	6	10 TYP	0.2	1

T I

TYP SLEW RATE (V/uS)	TYPICAL INPUT NOISE VOLTAGE (nV/√Hz)	MAXIMUM SUPPLY CURRENT (mA)	MAXIMUM SUPPLY VOLTAGE (±V)	OUTPUT CURRENT RATING (A)	UNIT PRICE ($)
0.5	50	0.25	18	0.001	1.56
0.5	50	0.5	18	0.001	2.53
3.5	42	0.25	18	0.001	1.27
3.5	42	0.25	18	0.001	.81
3.5	42	0.25	18	0.001	1.27
3.5	42	0.25	18	0.001	1.27
3.5	42	0.25	18	0.001	3.57
3.5	42	0.25	18	0.001	.81
3.5	42	0.25	18	0.001	1.27
3.5	42	0.25	18	0.001	3.80
3.5	42	0.5	18	0.001	1.84
3.5	42	0.5	18	0.001	5.52
3.5	42	0.5	18	0.001	1.38
3.5	42	0.5	18	0.001	1.84
3.5	42	0.5	18	0.001	6.30
3.5	42	1	18	0.001	2.65

OPERATIONAL AMPLIFIERS:

PART NUMBER	OP AMPS PER PKG	TEMP GRADE	MAXIMUM OFFSET VOLTAGE (mV)	MAXIMUM OFFSET DRIFT (uV/°C)	MAXIMUM BIAS CURRENT (nA)	TYP BAND-WIDTH (MHz)
TL064BC	4	COM	3	10 TYP	0.2	1
TL064C	4	COM	15	10 TYP	0.2	1
TL064I	4	IND	6	10 TYP	0.2	1
TL064M	4	MIL	9	10 TYP	0.2	1
TL066AC	1	COM	6	10 TYP	0.2	1
TL066BC	1	COM	3	10 TYP	0.2	1
TL066C	1	COM	15	10 TYP	0.4	1
TL066I	1	IND	6	10 TYP	0.2	1
TL070AC	1	COM	6	10 TYP	0.2	3
TL070C	1	COM	10	10 TYP	0.2	3
TL070I	1	IND	6	10 TYP	0.2	3
TL071AC	1	COM	6	10 TYP	0.2	3
TL071BC	1	COM	3	10 TYP	0.2	3
TL071C	1	COM	10	10 TYP	0.2	3
TL071I	1	IND	6	10 TYP	0.2	3
TL071M	1	MIL	6	10 TYP	0.2	3

T I

TYP SLEW RATE (V/uS)	TYPICAL INPUT NOISE VOLTAGE (nV/√Hz)	MAXIMUM SUPPLY CURRENT (mA)	MAXIMUM SUPPLY VOLTAGE (±V)	OUTPUT CURRENT RATING (A)	UNIT PRICE ($)
3.5	42	1	18	0.001	6.79
3.5	42	1	18	0.001	2.19
3.5	42	1	18	0.001	2.65
3.5	42	1	18	0.001	9.50
3.5	42	0.25	18	0.001	1.27
3.5	42	0.25	18	0.001	3.57
3.5	42	0.25	18	0.001	.81
3.5	42	0.25	18	0.001	1.27
13	18	2.5	18	0.005	1.10
13	18	2.5	18	0.005	.64
13	18	2.5	18	0.005	1.10
13	18	2.5	18	0.005	1.10
13	18	2.5	18	0.005	3.40
13	18	2.5	18	0.005	.64
13	18	2.5	18	0.005	1.10
13	18	2.5	18	0.005	3.50

PART NUMBER	OP AMPS PER PKG	TEMP GRADE	MAXIMUM OFFSET VOLTAGE (mV)	MAXIMUM OFFSET DRIFT (uV/°C)	MAXIMUM BIAS CURRENT (nA)	TYP BAND-WIDTH (MHz)
TL072AC	2	COM	6	10 TYP	0.2	3
TL072BC	2	COM	3	10 TYP	0.2	3
TL072C	2	COM	10	10 TYP	0.2	3
TL072I	2	IND	6	10 TYP	0.2	3
TL072M	2	MIL	6	10 TYP	0.2	3
TL074AC	4	COM	6	10 TYP	0.2	3
TL074BC	4	COM	3	10 TYP	0.2	3
TL074C	4	COM	10	10 TYP	0.2	3
TL074I	4	IND	6	10 TYP	0.2	3
TL074M	4	MIL	9	10 TYP	0.2	3
TL075C	4	COM	10	10 TYP	0.2	3
TL080AC	1	COM	6	10 TYP	0.2	3
TL080C	1	COM	15	10 TYP	0.4	3
TL080I	1	IND	6	10 TYP	0.2	3
TL081AC	1	COM	6	10 TYP	0.2	3
TL081BC	1	COM	3	10 TYP	0.2	3

T I

TYP SLEW RATE (V/uS)	TYPICAL INPUT NOISE VOLTAGE ($nV/\sqrt{Hz}$)	MAXIMUM SUPPLY CURRENT (mA)	MAXIMUM SUPPLY VOLTAGE (±V)	OUTPUT CURRENT RATING (A)	UNIT PRICE ($)
13	18	5	18	0.005	1.52
13	18	5	18	0.005	5.20
13	18	5	18	0.005	1.06
13	18	5	18	0.005	1.52
13	18	5	18	0.005	6.00
13	18	10	18	0.005	2.42
13	18	10	18	0.005	6.10
13	18	10	18	0.005	1.96
13	18	10	18	0.005	2.42
13	18	10	18	0.005	9.00
13	18	10	18	0.005	2.88
13	18	2.8	18	0.005	.97
13	18	2.8	18	0.005	.51
13	18	2.8	18	0.005	.97
13	18	2.8	18	0.005	.97
13	18	2.8	18	0.005	3.27

PART NUMBER	OP AMPS PER PKG	TEMP GRADE	MAXIMUM OFFSET VOLTAGE (mV)	MAXIMUM OFFSET DRIFT (uV/°C)	MAXIMUM BIAS CURRENT (nA)	TYP BAND-WIDTH (MHz)
TL081C	1	COM	15	10 TYP	0.4	3
TL081I	1	IND	6	10 TYP	0.2	3
TL081M	1	MIL	6	10 TYP	0.2	3
TL082AC	2	COM	6	10 TYP	0.2	3
TL082BC	2	COM	3	10 TYP	0.2	3
TL082C	2	COM	15	10 TYP	0.4	3
TL082I	2	IND	6	10 TYP	0.2	3
TL082M	2	MIL	6	10 TYP	0.2	3
TL083AC	2	COM	6	10 TYP	0.2	3
TL083C	2	COM	15	10 TYP	0.4	3
TL083I	2	IND	6	10 TYP	0.2	3
TL084AC	4	COM	6	10 TYP	0.2	3
TL084BC	4	COM	3	10 TYP	0.2	3
TL084C	4	COM	15	10 TYP	0.4	3
TL084I	4	IND	6	10 TYP	0.2	3
TL084M	4	MIL	9	10 TYP	0.2	3

T I

TYP SLEW RATE (V/uS)	TYPICAL INPUT NOISE VOLTAGE (nV/√Hz)	MAXIMUM SUPPLY CURRENT (mA)	MAXIMUM SUPPLY VOLTAGE (±V)	OUTPUT CURRENT RATING (A)	UNIT PRICE ($)
13	18	2.8	18	0.005	.51
13	18	2.8	18	0.005	.97
13	18	2.8	18	0.005	3.00
13	18	5.6	18	0.005	1.40
13	18	5.6	18	0.005	3.73
13	18	5.6	18	0.005	.97
13	18	5.6	18	0.005	1.40
13	18	5.6	18	0.005	4.90
13	18	5.6	18	0.005	1.96
13	18	5.6	18	0.005	1.50
13	18	5.6	18	0.005	1.96
13	18	11.2	18	0.005	2.30
13	18	11.2	18	0.005	5.98
13	18	11.2	18	0.005	1.84
13	18	11.2	18	0.005	2.30
13	18	11.2	18	0.005	7.40

PART NUMBER	OP AMPS PER PKG	TEMP GRADE	MAXIMUM OFFSET VOLTAGE (mV)	MAXIMUM OFFSET DRIFT (uV/°C)	MAXIMUM BIAS CURRENT (nA)	TYP BAND-WIDTH (MHz)
TL085C	4	COM	15	10 TYP	0.4	3
TL087C	1	COM	0.5	10 TYP	0.2	3
TL087I	1	IND	0.5	10 TYP	0.2	3
TL088C	1	COM	1	10 TYP	0.2	3
TL088I	1	IND	1	10 TYP	0.2	3
TL088M	1	MIL	3	10 TYP	0.4	3
TL136C	4	COM	6		500	3
TL287C	2	COM	0.5	10 TYP	0.2	3
TL287I	2	IND	0.5	10 TYP	0.2	3
TL288C	2	COM	1	10 TYP	0.2	3
TL288I	2	COM	1	10 TYP	0.2	3
TL288M	2	MIL	3	10 TYP	0.4	3
TL321C	1	COM	7		250	0.6
TL322C	2	COM	10	10 TYP	500	1
TL322I	2	IND	8	10 TYP	500	1
TLC251AC	1	COM	5	5 TYP	0.001 TYP	2.3

T I

TYP SLEW RATE (V/uS)	TYPICAL INPUT NOISE VOLTAGE (nV/√Hz)	MAXIMUM SUPPLY CURRENT (mA)	MAXIMUM SUPPLY VOLTAGE (±V)	OUTPUT CURRENT RATING (A)	UNIT PRICE ($)
13	18	11.2	18	0.005	1.96
13	18	2.8	18	0.005	6.90
13	18	2.8	18	0.005	11.50
13	18	2.8	18	0.005	4.03
13	18	2.8	18	0.005	5.75
13	18	2.8	18	0.005	14.00
2	7.5	11.3	18	0.005	1.50
13	18	5.6	18	0.005	11.50
13	18	5.6	18	0.005	34.50
13	18	5.6	18	0.005	6.90
13	18	5.6	18	0.005	10.35
13	18	5.6	18	0.005	24.00
0.3		1	16	0.01	.41
0.6		4	18	0.005	.74
0.6		4	18	0.005	1.15
4.5	30	2	8	0.0008	3.24

PART NUMBER	OP AMPS PER PKG	TEMP GRADE	MAXIMUM OFFSET VOLTAGE (mV)	MAXIMUM OFFSET DRIFT (uV/°C)	MAXIMUM BIAS CURRENT (nA)	TYP BAND-WIDTH (MHz)
TLC251BC	1	COM	2	5 TYP	0.001 TYP	2.3
TLC251C	1	COM	10	5 TYP	0.001 TYP	2.3
TLC252AC	2	COM	5	5 TYP	0.001 TYP	2.3
TLC252BC	2	COM	2	5 TYP	0.001 TYP	2.3
TLC252C	2	COM	10	5 TYP	0.001 TYP	2.3
TLC254AC	4	COM	5	5 TYP	0.001 TYP	2.3
TLC254BC	4	COM	2	5 TYP	0.001 TYP	2.3
TLC254C	4	COM	10	5 TYP	0.001 TYP	2.3
TLC25L2AC	2	COM	5	0.7 TYP	0.001 TYP	0.1
TLC25L2BC	2	COM	2	0.7 TYP	0.001 TYP	0.1
TLC25L2C	2	COM	10	0.7 TYP	0.001 TYP	0.1
TLC25L4AC	4	COM	5	0.7 TYP	0.001 TYP	0.1
TLC25L4BC	4	COM	2	0.7 TYP	0.001 TYP	0.1
TLC25L4C	4	COM	10	0.7 TYP	0.001 TYP	0.1
TLC25M2AC	2	COM	5	2 TYP	0.001 TYP	0.7
TLC25M2BC	2	COM	2	2 TYP	0.001 TYP	0.7

T I

TYP SLEW RATE (V/uS)	TYPICAL INPUT NOISE VOLTAGE (nV/$\sqrt{Hz}$)	MAXIMUM SUPPLY CURRENT (mA)	MAXIMUM SUPPLY VOLTAGE (±V)	OUTPUT CURRENT RATING (A)	UNIT PRICE ($)
4.5	30	2	8	0.0008	12.65
4.5	30	2	8	0.0008	1.93
4.5	30	4	8	0.0008	6.93
4.5	30	4	8	0.0008	27.06
4.5	30	4	8	0.0008	4.14
4.5	30	8	8	0.0008	11.37
4.5	30	8	8	0.0008	44.37
4.5	30	8	8	0.0008	6.78
0.04	70	0.04	8	8E-6	6.93
0.04	70	0.04	8	8E-6	27.06
0.04	70	0.04	8	8E-6	4.14
0.04	70	0.08	8	8E-6	11.37
0.04	70	0.08	8	8E-6	44.37
0.04	70	0.08	8	8E-6	6.78
0.6	38	0.6	8	8E-5	6.93
0.6	38	0.6	8	8E-5	27.06

PART NUMBER	OP AMPS PER PKG	TEMP GRADE	MAXIMUM OFFSET VOLTAGE (mV)	MAXIMUM OFFSET DRIFT (uV/°C)	MAXIMUM BIAS CURRENT (nA)	TYP BAND-WIDTH (MHz)
TLC25M2C	2	COM	10	2 TYP	0.001 TYP	0.7
TLC25M4AC	4	COM	5	2 TYP	0.001 TYP	0.7
TLC25M4BC	4	COM	2	2 TYP	0.001 TYP	0.7
TLC25M4C	4	COM	10	2 TYP	0.001 TYP	0.7
TLC271AC	1	COM	5	5 TYP	0.001 TYP	2.3
TLC271AM	1	MIL	5	5 TYP	0.001 TYP	2.3
TLC271BC	1	COM	2	5 TYP	0.001 TYP	2.3
TLC271BM	1	MIL	2	5 TYP	0.001 TYP	2.3
TLC271C	1	COM	10	5 TYP	0.001 TYP	2.3
TLC271M	1	MIL	10	5 TYP	0.001 TYP	2.3
TLC272AC	2	COM	5	5 TYP	0.001 TYP	2.3
TLC272AM	2	MIL	5	5 TYP	0.001 TYP	2.3
TLC272BC	2	COM	2	5 TYP	0.001 TYP	2.3
TLC272BM	2	MIL	2	5 TYP	0.001 TYP	2.3
TLC272C	2	COM	10	5 TYP	0.001 TYP	2.3
TLC272M	2	MIL	10	5 TYP	0.001 TYP	2.3

T I

TYP SLEW RATE (V/uS)	TYPICAL INPUT NOISE VOLTAGE (nV/√Hz)	MAXIMUM SUPPLY CURRENT (mA)	MAXIMUM SUPPLY VOLTAGE (±V)	OUTPUT CURRENT RATING (A)	UNIT PRICE ($)
0.6	38	0.6	8	8E-5	4.14
0.6	38	1.2	8	8E-5	11.37
0.6	38	1.2	8	8E-5	44.37
0.6	38	1.2	8	8E-5	6.78
4.5	30	2	8	0.0008	.81
4.5	30	2	8	0.0008	4.50
4.5	30	2	8	0.0008	5.18
4.5	30	2	8	0.0008	18.00
4.5	30	2	8	0.0008	.64
4.5	30	2	8	0.0008	4.00
4.5	30	4	8	0.0008	1.71
4.5	30	4	8	0.0008	6.70
4.5	30	4	8	0.0008	11.10
4.5	30	4	8	0.0008	23.00
4.5	30	4	8	0.0008	1.06
4.5	30	4	8	0.0008	6.00

PART NUMBER	OP AMPS PER PKG	TEMP GRADE	MAXIMUM OFFSET VOLTAGE (mV)	MAXIMUM OFFSET DRIFT (uV/°C)	MAXIMUM BIAS CURRENT (nA)	TYP BAND-WIDTH (MHz)
TLC274AC	4	COM	5	5 TYP	0.001 TYP	2.3
TLC274AM	4	MIL	5	5 TYP	0.001 TYP	2.3
TLC274BC	4	COM	2	5 TYP	0.001 TYP	2.3
TLC274BM	4	MIL	2	5 TYP	0.001 TYP	2.3
TLC274C	4	COM	10	5 TYP	0.001 TYP	2.3
TLC274M	4	MIL	10	5 TYP	0.001 TYP	2.3
TLC27L2AC	2	COM	5	0.7 TYP	0.001 TYP	0.1
TLC27L2AM	2	MIL	5	0.7 TYP	0.001 TYP	0.1
TLC27L2BC	2	COM	2	0.7 TYP	0.001 TYP	0.1
TLC27L2BM	2	MIL	2	0.7 TYP	0.001 TYP	0.1
TLC27L2C	2	COM	10	0.7 TYP	0.001 TYP	0.1
TLC27L2M	2	MIL	10	0.7 TYP	0.001 TYP	0.1
TLC27L4AC	4	COM	5	0.7 TYP	0.001 TYP	0.1
TLC27L4AM	4	MIL	5	0.7 TYP	0.001 TYP	0.1
TLC27L4BC	4	COM	2	0.7 TYP	0.001 TYP	0.1
TLC27L4BM	4	MIL	2	0.7 TYP	0.001 TYP	0.1

T I

TYP SLEW RATE (V/uS)	TYPICAL INPUT NOISE VOLTAGE (nV/√Hz)	MAXIMUM SUPPLY CURRENT (mA)	MAXIMUM SUPPLY VOLTAGE (±V)	OUTPUT CURRENT RATING (A)	UNIT PRICE ($)
4.5	30	8	8	0.0008	2.79
4.5	30	8	8	0.0008	10.00
4.5	30	8	8	0.0008	18.21
4.5	30	8	8	0.0008	29.00
4.5	30	8	8	0.0008	2.25
4.5	30	8	8	0.0008	9.00
0.04	70	0.04	8	8E-6	1.71
0.04	70	0.04	8	8E-6	6.70
0.04	70	0.04	8	8E-6	11.10
0.04	70	0.04	8	8E-6	23.00
0.04	70	0.04	8	8E-6	1.06
0.04	70	0.04	8	8E-6	6.00
0.04	70	0.08	8	8E-6	2.79
0.04	70	0.08	8	8E-6	10.00
0.04	70	0.08	8	8E-6	18.21
0.04	70	0.08	8	8E-6	29.00

PART NUMBER	OP AMPS PER PKG	TEMP GRADE	MAXIMUM OFFSET VOLTAGE (mV)	MAXIMUM OFFSET DRIFT (uV/°C)	MAXIMUM BIAS CURRENT (nA)	TYP BAND-WIDTH (MHz)
TLC27L4C	4	COM	10	0.7 TYP	0.001 TYP	0.1
TLC27L4M	4	MIL	10	0.7 TYP	0.001 TYP	0.1
TLC27M2AC	2	COM	5	2 TYP	0.001 TYP	0.7
TLC27M2AM	2	MIL	5	2 TYP	0.001 TYP	0.7
TLC27M2BC	2	COM	2	2 TYP	0.001 TYP	0.7
TLC27M2BM	2	MIL	2	2 TYP	0.001 TYP	0.7
TLC27M2C	2	COM	10	2 TYP	0.001 TYP	0.7
TLC27M2M	2	MIL	10	2 TYP	0.001 TYP	0.7
TLC27M4AC	4	COM	5	2 TYP	0.001 TYP	0.7
TLC27M4AM	4	MIL	5	2 TYP	0.001 TYP	0.7
TLC27M4BC	4	COM	2	2 TYP	0.001 TYP	0.7
TLC27M4BM	4	MIL	2	2 TYP	0.001 TYP	0.7
TLC27M4C	4	COM	10	2 TYP	0.001 TYP	0.7
TLC27M4M	4	MIL	10	2 TYP	0.001 TYP	0.7

T I

TYP SLEW RATE (V/uS)	TYPICAL INPUT NOISE VOLTAGE ($nV/\sqrt{Hz}$)	MAXIMUM SUPPLY CURRENT (mA)	MAXIMUM SUPPLY VOLTAGE (±V)	OUTPUT CURRENT RATING (A)	UNIT PRICE ($)
0.04	70	0.08	8	8E-6	2.25
0.04	70	0.08	8	8E-6	9.00
0.6	38	0.6	8	8E-5	1.71
0.6	38	0.6	8	8E-5	6.70
0.6	38	0.6	8	8E-5	11.10
0.6	38	0.6	8	8E-5	23.00
0.6	38	0.6	8	8E-5	1.06
0.6	38	0.6	8	8E-5	6.00
0.6	38	1.2	8	8E-5	2.79
0.6	38	1.2	8	8E-5	10.00
0.6	38	1.2	8	8E-5	18.21
0.6	38	1.2	8	8E-5	29.00
0.6	38	1.2	8	8E-5	2.25
0.6	38	1.2	8	8E-5	9.00

Appendix B
Comparators

PART NUMBER	CMP PER PKG	OPERATING TEMP (C) MIN	MAX	MAXIMUM OFFSET VOLTAGE (mV)	MAXIMUM BIAS CURRENT (nA)
Am685L	1	-30	85	2	10,000
Am685M	1	-55	125	2	10,000
Am686C	1	0	70	3	10,000
Am686C-1	1	0	70	6	12,000
Am686M	1	-55	125	2	10,000
Am687AL	2	-30	85	3	10,000
Am687AM	2	-55	125	2	10,000
Am687L	2	-30	85	3	10,000
Am687M	2	-55	125	2	10,000

ADVANCED MICRO DEVICES (AMD)

OUTPUT TYPE	TYPICAL RESPONSE TIME (nS)	TYPICAL SUPPLY VOLTAGES (V,V)	MAXIMUM VCC CURRENT (mA)	MAXIMUM VEE CURRENT (mA)	UNIT PRICE ($)
ECL	5.5	+6,-5.2	22	26	13.40
ECL	5.5	+6,-5.2	22	26	30.00
TTL	9	+5,-6	42	34	5.25
TTL	9	+5,-6	50	40	3.00
TTL	9	+5,-6	40	32	30.00
ECL	7	+5,-5.2	35	48	28.50
ECL	7	+5,-5.2	32	44	66.00
ECL	7	+5,-5.2	35	48	22.50
ECL	7	+5,-5.2	32	44	42.00

PART NUMBER	CMP PER PKG	OPERATING TEMP (C) MIN	MAX	MAXIMUM OFFSET VOLTAGE (mV)	MAXIMUM BIAS CURRENT (nA)
uA111	1	-55	125	3	100
uA139	4	-55	125	5	100
uA139A	4	-55	125	2	100
uA239	4	-25	85	5	250
uA239A	4	-25	85	2	250
uA311	1	0	70	7.5	250
uA339	4	0	70	5	250
uA339A	4	0	70	2	250
uA710	1	-55	125	2	20,000
uA710C	1	0	70	5	25,000
uA711	2	-55	125	3.5	75,000
uA711C	2	0	70	5	100,000
uA760	1	-55	125	6	60,000
uA760C	1	0	70	6	60,000
uA2901	4	-40	85	7	250
uA3302	4	-40	85	20	500

FAIRCHILD

OUTPUT TYPE	TYPICAL RESPONSE TIME (nS)	TYPICAL SUPPLY VOLTAGES (V,V)	MAXIMUM VCC CURRENT (mA)	MAXIMUM VEE CURRENT (mA)	UNIT PRICE ($)
O.C.	200	+15,-15	6	5	2.57
O.C.	1300	+5,-0	2	0	3.23
O.C.	1300	+5,-0	2	0	7.60
O.C.	1300	+5,-0	2	0	2.22
O.C.	1300	+5,-0	2	0	5.13
O.C.	200	+15,-15	7.5	5	0.57
O.C.	1300	+5,-0	2	0	0.67
O.C.	1300	+5,-0	2	0	1.65
TTL	40	+12,-6	9	7	1.90
TTL	40	+12,-6	9	7	0.74
TTL	40	+12,-6	12	6	2.19
TTL	40	+12,-6	12	6	1.52
TTL	16	+5,-5	32	16	15.20
TTL	16	+5,-5	34	16	14.20
O.C.	1300	+5,-0	2	0	1.77
O.C.	1300	+5,-0	2	0	0.51

COMPARATORS:

PART NUMBER	CMP PER PKG	OPERATING TEMP (C) MIN	MAX	MAXIMUM OFFSET VOLTAGE (mV)	MAXIMUM BIAS CURRENT (nA)
MC1414	2	0	75	5	25,000
MC1514	2	-55	125	2	20,000
MC3302	4	-40	85	20	100

COMPARATORS:

PART NUMBER	CMP PER PKG	OPERATING TEMP (C) MIN	MAX	MAXIMUM OFFSET VOLTAGE (mV)	MAXIMUM BIAS CURRENT (nA)
LF111	1	-55	125	4	0.05
LF211	1	-25	85	4	0.05
LF311	1	0	70	10	0.15
LH2111	2	-55	125	3	100
LH2211	2	-25	85	3	100
LF2311	2	0	70	7.5	250
LM106	1	-55	125	2	20,000
LM111	1	-55	125	3	100
LM119	2	-55	125	4	500
LM139	4	-55	125	5	100
LM139A	4	-55	125	2	100
LM160	1	-55	125	5	20,000
LM193	2	-55	125	5	100
LM193A	2	-55	125	2	100
LM206	1	-25	85	2	20,000
LM211	1	-25	85	3	100

MOTOROLA

OUTPUT TYPE	TYPICAL RESPONSE TIME (nS)	TYPICAL SUPPLY VOLTAGES (V,V)	MAXIMUM VCC CURRENT (mA)	MAXIMUM VEE CURRENT (mA)	UNIT PRICE ($)
TTL	40	+12,-6	18	14	1.83
TTL	40	+12,-6	18	14	4.92
O.C.	1300	+5,-0	2	0	0.73

NATIONAL

OUTPUT TYPE	TYPICAL RESPONSE TIME (nS)	TYPICAL SUPPLY VOLTAGES (V,V)	MAXIMUM VCC CURRENT (mA)	MAXIMUM VEE CURRENT (mA)	UNIT PRICE ($)
O.C.	200	+15,-15	6	5	25.50
O.C.	200	+15,-15	6	5	14.25
O.C.	200	+15,-15	7.5	5	6.40
O.C.	200	+15,-15	6	5	17.25
O.C.	200	+15,-15	6	5	12.00
O.C.	200	+15,-15	7.5	5	10.00
TTL	28	+12,-6	10	3.6	27.00
O.C.	200	+15,-15	6	5	3.60
O.C.	80	+15,-15	11.5	4.5	8.50
O.C.	1300	+5,-0	2	0	2.90
O.C.	1300	+5,-0	2	0	10.70
TTL	14	+5,-5	32	16	14.00
O.C.	1300	+5,-0	1	0	5.50
O.C.	1300	+5,-0	1	0	9.00
TTL	28	+12,-6	10	3.6	17.20
O.C.	200	+15,-15	6	5	2.70

COMPARATORS:

PART NUMBER	CMP PER PKG	OPERATING TEMP (C) MIN	MAX	MAXIMUM OFFSET VOLTAGE (mV)	MAXIMUM BIAS CURRENT (nA)
LM219	2	-25	85	4	500
LM239	4	-25	85	5	250
LM239A	4	-25	85	2	250
LM260	1	-25	85	5	20,000
LM293	2	-25	85	5	250
LM293A	2	-25	85	2	250
LM306	1	0	70	5	25,000
LM311	1	0	70	7.5	250
LM319	2	0	70	8	1,000
LM339	4	0	70	5	250
LM339A	4	0	70	2	250
LM360	1	0	70	5	20,000
LM393	2	0	70	5	250
LM393A	2	0	70	2	250
LM710	1	-55	125	2	20,000
LM710C	1	0	70	5	25,000

NATIONAL

OUTPUT TYPE	TYPICAL RESPONSE TIME (nS)	TYPICAL SUPPLY VOLTAGES (V,V)	MAXIMUM VCC CURRENT (mA)	MAXIMUM VEE CURRENT (mA)	UNIT PRICE ($)
O.C.	80	+15,-15	11.5	4.5	7.70
O.C.	1300	+5,-0	2	0	2.75
O.C.	1300	+5,-0	2	0	9.98
TTL	14	+5,-5	32	16	7.49
O.C.	1300	+5,-0	1	0	4.10
O.C.	1300	+5,-0	1	0	6.15
TTL	28	+12,-6	10	3.6	10.20
O.C	200	+15,-15	7.5	5	0.65
O.C.	80	+15,-15	12.5	5	2.40
O.C.	1300	+5,-0	2	0	0.75
O.C.	1300	+5,-0	2	0	1.50
TTL	14	+5,-5	32	16	4.50
O.C.	1300	+5,-0	1	0	0.60
O.C.	1300	+5,-0	1	0	1.80
TTL	40	+12,-6	9	7	2.20
TTL	40	+12,-6	9	7	0.70

COMPARATORS:

PART NUMBER	CMP PER PKG	OPERATING TEMP (C) MIN	MAX	MAXIMUM OFFSET VOLTAGE (mV)	MAXIMUM BIAS CURRENT (nA)
LM711	2	-55	125	3.5	75,000
LM711C	2	0	70	5	100,000
LM1414	2	0	70	5	25,000
LM1514	2	-55	125	2	20,000
LM2901	4	-40	85	7	250
LM2903	2	-40	85	7	250
LM3302	4	-40	85	20	500

NATIONAL

OUTPUT TYPE	TYPICAL RESPONSE TIME (nS)	TYPICAL SUPPLY VOLTAGES (V,V)	MAXIMUM VCC CURRENT (mA)	MAXIMUM VEE CURRENT (mA)	UNIT PRICE ($)
TTL	40	+12,-6	13	6	2.50
TTL	40	+12,-6	13	6	0.90
TTL	30	+12,-6	18	14	1.49
TTL	30	+12,-6	18	14	4.98
O.C.	1300	+5,-0	2	0	1.10
O.C.	1300	+5,-0	1	0	1.15
O.C.	1300	+5,-0	2	0	0.67

COMPARATORS:

PART NUMBER	CMP PER PKG	OPERATING TEMP (C) MIN	MAX	MAXIMUM OFFSET VOLTAGE (mV)	MAXIMUM BIAS CURRENT (nA)
CMP-01	1	-55	125	0.8	600
CMP-01C	1	0	70	2.8	900
CMP-01E	1	0	70	0.8	600
CMP-02	1	-55	125	0.8	50
CMP-02C	1	0	70	2.8	100
CMP-02E	1	0	70	0.8	50
CMP-04B	4	-55	125	1	100
CMP-04FP	4	0	70	1	100
CMP-04FY	4	-25	85	1	100
CMP-05A	1	-55	125	0.25	1,200
CMP-05B	1	-55	125	0.6	1,800
CMP-05EP	1	0	70	0.25	1,200
CMP-05EZ	1	-25	85	0.25	1,200
CMP-05FP	1	0	70	0.6	1,800
CMP-05FZ	1	-25	85	0.6	1,800

PRECISION MONOLITHICS INCORPORATED (PMI)

OUTPUT TYPE	TYPICAL RESPONSE TIME (nS)	TYPICAL SUPPLY VOLTAGES (V,V)	MAXIMUM VCC CURRENT (mA)	MAXIMUM VEE CURRENT (mA)	UNIT PRICE ($)
TTL	110	+15,-15	8	2.2	20.40
TTL	110	+15,-15	8.5	2.2	3.75
TTL	110	+15,-15	8	2.2	5.63
TTL	190	+15,-15	8	2.2	20.40
TTL	190	+15,-15	8.5	2.2	3.75
TTL	190	+15,-15	8	2.2	5.63
O.C.	1300	+5,-0	2	0	14.93
O.C.	1300	+5,-0	2	0	7.80
O.C.	1300	+5,-0	2	0	9.75
TTL	37	+5,-5	15	16	21.68
TTL	37	+5,-5	16	18	12.75
TTL	37	+5,-5	15	16	9.15
TTL	37	+5,-5	15	16	11.40
TTL	37	+5,-5	16	18	6.00
TTL	37	+5,-5	16	18	7.50

COMPARATORS:

PART NUMBER	CMP PER PKG	OPERATING TEMP (C) MIN	MAX	MAXIMUM OFFSET VOLTAGE (mV)	MAXIMUM BIAS CURRENT (nA)
CA139	4	-55	125	5	100
CA139A	4	-55	125	2	100
CA239	4	-25	85	5	250
CA239A	4	-25	85	2	250
CA311	1	0	70	7.5	250
CA339	4	0	70	5	250
CA339A	4	0	70	2	250
CA3290	2	-55	125	20	0.05
CA3290A	2	-55	125	10	0.04
CA3290B	2	-55	125	6	0.03

RCA

OUTPUT TYPE	TYPICAL RESPONSE TIME (nS)	TYPICAL SUPPLY VOLTAGES (V,V)	MAXIMUM VCC CURRENT (mA)	MAXIMUM VEE CURRENT (mA)	UNIT PRICE ($)
O.C.	1300	+5,-0	2	0	1.30
O.C.	1300	+5,-0	2	0	4.42
O.C.	1300	+5,-0	2	0	0.94
O.C.	1300	+5,-0	2	0	2.30
O.C.	200	+15,-15	7.5	5	0.60
O.C.	1300	+5,-0	2	0	0.66
O.C.	1300	+5,-0	2	0	1.72
O.C.	1200	+5,-0	1.4	0	0.90
O.C.	1200	+5,-0	1.4	0	1.26
O.C.	1200	+5,-0	1.4	0	3.20

COMPARATORS:

PART NUMBER	CMP PER PKG	OPERATING TEMP (C) MIN	MAX	MAXIMUM OFFSET VOLTAGE (mV)	MAXIMUM BIAS CURRENT (nA)
TL331C	1	0	70	5	250
TL331I	1	-25	85	5	100
TL506C	2	0	70	5	25,000
TL506M	2	-55	125	2	20,000
TL514C	2	0	70	3.5	20,000
TL514M	2	-55	125	2	15,000
TL710C	1	0	70	7.5	100,000
TL710M	1	-55	125	5	75,000
TL810C	1	0	70	3.5	20,000
TL811C	2	0	70	5	30,000
TL811M	2	-55	125	3.5	20,000
TL820C	2	0	70	3.5	20,000
TL820M	2	-55	125	2	15,000

TEXAS INSTRUMENTS (TI)

OUTPUT TYPE	TYPICAL RESPONSE TIME (nS)	TYPICAL SUPPLY VOLTAGES (V,V)	MAXIMUM VCC CURRENT (mA)	MAXIMUM VEE CURRENT (mA)	UNIT PRICE ($)
O.C.	1300	+5,-0	0.8	0	0.58
O.C.	1300	+5,-0	0.8	0	0.92
TTL	28	+12,-6	20	7.2	2.81
TTL	28	+12,-6	20	7.2	4.40
TTL	30	+12,-6	9	7	1.33
TTL	30	+12,-6	9	7	4.40
TTL	40	+12,-6	10.1	8.9	0.53
TTL	40	+12,-6	10.1	8.9	1.70
TTL	30	+12,-6	9	7	1.01
TTL	33	+12,-6	10	6	1.31
TTL	33	+12,-6	10	6	2.20
TTL	30	+12,-6	9	7	1.68
TTL	30	+12,-6	9	7	2.20

Appendix C
Voltage References

VOLTAGE REFERENCES:

PART NUMBER	OPERATING TEMP (C)		OUTPUT VOLTAGE		
	MIN	MAX	TYP (V)	TOL (%)	TEMPCO (PPM/C)
AD580J	0	70	2.5	3	85
AD580K	0	70	2.5	1	40
AD580L	0	70	2.5	0.4	25
AD580M	0	70	2.5	0.4	10
AD580S	-55	125	2.5	1	55
AD580T	-55	125	2.5	0.4	25
AD580U	-55	125	2.5	0.4	10
AD581J	0	70	10	0.3	30
AD581K	0	70	10	0.1	15
AD581L	0	70	10	0.05	5
AD581S	-55	125	10	0.3	30
AD581T	-55	125	10	0.1	15
AD581U	-55	125	10	0.05	10

ANALOG DEVICES

MAXIMUM OUTPUT CURRENT (mA)	MAX REGULATION		MAXIMUM INPUT VOLTAGE (V)	UNIT PRICE ($)
	LINE (%/V)	LOAD (%/mA)		
10	0.05	0.04	30	3.90
10	0.03	0.04	30	8.00
10	0.02	0.04	30	11.00
10	0.02	0.04	30	17.00
10	0.05	0.04	30	16.00
10	0.02	0.04	30	25.10
10	0.02	0.04	30	53.80
5	0.005	0.005	30	5.60
5	0.005	0.005	30	9.20
5	0.005	0.005	30	21.80
5	0.005	0.005	30	17.20
5	0.005	0.005	30	26.00
5	0.005	0.005	30	44.80

VOLTAGE REFERENCES:

PART NUMBER	OPERATING TEMP (C) MIN	MAX	OUTPUT VOLTAGE TYP (V)	TOL (%)	TEMPCO (PPM/C)
AD584J	0	70	10	0.3	30
			7.5	0.27	30
			5.0	0.3	30
			2.5	0.3	30
AD584K	0	70	10	0.1	15
			7.5	0.11	15
			5.0	0.12	15
			2.5	0.14	15
AD584L	0	70	10	0.05	5
			7.5	0.05	5
			5.0	0.06	5
			2.5	0.1	10
AD584S	-55	125	10	0.3	30
			7.5	0.27	30
			5.0	0.3	30
			2.5	0.3	30
AD584T	-55	125	10	0.1	15
			7.5	0.11	15
			5.0	0.12	15
			2.5	0.14	20

VOLTAGE REFERENCES:

PART NUMBER	OPERATING TEMP (C) MIN	MAX	BREAKDOWN VOLTAGE (V) MIN	TYP	MAX
AD589J	0	70	1.200	1.235	1.250
AD589K	0	70	1.200	1.235	1.250
AD589L	0	70	1.200	1.235	1.250
AD589M	0	70	1.200	1.235	1.250
AD589S	-55	125	1.200	1.235	1.250
AD589T	-55	125	1.200	1.235	1.250
AD589U	-55	125	1.200	1.235	1.250

ANALOG DEVICES

MAXIMUM OUTPUT CURRENT (mA)	MAX REGULATION		MAXIMUM INPUT VOLTAGE (V)	UNIT PRICE ($)
	LINE (%/V)	LOAD (%/mA)		
5	0.005	0.005	30	5.70
5	0.005	0.005	30	12.20
5	0.005	0.005	30	21.80
5	0.005	0.005	30	17.10
5	0.005	0.005	30	25.30

ANALOG DEVICES

TEMPERATURE COEFFICIENT (PPM/C)	OPERATING CURRENT (mA)		MAXIMUM DYNAMIC RESISTANCE (OHMS)	UNIT PRICE ($)
	MIN	MAX		
100	0.05	5	2	2.20
50	0.05	5	2	2.90
25	0.05	5	2	8.10
10	0.05	5	2	18.10
100	0.05	5	2	4.40
50	0.05	5	2	6.50
25	0.05	5	2	18.40

VOLTAGE REFERENCES:

PART NUMBER	OPERATING TEMP (C)		OUTPUT VOLTAGE		
	MIN	MAX	TYP (V)	TOL (%)	TEMPCO (PPM/C)
AD1403	0	70	2.5	1	40
AD1403A	0	70	2.5	0.4	25
AD2700J	-25	85	10	0.05	10
AD2700L	-25	85	10	0.025	3
AD2700S	-55	125	10	0.05	3
AD2700U	-55	125	10	0.025	3
AD2701J	-25	85	-10	0.05	10
AD2701L	-25	85	-10	0.025	3
AD2701S	-55	125	-10	0.05	3
AD2701U	-55	125	-10	0.025	3
AD2702J	-25	85	10,-10	0.05	10
AD2702L	-25	85	10,-10	0.025	5
AD2702S	-55	125	10,-10	0.05	5
AD2702U	-55	125	10,-10	0.025	3

* = TYPICAL

ANALOG DEVICES

MAXIMUM OUTPUT CURRENT (mA)	MAX REGULATION LINE (%/V)	LOAD (%/mA)	MAXIMUM INPUT VOLTAGE (V)	UNIT PRICE ($)
10	0.01	0.04	40	2.50
10	0.01	0.04	40	6.00
10	0.003*	0.0005*	16.5	42.50
10	0.003*	0.0005*	16.5	70.50
10	0.003*	0.0005*	16.5	53.00
10	0.003*	0.0005*	16.5	83.00
10	0.003*	0.0005*	16.5	50.00
10	0.003*	0.0005*	16.5	79.75
10	0.003*	0.0005*	16.5	58.25
10	0.000*	0.0005*	16.5	93.25
10	0.003*	0.0005*	16.5	48.25
10	0.003*	0.0005*	16.5	77.50
10	0.003*	0.0005*	16.5	56.75
10	0.003*	0.0005*	16.5	90.75

VOLTAGE REFERENCES:

PART NUMBER	OPERATING TEMP (C) MIN	MAX	OUTPUT VOLTAGE TYP (V)	TOL (%)	TEMPCO (PPM/C)
AD2710K	0	70	10	0.01	5
	25	70	10	0.01	2
AD2710L	0	70	10	0.01	5
	25	70	10	0.01	1
AD2712K	0	70	10	0.01	5
	0	70	-10	0.01	5
	25	70	10	0.01	2
	25	70	-10	0.01	3
AD2712L	0	70	10	0.01	5
	0	70	-10	0.01	5
	25	70	10	0.01	1
	25	70	-10	0.01	2

VOLTAGE REFERENCES:

PART NUMBER	OPERATING TEMP (C) MIN	MAX	BREAKDOWN VOLTAGE (V) MIN	TYP	MAX
ICL8069A	0	70	1.20	1.23	1.25
ICL8069B	0	70	1.20	1.23	1.25
ICL8069CC	0	70	1.20	1.23	1.25
ICL8069CM	-55	125	1.20	1.23	1.25
ICL8069DC	0	70	1.20	1.23	1.25
ICL8069DM	-55	125	1.20	1.23	1.25

ANALOG DEVICES

MAXIMUM OUTPUT CURRENT (mA)	MAX REGULATION		MAXIMUM INPUT VOLTAGE (V)	UNIT PRICE ($)
	LINE (%/V)	LOAD (%/mA)		
5	0.002	0.001	16.5	35.75
5	0.002	0.001	16.5	45.25
5	0.002	0.001	16.5	41.00
5	0.002	0.001	16.5	48.25

INTERSIL

TEMPERATURE COEFFICIENT (PPM/C)	OPERATING CURRENT (mA)		MAXIMUM DYNAMIC RESISTANCE (OHMS)	UNIT PRICE ($)
	MIN	MAX		
10	0.05	5	2	15.85
25	0.05	5	2	7.15
50	0.05	5	2	1.60
50	0.05	5	2	5.55
100	0.05	5	2	1.15
100	0.05	5	2	4.65

VOLTAGE REFERENCES:

PART NUMBER	OPERATING TEMP (C)		OUTPUT VOLTAGE		
	MIN	MAX	TYP (V)	TOL (%)	TEMPCO (PPM/C)
MC1400AG2	0	70	2.5	0.2	10
MC1400AG5	0	70	5	0.2	10
MC1400AG6	0	70	6.25	0.2	10
MC1400AG10	0	70	10	0.2	10
MC1400G2	0	70	2.5	0.2	25
MC1400G5	0	70	5	0.2	25
MC1400G6	0	70	6.25	0.2	25
MC1400G10	0	70	10	0.2	25
MC1403	0	70	2.5	1	40
MC1403A	0	70	2.5	1	25
MC1404AU5	0	70	5	1	25
MC1404AU6	0	70	6.25	1	25
MC1404AU10	0	70	10	1	25
MC1404U5	0	70	5	1	40
MC1404U6	0	70	6.25	1	40
MC1404U10	0	70	10	1	40

MOTOROLA

MAXIMUM OUTPUT CURRENT (mA)	MAX REGULATION		MAXIMUM INPUT VOLTAGE (V)	UNIT PRICE ($)
	LINE (%/V)	LOAD (%/mA)		
10	0.0033	0.04	40	10.43
10	0.0024	0.04	40	10.43
10	0.0020	0.032	40	10.43
10	0.0014	0.02	40	10.43
10	0.0033	0.04	40	6.38
10	0.0024	0.04	40	6.38
10	0.0020	0.032	40	6.38
10	0.0014	0.02	40	6.38
10	0.01	0.04	40	2.71
10	0.01	0.04	40	6.24
10	0.0037	0.02	40	3.85
10	0.0031	0.016	40	3.85
10	0.0022	0.01	40	3.85
10	0.0037	0.02	40	2.73
10	0.0031	0.016	40	2.73
10	0.0022	0.01	40	2.73

VOLTAGE REFERENCES:

PART NUMBER	OPERATING TEMP (C)		OUTPUT VOLTAGE		
	MIN	MAX	TYP (V)	TOL (%)	TEMPCO (PPM/C)
MC1500AG2	-55	125	2.5	0.2	10
MC1500AG5	-55	125	5	0.2	10
MC1500AG6	-55	125	6.25	0.2	10
MC1500AG10	-55	125	10	0.2	10
MC1500G2	-55	125	2.5	0.2	40
MC1500G5	-55	125	5	0.2	40
MC1500G6	-55	125	6.25	0.2	40
MC1500G10	-55	125	10	0.2	40
MC1503	-55	125	2.5	1	55
MC1503A	-55	125	2.5	1	25
MC1504AU5	-55	125	5	1	25
MC1504AU6	-55	125	6.25	1	25
MC1504AU10	-55	125	10	1	25
MC1504U5	-55	125	5	1	55
MC1504U6	-55	125	6.25	1	55
MC1504U10	-55	125	10	1	55

MOTOROLA

MAXIMUM OUTPUT CURRENT (mA)	MAX REGULATION		MAXIMUM INPUT VOLTAGE (V)	UNIT PRICE ($)
	LINE (%/V)	LOAD (%/mA)		
10	0.0033	0.04	40	36.75
10	0.0024	0.04	40	36.75
10	0.0020	0.032	40	36.75
10	0.0014	0.02	40	36.75
10	0.0033	0.04	40	17.63
10	0.0024	0.04	40	17.63
10	0.0020	0.032	40	17.63
10	0.0014	0.02	40	17.63
10	0.01	0.04	40	7.61
10	0.01	0.04	40	14.25
10	0.0037	0.02	40	14.38
10	0.0031	0.016	40	14.38
10	0.0022	0.01	40	14.38
10	0.0037	0.02	40	9.12
10	0.0031	0.016	40	9.12
10	0.0022	0.01	40	9.12

VOLTAGE REFERENCES:

PART NUMBER	OPERATING TEMP (C)		OUTPUT VOLTAGE		
	MIN	MAX	TYP (V)	TOL (%)	TEMPCO (PPM/C)
LH0070-0	-55	125	10	0.1	40
LH0070-1	-55	125	10	0.1	20
LH0070-2	-55	125	10	0.05	8
LH0071-0	-55	125	10.24	0.1	40
LH0071-1	-55	125	10.24	0.1	20
LH0071-2	-55	125	10.24	0.1	8

VOLTAGE REFERENCES:

PART NUMBER	OPERATING TEMP (C)		BREAKDOWN VOLTAGE	
	MIN	MAX	TYP(V)	TOL(%)
LM103H-1.8	-55	125	1.8	10
LM103H-2.0	-55	125	2.0	10
LM103H-2.2	-55	125	2.2	10
LM103H-2.4	-55	125	2.4	10
LM103H-2.7	-55	125	2.7	10
LM103H-3.0	-55	125	3.0	10
LM103H-3.3	-55	125	3.3	10
LM103H-3.6	-55	125	3.6	10
LM103H-3.9	-55	125	3.9	10
LM103H-4.3	-55	125	4.3	10
LM103H-4.7	-55	125	4.7	10
LM103H-5.1	-55	125	5.1	10
LM103H-5.6	-55	125	5.6	10

NATIONAL

MAXIMUM OUTPUT CURRENT (mA)	MAX REGULATION LINE (%/V)	LOAD (%/mA)	MAXIMUM INPUT VOLTAGE (V)	UNIT PRICE ($)
5	0.005	0.006	40	6.60
5	0.005	0.006	40	10.20
5	0.0015	0.006	40	18.50
5	0.005	0.006	40	7.10
5	0.005	0.006	40	12.55
5	0.0015	0.006	40	18.10

NATIONAL

TYPICAL TEMPERATURE COEFFICIENT (mV/C)	OPERATING CURRENT (mA) MIN	MAX	MAXIMUM DYNAMIC RESISTANCE (OHMS)	UNIT PRICE ($)
-5	0.01	10	25	9.00
-5	0.01	10	25	9.00
-5	0.01	10	25	9.00
-5	0.01	10	25	9.00
-5	0.01	10	25	9.00
-5	0.01	10	25	9.00
-5	0.01	10	25	9.00
-5	0.01	10	25	9.00
-5	0.01	10	25	9.00
-5	0.01	10	25	9.00
-5	0.01	10	25	9.00
-5	0.01	10	25	9.00
-5	0.01	10	25	9.00

VOLTAGE REFERENCES:

PART NUMBER	OPERATING TEMP (C)		BREAKDOWN VOLTAGE (V)		
	MIN	MAX	MIN	TYP	MAX
LM113	-55	125	1.16	1.22	1.28
LM113-1	-55	125	1.21	1.22	1.232
LM113-2	-55	125	1.195	1.22	1.245
LM129A	-55	125	6.7	6.9	7.2
LM129B	-55	125	6.7	6.9	7.2
LM129C	-55	125	6.7	6.9	7.2
LM136-2.5	-55	125	2.44	2.49	2.54
LM136-5.0	-55	125	4.9	5.0	5.1
LM136A-2.5	-55	125	2.465	2.49	2.515
LM136A-5.0	-55	125	4.95	5.0	5.05
LM185-1.2	-55	125	1.223	1.235	1.247
LM185-2.5	-55	125	2.462	2.5	2.538
LM199	-55	125	6.8	6.95	7.1
LM199A	-55	125	6.8	6.95	7.1

* = TYPICAL

NATIONAL

MAXIMUM TEMPERATURE COEFFICIENT (PPM/C)	OPERATING CURRENT (mA)		MAXIMUM DYNAMIC RESISTANCE (OHMS)	UNIT PRICE ($)
	MIN	MAX		
100*	0.5	20	1	11.25
100*	0.5	20	1	22.50
100*	0.5	20	1	15.00
10	0.6	15	1	12.95
20	0.6	15	1	8.95
50	0.6	15	1	5.95
72	0.4	10	0.6	8.64
144	0.6	10	1.2	8.64
72	0.4	10	0.6	12.50
144	0.6	10	1.2	12.50
20*	0.01	20	0.6	17.70
20*	0.02	20	0.6	18.45
15	0.5	10	1	29.50
10	0.5	10	1	44.00

VOLTAGE REFERENCES:

PART NUMBER	OPERATING TEMP (C)		BREAKDOWN VOLTAGE (V)		
	MIN	MAX	MIN	TYP	MAX
LM236-2.5	-25	85	2.44	2.49	2.54
LM236-5.0	-25	85	4.9	5.0	5.1
LM236A-2.5	-25	85	2.465	2.49	2.515
LM236A-5.0	-25	85	4.95	5.0	5.05
LM285-1.2	-25	85	1.223	1.235	1.247
LM285-2.5	-25	85	2.462	2.5	2.538
LM299	-25	85	6.8	6.95	7.1
LM299A	-25	85	6.8	6.95	7.1

* = TYPICAL

NATIONAL

MAXIMUM TEMPERATURE COEFFICIENT (PPM/C)	OPERATING CURRENT (mA)		MAXIMUM DYNAMIC RESISTANCE (OHMS)	UNIT PRICE ($)
	MIN	MAX		
36	0.4	10	0.6	5.75
72	0.6	10	1.2	5.75
36	0.4	10	0.6	8.65
72	0.6	10	1.2	8.65
20*	0.01	20	0.6	15.55
20*	0.02	20	0.6	15.55
1	0.5	10	1	14.40
0.5	0.5	10	1	19.50

VOLTAGE REFERENCES:

PART NUMBER	OPERATING TEMP (C)		BREAKDOWN VOLTAGE (V)		
	MIN	MAX	MIN	TYP	MAX
LM313	0	70	1.16	1.22	1.28
LM329A	0	70	6.7	6.9	7.2
LM329B	0	70	6.7	6.9	7.2
LM329C	0	70	6.7	6.9	7.2
LM329D	0	70	6.7	6.9	7.2
LM336-2.5	0	70	2.39	2.49	2.59
LM336-5.0	0	70	4.8	5.0	5.2
LM336B-2.5	0	70	2.44	2.49	2.54
LM336B-5.0	0	70	4.9	5.0	5.1
LM385-1.2	0	70	1.205	1.235	1.260
LM385B-1.2	0	70	1.223	1.235	1.247
LM385-2.5	0	70	2.425	2.5	2.575
LM385B-2.5	0	70	2.462	2.5	2.538
LM399	0	70	6.6	6.95	7.3
LM399A	0	70	6.6	6.95	7.3
LM3999	0	70	6.6	6.95	7.3

* = TYPICAL

NATIONAL

MAXIMUM TEMPERATURE COEFFICIENT (PPM/C)	OPERATING CURRENT (mA)		MAXIMUM DYNAMIC RESISTANCE (OHMS)	UNIT PRICE ($)
	MIN	MAX		
100*	0.5	20	1	6.75
10	0.6	15	2	5.49
20	0.6	15	2	1.99
50	0.6	15	2	1.49
100	0.6	15	2	0.65
24	0.4	10	1	1.10
48	0.6	10	2	1.10
24	0.4	10	1	2.00
48	0.6	10	2	2.00
20*	0.01	20	1	1.75
20*	0.01	20	1	2.45
20*	0.02	20	1	1.75
20*	0.02	20	1	2.25
2	0.5	10	1.5	5.45
1	0.5	10	1.5	7.90
5	0.6	10	2.2	2.75

VOLTAGE REFERENCES:

PART NUMBER	OPERATING TEMP (C)		OUTPUT VOLTAGE		
	MIN	MAX	TYP (V)	TOL (%)	TEMPCO (PPM/C)
REF-01	-55	125	10	0.5	25
REF-01A	-55	125	10	0.3	8.5
REF-01C	0	70	10	1	65
REF-01E	0	70	10	0.3	8.5
REF-01H	0	70	10	0.5	25
REF-02	-55	125	5	0.5	25
REF-02A	-55	125	5	0.3	8.5
REF-02C	0	70	5	1	65
REF-02D	0	70	5	2	250
REF-02E	0	70	5	0.3	8.5
REF-02H	0	70	5	0.5	25
REF-05A	-55	125	5	0.3	8.5
REF-05B	-55	125	5	0.5	25
REF-10A	-55	125	10	0.3	8.5
REF-10B	-55	125	10	0.5	25

PRECISION MONOLITHICS INCORPORATED (PMI)

MAXIMUM OUTPUT CURRENT (mA)	MAX REGULATION		MAXIMUM INPUT VOLTAGE (V)	UNIT PRICE ($)
	LINE (%/V)	LOAD (%/mA)		
10	0.01	0.01	40	19.13
10	0.01	0.008	40	38.25
8	0.015	0.015	30	3.75
10	0.01	0.008	40	9.60
10	0.01	0.01	40	4.88
10	0.01	0.01	40	19.13
10	0.01	0.01	40	38.25
8	0.015	0.015	30	3.75
4	0.04	0.04	30	2.10
10	0.01	0.01	40	9.60
10	0.01	0.01	40	4.88
10	0.01	0.01	40	75.00
10	0.01	0.01	40	37.50
10	0.01	0.01	40	75.00
10	0.01	0.01	40	37.50

Appendix D
Voltage Regulators

PART NUMBER	OPERATING TEMP (C)		OUTPUT VOLTAGE (V)		
	MIN	MAX	MIN	TYP	MAX
UA105	-55	125	4.5	ADJ	40
UA109	-55	150	4.7	5.05	5.3
UA117	-55	150	1.2	ADJ	37
UA305	0	70	4.5	ADJ	40
UA305A	0	70	4.5	ADJ	40
UA317	0	125	1.2	ADJ	37
UA376	0	70	5	ADJ	37
UA309	0	125	4.8	5.05	5.2
UA723	-55	125	2	ADJ	37
UA723C	0	70	2	ADJ	37
UA78G	-55	150	5	ADJ	30
UA78GC	0	125	5	ADJ	30
UA78HGA	-55	150	5	ADJ	24
UA78HGAC	0	125	5	ADJ	24
UA78H05	-55	150	4.85	5.0	5.25
UA78H05A	-55	150	4.85	5.0	5.25
UA78H05AC	0	150	4.85	5.0	5.25
UA78H05C	0	150	4.85	5.0	5.25

FAIRCHILD

MAXIMUM OUTPUT CURRENT (A)	MAX REGULATION		MAXIMUM INPUT VOLTAGE (V)	UNIT PRICE ($)
	LINE (%/V)	LOAD (%/A)		
0.012	0.03	4.2	50	2.00
1.5	0.055	1.32	25	6.65
1.5	0.05	0.2		7.98
0.012	0.03	4.2	50	0.76
0.045	0.03	4.4	50	1.79
1.5	0.07	0.33		1.52
0.025	0.03	8	40	0.57
1.5	0.055	1.32	25	2.09
0.05	0.033	3	40	1.71
0.05	0.033	4	40	0.49
1.5	0.167	2	40	5.70
1.5	0.167	2	40	1.14
5	0.057	0.2	27.5	47.92
5	0.057	0.2	27.5	9.90
5	0.061	0.2	25	43.23
5	0.057	0.2	25	47.92
5	0.057	0.2	25	8.23
5	0.061	0.2	25	7.81

VOLTAGE REGULATORS:

PART NUMBER	OPERATING TEMP (C) MIN	MAX	OUTPUT VOLTAGE (V) MIN	TYP	MAX
UA78H12A	-55	150	11.5	12	12.5
UA78H12AC	0	150	11.5	12	12.5
UA78L05C	0	125	4.8	5.0	5.2
UA78L09C	0	125	8.64	9.0	9.36
UA78L12C	0	125	11.5	12.0	12.5
UA78L15C	0	125	14.4	15.0	15.6
UA78L62C	0	125	5.95	6.2	6.45
UA78L82C	0	125	7.87	8.2	8.53
UA78MGC	0	125	5	ADJ	30
UA78M05	-55	150	4.8	5.0	5.2
UA78M05C	0	125	4.8	5.0	5.2
UA78M06	-55	150	5.75	6.0	6.25
UA78M06C	0	125	5.75	6.0	6.25
UA78M08	-55	150	7.7	8.0	8.3
UA78M08C	0	125	7.7	8.0	8.3
UA78M12	-55	150	11.5	12.0	12.5
UA78M12C	0	125	11.5	12.0	12.5

FAIRCHILD

MAXIMUM OUTPUT CURRENT (A)	MAX REGULATION LINE (%/V)	LOAD (%/A)	MAXIMUM INPUT VOLTAGE (V)	UNIT PRICE ($)
5	0.111	0.2	25	47.92
5	0.111	0.2	25	8.75
0.1	0.167	15	20	0.40
0.1	0.152	12.5	24	0.40
0.1	0.152	10.4	27	0.40
0.1	0.167	12.5	30	0.40
0.1	0.183	16.1	20	0.40
0.1	0.139	12.2	23	0.40
0.5	0.167	2	40	0.95
0.5	0.042	2.5	25	3.80
0.5	0.059	5	25	0.76
0.5	0.045	2.5	25	3.80
0.5	0.052	5	25	0.76
0.5	0.042	2.5	25	3.80
0.5	0.044	5	25	0.76
0.5	0.028	2.5	30	3.80
0.5	0.030	5	30	0.76

VOLTAGE REGULATORS:

PART NUMBER	OPERATING TEMP (C) MIN	MAX	OUTPUT VOLTAGE (V) MIN	TYP	MAX
UA78M15	-55	150	14.4	15.0	15.6
UA78M15C	0	125	14.4	15.0	15.6
UA78M24	-55	150	23	24	25
UA78M24C	0	125	23	24	25
UA78P05	-55	150	4.85	5.0	5.25
UA78P05C	0	150	4.85	5.0	5.25
UA7805	-55	150	4.8	5.0	5.2
UA7805C	0	125	4.8	5.0	5.2
UA7806C	0	125	5.75	6.0	6.25
UA7808	-55	150	7.7	8.0	8.3
UA7808C	0	125	7.7	8.0	8.3
UA7812	-55	150	11.5	12.0	12.5
UA7812C	0	125	11.5	12.0	12.5
UA7815	-55	150	14.4	15.0	15.6
UA7815C	0	125	14.4	15.0	15.6
UA7818	-55	150	17.3	18.0	18.7
UA7818C	0	125	17.3	18.0	18.7

FAIRCHILD

MAXIMUM OUTPUT CURRENT (A)	MAX REGULATION LINE (%/V)	LOAD (%/A)	MAXIMUM INPUT VOLTAGE (V)	UNIT PRICE ($)
0.5	0.025	2.5	30	3.80
0.5	0.033	5	30	0.76
0.5	0.021	2.5	38	5.70
0.5	0.021	5	38	0.76
10	0.059	0.16	25	58.33
10	0.059	0.16	25	14.38
1.5	0.125	1	25	5.70
1.5	0.25	2	25	0.86
1.5	0.25	2	25	0.86
1.5	0.083	1	25	5.70
1.5	0.167	2	25	0.86
1.5	0.083	1	30	5.70
1.5	0.167	2	30	0.86
1.5	0.083	1	30	5.70
1.5	0.167	2	30	0.86
1.5	0.083	1	33	5.70
1.5	0.167	2	33	0.86

VOLTAGE REGULATORS:

PART NUMBER	OPERATING TEMP (C) MIN	MAX	OUTPUT VOLTAGE (V) MIN	TYP	MAX
UA7824	-55	150	23	24	25
UA7824C	0	125	23	24	25
UA7885C	0	125	8.15	8.5	8.85
UA79G	-55	150	-2.23	ADJ	-30
UA79GC	0	125	-2.23	ADJ	-30
UA79HG	-55	150	-2.11	ADJ	-24
UA79HGC	0	150	-2.11	ADJ	-24
UA79MGC	0	125	-2.23	ADJ	-30
UA79M05	-55	150	-4.8	-5.0	-5.2
UA79M05C	0	125	-4.8	-5.0	-5.2
UA79M08	-55	150	-7.7	-8.0	-8.3
UA79M08C	0	125	-7.7	-8.0	-8.3
UA79M12	-55	150	-11.5	-12.0	-12.5
UA79M12C	0	125	-11.5	-12.0	-12.5
UA79M15	-55	150	-14.4	-15.0	-15.6
UA79M15C	0	125	-14.4	-15.0	-15.6

FAIRCHILD

MAXIMUM OUTPUT CURRENT (A)	MAX REGULATION LINE (%/V)	LOAD (%/A)	MAXIMUM INPUT VOLTAGE (V)	UNIT PRICE ($)
1.5	0.083	1	38	5.70
1.5	0.167	2	38	0.86
1.5	0.167	2	25	1.52
1.5	0.167	2	-40	5.70
1.5	0.167	2	-40	1.14
5	0.03	0.2	-40	58.33
5	0.03	0.2	-40	16.25
0.5	0.167	2	-40	0.95
0.5	0.060	4	-25	3.80
0.5	0.060	4	-25	0.76
0.5	0.063	4	-25	3.80
0.5	0.063	4	-25	0.76
0.5	0.042	4	-30	3.80
0.5	0.042	4	-30	0.76
0.5	0.033	3.2	-30	3.80
0.5	0.033	3.2	-30	0.76

VOLTAGE REGULATORS:

PART NUMBER	OPERATING TEMP (C)		OUTPUT VOLTAGE (V)		
	MIN	MAX	MIN	TYP	MAX
UA7905	-55	150	-4.8	-5.0	-5.2
UA7905C	0	125	-4.8	-5.0	-5.2
UA7908	-55	150	-7.7	-8.0	-8.3
UA7908C	0	125	-7.7	-8.0	-8.3
UA7912	-55	150	-11.5	-12.0	-12.5
UA7912C	0	125	-11.5	-12.0	-12.5
UA7915	-55	150	-14.4	-15.0	-15.6
UA7915C	0	125	-14.4	-15.0	-15.6

FAIRCHILD

MAXIMUM OUTPUT CURRENT (A)	MAX REGULATION LINE (%/V)	LOAD (%/A)	MAXIMUM INPUT VOLTAGE (V)	UNIT PRICE ($)
1.5	0.125	1	-25	5.70
1.5	0.25	2	-25	0.86
1.5	0.083	1	-25	5.70
1.5	0.167	2	-30	0.86
1.5	0.083	1	-30	5.70
1.5	0.167	2	-30	0.86
1.5	0.083	1	-30	5.70
1.5	0.167	2	-30	0.86

PART NUMBER	OPERATING TEMP (C) MIN	MAX	OUTPUT VOLTAGE (V) MIN	TYP	MAX
MC1463G	0	70	-3.8	ADJ	-32
MC1463R	0	70	-3.8	ADJ	-32
MC1468	0	70	±14.5	±15	±15.5
MC1469G	0	70	2.5	ADJ	32
MC1469R	0	70	2.5	ADJ	32
MC1563G	-55	125	-3.6	ADJ	-37
MC1563R	-55	125	-3.6	ADJ	-37
MC1568	-55	125	±14.8	±15	±15.2
MC1569G	-55	125	2.5	ADJ	37
MC1569R	-55	125	2.5	ADJ	37
MC1723	-55	125	2	ADJ	37
MC1723C	0	70	2	ADJ	37
MC78L05AC	0	125	4.8	5.0	5.2
MC78L05C	0	125	4.6	5.0	5.4
MC78L08AC	0	125	7.7	8.0	8.3
MC78L08C	0	125	7.36	8.0	8.64
MC78L12AC	0	125	11.5	12	12.5

MOTOROLA

MAXIMUM OUTPUT CURRENT (A)	MAX REGULATION		MAXIMUM INPUT VOLTAGE (V)	UNIT PRICE ($)
	LINE (%/V)	LOAD (%/A)		
0.2	0.03	2.6	−35	4.32
0.5	0.03	1	−35	7.70
0.05	0.006	0.013	±30	5.33
0.2	0.03	2.6	35	4.30
0.5	0.03	1	35	5.65
0.2	0.015	2.6	−40	8.08
0.5	0.015	1	−40	13.61
0.05	0.006	0.013	±30	6.08
0.2	0.015	2.6	40	6.88
0.5	0.015	1	40	10.49
0.05	0.033	3	40	2.00
0.05	0.033	4	40	0.54
0.1	0.167	15	20	0.56
0.1	0.25	15	20	0.52
0.1	0.13	12.5	23	0.56
0.1	0.156	12.5	23	0.52
0.1	0.152	10.4	27	0.56

VOLTAGE REGULATORS:

PART NUMBER	OPERATING TEMP (C)		OUTPUT VOLTAGE (V)		
	MIN	MAX	MIN	TYP	MAX
MC78L12C	0	125	11.1	12	12.9
MC78L15AC	0	125	14.4	15	15.6
MC78L15C	0	125	13.8	15	16.2
MC78L18AC	0	125	17.3	18	18.7
MC78L18C	0	125	16.6	18	19.4
MC78L24AC	0	125	23	24	25
MC78L24C	0	125	22.1	24	25.9
MC78M05B	-40	125	4.8	5.0	5.2
MC78M05C	0	125	4.8	5.0	5.2
MC78M06B	-40	125	5.75	6.0	6.25
MC78M06C	0	125	5.75	6.0	6.25
MC78M08B	-40	125	7.7	8.0	8.3
MC78M08C	0	125	7.7	8.0	8.3
MC78M12B	-40	125	11.5	12	12.5
MC78M12C	0	125	11.5	12	12.5
MC78M15B	-40	125	14.4	15	15.6
MC78M15C	0	125	14.4	15	15.6

MOTOROLA

MAXIMUM OUTPUT CURRENT (A)	MAX REGULATION		MAXIMUM INPUT VOLTAGE (V)	UNIT PRICE ($)
	LINE (%/V)	LOAD (%/A)		
0.1	0.152	10.4	27	0.52
0.1	0.167	12.5	30	0.56
0.1	0.167	12.5	30	0.52
0.1	0.127	11.8	33	0.56
0.1	0.139	11.8	33	0.52
0.1	0.125	10.4	38	0.56
0.1	0.125	10.4	38	0.52
0.5	0.059	5	25	1.23
0.5	0.059	5	25	0.86
0.5	0.052	5	25	1.23
0.5	0.052	5	25	0.86
0.5	0.045	5	25	1.23
0.5	0.045	5	25	0.86
0.5	0.069	5	30	1.23
0.5	0.069	5	30	0.86
0.5	0.033	5	30	1.23
0.5	0.033	5	30	0.86

PART NUMBER	OPERATING TEMP (C) MIN	MAX	OUTPUT VOLTAGE (V) MIN	TYP	MAX
MC78M18B	-40	125	17.3	18	18.7
MC78M18C	0	125	17.3	18	18.7
MC78M20B	-40	125	19.2	20	20.8
MC78M20C	0	125	19.2	20	20.8
MC78M24B	-40	125	23	24	25
MC78M24C	0	125	23	24	25
MC78T05	-55	150	4.8	5.0	5.2
MC78T05A	-55	150	4.9	5.0	5.1
MC78T05AC	0	125	4.9	5.0	5.1
MC78T05C	0	125	4.8	5.0	5.2
MC78T06	-55	150	5.75	6.0	6.25
MC78T06C	0	125	5.75	6.0	6.25
MC78T08	-55	150	7.7	8.0	8.3
MC78T08C	0	125	7.7	8.0	8.3
MC78T12	-55	150	11.5	12	12.5
MC78T12A	-55	150	11.75	12	12.25
MC78T12AC	0	125	11.75	12	12.25
MC78T12C	0	125	11.5	12	12.5

MOTOROLA

MAXIMUM OUTPUT CURRENT (A)	MAX REGULATION		MAXIMUM INPUT VOLTAGE (V)	UNIT PRICE ($)
	LINE (%/V)	LOAD (%/A)		
0.5	0.031	5	33	1.23
0.5	0.031	5	33	0.86
0.5	0.023	5	35	1.23
0.5	0.023	5	35	0.86
0.5	0.021	5	38	1.23
0.5	0.021	5	38	0.86
3	0.125	0.2	35	26.50
3	0.05	0.17	35	31.25
3	0.05	0.17	35	5.70
3	0.125	0.2	35	5.15
3	0.125	0.17	35	26.50
3	0.125	0.17	35	6.85
3	0.073	0.13	35	26.50
3	0.073	0.13	35	5.15
3	0.063	0.08	35	26.50
3	0.025	0.07	35	31.25
3	0.025	0.07	35	5.70
3	0.063	0.08	35	5.15

VOLTAGE REGULATORS:

PART NUMBER	OPERATING TEMP (C)		OUTPUT VOLTAGE (V)		
	MIN	MAX	MIN	TYP	MAX
MC78T15	-55	150	14.4	15	15.6
MC78T15A	-55	150	14.7	15	15.3
MC78T15AC	0	125	14.7	15	15.3
MC78T15C	0	125	14.4	15	15.6
MC78T18	-55	150	17.3	18	18.7
MC78T18C	0	125	17.3	18	18.7
MC78T24	-55	150	23	24	25
MC78T24C	0	125	23	24	25
MC7805	-55	150	4.8	5.0	5.2
MC7805A	-55	150	4.9	5.0	5.1
MC7805AC	0	125	4.9	5.0	5.1
MC7805B	-40	125	4.8	5.0	5.2
MC7805C	0	125	4.8	5.0	5.2
MC7806	-55	150	5.75	6.0	6.25
MC7806A	-55	150	5.88	6.0	6.12
MC7806AC	0	125	5.88	6.0	6.12
MC7806B	-40	125	5.75	6.0	6.25
MC7806C	0	125	5.75	6.0	6.25

MOTOROLA

MAXIMUM OUTPUT CURRENT (A)	MAX REGULATION		MAXIMUM INPUT VOLTAGE (V)	UNIT PRICE ($)
	LINE (%/V)	LOAD (%/A)		
3	0.061	0.07	40	26.50
3	0.024	0.06	40	31.25
3	0.024	0.06	40	5.70
3	0.061	0.07	40	5.15
3	0.074	0.06	40	26.50
3	0.074	0.06	40	5.15
3	0.063	0.04	40	26.50
3	0.063	0.04	40	5.70
1.5	0.125	1	25	11.69
1.5	0.02	0.6	25	13.58
1.5	0.125	2	25	1.14
1.5	0.25	2	25	1.27
1.5	0.25	2	25	0.95
1.5	0.125	1	25	11.69
1.5	0.02	0.6	25	13.58
1.5	0.125	2	25	1.14
1.5	0.25	2	25	1.27
1.5	0.25	2	25	0.95

PART NUMBER	OPERATING TEMP (C) MIN	MAX	OUTPUT VOLTAGE (V) MIN	TYP	MAX
MC7808	-55	150	7.7	8.0	8.3
MC7808A	-55	150	7.84	8.0	8.16
MC7808AC	0	125	7.84	8.0	8.16
MC7808B	-40	125	7.7	8.0	8.3
MC7808C	0	125	7.7	8.0	8.3
MC7812	-55	150	11.5	12.0	12.5
MC7812A	-55	150	11.75	12.0	12.25
MC7812AC	0	125	11.75	12.0	12.25
MC7812B	-40	125	11.5	12.0	12.5
MC7812C	0	125	11.5	12.0	12.5
MC7815	-55	150	14.4	15.0	15.6
MC7815A	-55	150	14.7	15.0	15.3
MC7815AC	0	125	14.7	15.0	15.3
MC7815B	-40	125	14.4	15.0	15.6
MC7815C	0	125	14.4	15.0	15.6
MC7818	-55	150	17.3	18.0	18.7
MC7818A	-55	150	17.64	18.0	18.36
MC7818AC	0	125	17.64	18.0	18.36

MOTOROLA

MAXIMUM OUTPUT CURRENT (A)	MAX REGULATION		MAXIMUM INPUT VOLTAGE (V)	UNIT PRICE ($)
	LINE (%/V)	LOAD (%/A)		
1.5	0.083	1	25	11.69
1.5	0.013	0.38	25	13.58
1.5	0.083	1.25	25	1.14
1.5	0.167	2	25	1.27
1.5	0.167	2	25	0.95
1.5	0.083	1	30	11.69
1.5	0.013	0.25	30	13.58
1.5	0.083	0.83	30	1.14
1.5	0.167	2	30	1.27
1.5	0.167	2	30	0.95
1.5	0.083	1	30	11.69
1.5	0.011	0.2	30	13.58
1.5	0.083	0.67	30	1.14
1.5	0.167	2	30	1.27
1.5	0.167	2	30	0.95
1.5	0.083	1	33	11.69
1.5	0.014	0.17	33	13.58
1.5	0.083	0.56	33	1.14

PART NUMBER	OPERATING TEMP (C)		OUTPUT VOLTAGE (V)		
	MIN	MAX	MIN	TYP	MAX
MC7818B	-40	125	17.3	18.0	18.7
MC7818C	0	125	17.3	18.0	18.7
MC7824	-55	150	23	24	25
MC7824A	-55	150	23.5	24.0	24.5
MC7824AC	0	125	23.5	24.0	24.5
MC7824B	-40	125	23	24	25
MC7824C	0	125	23	24	25
MC79L03AC	0	125	-2.88	-3.0	-3.12
MC79L03C	0	125	-2.76	-3.0	-3.24
MC79L05AC	0	125	-4.8	-5.0	-5.2
MC79L05C	0	125	-4.6	-5.0	-5.4
MC79L12AC	0	125	-11.5	-12	-12.5
MC79L12C	0	125	-11.1	-12	-12.9
MC79L15AC	0	125	-14.4	-15	-15.6
MC79L15C	0	125	-13.8	-15	-16.2

MOTOROLA

MAXIMUM OUTPUT CURRENT (A)	MAX REGULATION		MAXIMUM INPUT VOLTAGE (V)	UNIT PRICE ($)
	LINE (%/V)	LOAD (%/A)		
1.5	0.167	2	33	1.27
1.5	0.167	2	33	0.95
1.5	0.083	1	38	11.69
1.5	0.013	0.13	38	13.58
1.5	0.083	0.42	38	1.14
1.5	0.167	2	38	1.27
1.5	0.167	2	38	0.95
0.1	0.111	30	-20	0.99
0.1	0.167	30	-20	0.80
0.1	0.167	15	-20	0.82
0.1	0.25	15	-20	0.73
0.1	0.152	10.4	-27	0.82
0.1	0.152	10.4	-27	0.73
0.1	0.167	12.5	-30	0.82
0.1	0.167	12.5	-30	0.73

PART NUMBER	OPERATING TEMP (C)		OUTPUT VOLTAGE (V)		
	MIN	MAX	MIN	TYP	MAX
MC79L18AC	0	125	-17.3	-18	-18.7
MC79L18C	0	125	-16.6	-18	-19.4
MC79L24AC	0	125	-23	-24	-25
MC79L24C	0	125	-22.1	-24	-25.9
MC7902C	0	125	-1.92	-2.0	-2.08
MC7905C	0	125	-4.8	-5.0	-5.2
MC7905.2C	0	125	-5.0	-5.2	-5.4
MC7906C	0	125	-5.75	-6.0	-6.25
MC7908C	0	125	-7.7	-8.0	-8.3
MC7912C	0	125	-11.5	-12	-12.5
MC7915C	0	125	-14.4	-15	-15.6
MC7918C	0	125	-17.3	-18	-18.7
MC7924C	0	125	-23	-24	-25

MOTOROLA

MAXIMUM OUTPUT CURRENT (A)	MAX REGULATION		MAXIMUM INPUT VOLTAGE (V)	UNIT PRICE ($)
	LINE (%/V)	LOAD (%/A)		
0.1	0.127	11.8	-33	0.88
0.1	0.139	11.8	-33	0.80
0.1	0.125	10.4	-38	0.88
0.1	0.125	10.4	-38	0.80
1.5	0.25	6	-25	1.50
1.5	0.25	2	-25	0.95
1.5	0.25	2	-25	1.68
1.5	0.25	2	-25	0.95
1.5	0.167	2	-25	0.95
1.5	0.167	2	-30	0.95
1.5	0.167	2	-30	0.95
1.5	0.167	2	-33	0.95
1.5	0.167	2	-38	0.95

PART NUMBER	OPERATING TEMP (C) MIN	MAX	OUTPUT VOLTAGE (V) MIN	TYP	MAX
LM104	-55	125	-0.015	ADJ	-40
LM105	-55	125	4.5	ADJ	40
LM109H	-55	150	4.7	5.05	5.3
LM109K	-55	150	4.7	5.05	5.3
LM117H	-55	150	1.2	ADJ	37
LM117HVH	-55	150	1.2	ADJ	57
LM117HVK	-55	150	1.2	ADJ	57
LM117K	-55	150	1.2	ADJ	37
LM120H-5	-55	150	-4.9	-5.0	-5.1
LM120H-12	-55	150	-11.7	-12.0	-12.3
LM120H-15	-55	150	-14.7	-15.0	-15.3
LM120K-5	-55	150	-4.9	-5.0	-5.1
LM120K-12	-55	150	-11.7	-12.0	-12.3
LM120K-15	-55	150	-14.7	-15.0	-15.3
LM123	-55	150	4.7	5	5.3
LM125	-55	125	±14.8	±15	±15.2
LM126	-55	125	±11.8	±12	±12.2

NATIONAL

MAXIMUM OUTPUT CURRENT (A)	MAX REGULATION LINE (%/V)	LOAD (%/A)	MAXIMUM INPUT VOLTAGE (V)	UNIT PRICE ($)
0.02			-50	9.65
0.012	0.03	4.2	50	3.85
0.5	0.055	2	25	13.00
1.5	0.055	1.32	25	15.40
0.5	0.02	0.6		15.00
0.5	0.02	0.6		30.00
1.5	0.02	0.2		30.00
1.5	0.02	0.2		16.00
0.5	0.028	2	-25	14.20
0.5	0.005	0.42	-32	14.20
0.5	0.004	0.33	-35	14.20
1.5	0.028	1	-25	14.95
1.5	0.005	0.44	-32	14.95
1.5	0.004	0.36	-35	14.95
3	0.067	0.67	15	30.75
0.05	0.006	1.33	±30	10.45
0.05	0.006	1.67	±30	10.45

PART NUMBER	OPERATING TEMP (C)		OUTPUT VOLTAGE (V)		
	MIN	MAX	MIN	TYP	MAX
LM137H	-55	150	-1.2	ADJ	-37
LM137HVH	-55	150	-1.2	ADJ	-47
LM137HVK	-55	150	-1.2	ADJ	-47
LM137K	-55	150	-1.2	ADJ	-37
LM138	-55	150	1.2	ADJ	32
LM140-5	-55	125	4.8	5	5.2
LM140-12	-55	125	11.5	12	12.5
LM140-15	-55	125	14.4	15	15.6
LM140A-5	-55	125	4.9	5	5.1
LM140A-12	-55	125	11.75	12	12.25
LM140A-15	-55	125	14.7	15	15.3
LM140L-5	-55	125	4.9	5	5.1
LM140L-12	-55	125	11.75	12	12.25
LM140L-15	-55	125	14.7	15	15.3
LM145-5	-55	150	-4.9	-5	-5.1
LM145-5.2	-55	150	-5.1	-5.2	-5.3
LM150	-55	150	1.2	ADJ	33

NATIONAL

MAXIMUM OUTPUT CURRENT (A)	MAX REGULATION		MAXIMUM INPUT VOLTAGE (V)	UNIT PRICE ($)
	LINE (%/V)	LOAD (%/A)		
0.5	0.02	1		18.00
0.5	0.02	1		28.85
1.5	0.02	0.33		33.00
1.5	0.02	0.33		19.60
5	0.01	0.06		48.75
1.5	0.079	0.67	20	8.45
1.5	0.081	0.67	27	8.45
1.5	0.081	0.67	30	8.45
1.5	0.016	0.33	20	13.45
1.5	0.012	0.18	27	13.45
1.5	0.012	0.16	30	13.45
0.1	0.034	8	25	3.00
0.1	0.035	6.7	30	3.00
0.1	0.037	6.7	30	3.00
3	0.024	0.5	-20	52.50
3	0.023	0.48	-20	52.50
3	0.01	0.1		34.60

VOLTAGE REGULATORS:

PART NUMBER	OPERATING TEMP (C)		OUTPUT VOLTAGE (V)		
	MIN	MAX	MIN	TYP	MAX
LM196	-55	150	1.25	ADJ	15
LM204	-25	85	-0.015	ADJ	-40
LM205	-25	85	4.5	ADJ	40
LM209H	-25	150	4.7	5.05	5.3
LM209K	-25	150	4.7	5.05	5.3
LM217H	-25	150	1.2	ADJ	37
LM217HVK	-25	150	1.2	ADJ	57
LM217K	-25	150	1.2	ADJ	37
LM223	-25	150	4.7	5	5.3
LM237H	-25	150	-1.2	ADJ	-37
LM237HVK	-25	150	-1.2	ADJ	-47
LM237K	-25	150	-1.2	ADJ	-37
LM238	-25	150	1.2	ADJ	32
LM250	-25	150	1.2	ADJ	33
LM304	0	70	-0.035	ADJ	-30
LM305	0	70	4.5	ADJ	30
LM305A	0	70	4.5	ADJ	40

NATIONAL

MAXIMUM OUTPUT CURRENT (A)	MAX REGULATION LINE (%/V)	LOAD (%/A)	MAXIMUM INPUT VOLTAGE (V)	UNIT PRICE ($)
10	0.01	0.1		90.00
0.02			-50	6.90
0.012	0.03	4.2	50	4.70
0.5	0.055	2	25	11.25
1.5	0.055	1.32	25	5.75
0.5	0.02	0.6		11.00
1.5	0.02	0.2		20.49
1.5	0.02	0.2		11.50
3	0.067	0.67	15	15.35
0.5	0.02	1		12.45
1.5	0.02	0.33		19.85
1.5	0.02	0.33		13.85
5	0.01	0.06		22.75
3	0.01	0.1		17.28
0.02			-40	3.50
0.012	0.03	4.2	40	1.00
0.045	0.03	4.4	50	3.40

PART NUMBER	OPERATING TEMP (C) MIN	MAX	OUTPUT VOLTAGE (V) MIN	TYP	MAX
LM309H	0	125	4.8	5.05	5.2
LM309K	0	125	4.8	5.05	5.2
LM317H	0	125	1.2	ADJ	37
LM317HVH	0	125	1.2	ADJ	57
LM317HVK	0	125	1.2	ADJ	57
LM317K	0	125	1.2	ADJ	37
LM317LZ	0	125	1.2	ADJ	37
LM317M	0	125	1.2	ADJ	37
LM317T	0	125	1.2	ADJ	37
LM320H-5	0	125	-4.8	-5.0	-5.2
LM320H-12	0	125	-11.6	-12.0	-12.4
LM320H-15	0	125	-14.6	-15.0	-15.4
LM320K-5	0	125	-4.8	-5.0	-5.2
LM320K-12	0	125	-11.6	-12.0	-12.4
LM320K-15	0	125	-14.6	-15.0	-15.4
LM320L-5	0	70	-4.8	-5	-5.2
LM320L-12	0	70	-11.5	-12	-12.5

NATIONAL

MAXIMUM OUTPUT CURRENT (A)	MAX REGULATION		MAXIMUM INPUT VOLTAGE (V)	UNIT PRICE ($)
	LINE (%/V)	LOAD (%/A)		
0.5	0.055	2	25	3.10
1.5	0.055	1.32	25	2.20
0.5	0.04	1		4.18
0.5	0.04	1		6.20
1.5	0.04	0.33		6.20
1.5	0.04	0.33		2.88
0.1	0.04	5		0.75
0.5	0.04	1		1.05
1.5	0.04	0.33		1.50
0.5	0.044	2	-25	5.50
0.5	0.009	0.67	-32	5.50
0.5	0.007	0.53	-35	5.50
1.5	0.044	1.33	-25	3.55
1.5	0.009	0.44	-32	3.55
1.5	0.007	0.36	-35	3.55
0.1	0.094	10	-20	1.00
0.1	0.030	8.3	-27	1.00

VOLTAGE REGULATORS:

PART NUMBER	OPERATING TEMP (C)		OUTPUT VOLTAGE (V)		
	MIN	MAX	MIN	TYP	MAX
LM320L-15	0	70	-14.4	-15	-15.6
LM320ML-5	0	70	-4.8	-5	-5.2
LM320ML-12	0	70	-11.5	-12	-12.5
LM320ML-15	0	70	-14.4	-15	-15.6
LM320MP-5	0	125	-4.8	-5.0	-5.2
LM320MP-12	0	125	-11.5	-12.0	-12.5
LM320MP-15	0	125	-14.4	-15.0	-15.6
LM320T-5	0	125	-4.8	-5.0	-5.2
LM320T-12	0	125	-11.6	-12.0	-12.4
LM320T-15	0	125	-14.5	-15.0	-15.5
LM323	0	125	4.7	5	5.3
LM325	0	70	±14.5	±15	±15.5
LM325A	0	70	±14.8	±15	±15.2
LM326	0	70	±11.5	±12	±12.5
LM330	0	70	4.8	5	5.2
LM337H	0	125	-1.2	ADJ	-37
LM337HVH	0	125	-1.2	ADJ	-47

NATIONAL

MAXIMUM OUTPUT CURRENT (A)	MAX REGULATION LINE (%/V)	LOAD (%/A)	MAXIMUM INPUT VOLTAGE (V)	UNIT PRICE ($)
0.1	0.024	8.3	-30	1.00
0.25	0.056	4	-25	1.20
0.25	0.022	4	-30	1.20
0.25	0.022	4	-30	1.20
0.5	0.046	4	-25	1.85
0.5	0.011	1.67	-32	1.85
0.5	0.011	1.33	-35	1.85
1.5	0.046	1.33	-25	2.49
1.5	0.010	0.44	-32	2.49
1.5	0.008	0.36	-35	2.49
3	0.067	0.67	15	6.95
0.05	0.006	1.33	±30	3.30
0.05	0.006	1.33	±30	5.20
0.05	0.006	1.67	±30	3.30
0.15	0.056	6.7	26	1.00
0.5	0.04	2		4.50
0.5	0.04	2		7.95

PART NUMBER	OPERATING TEMP (C) MIN	MAX	OUTPUT VOLTAGE (V) MIN	TYP	MAX
LM337HVK	0	125	-1.2	ADJ	-47
LM337K	0	125	-1.2	ADJ	-37
LM337LZ	0	125	-1.2	ADJ	-37
LM337M	0	125	-1.2	ADJ	-37
LM337T	0	125	-1.2	ADJ	-37
LM338	0	125	1.2	ADJ	32
LM340-5	0	70	4.8	5	5.2
LM340-12	0	70	11.5	12	12.5
LM340-15	0	70	14.4	15	15.6
LM340A-5	0	70	4.9	5	5.1
LM340A-12	0	70	11.75	12	12.25
LM340A-15	0	70	14.7	15	15.3
LM340AT-5	0	70	4.9	5	5.1
LM340AT-12	0	70	11.75	12	12.25
LM340AT-15	0	70	14.7	15	15.3
LM340L-5	0	70	4.9	5	5.1
LM340L-12	0	70	11.75	12	12.25

NATIONAL

MAXIMUM OUTPUT CURRENT (A)	MAX REGULATION LINE (%/V)	LOAD (%/A)	MAXIMUM INPUT VOLTAGE (V)	UNIT PRICE ($)
1.5	0.04	0.67		10.00
1.5	0.04	0.67		5.20
0.1	0.04	10		1.05
0.5	0.04	2		1.15
1.5	0.04	0.67		1.95
5	0.03	0.1		7.60
1.5	0.079	0.67	20	2.10
1.5	0.081	0.67	27	2.10
1.5	0.081	0.67	30	2.10
1.5	0.016	0.33	20	3.65
1.5	0.012	0.18	27	3.65
1.5	0.012	0.16	30	3.65
1.5	0.016	0.33	20	1.50
1.5	0.012	0.18	27	1.50
1.5	0.012	0.16	30	1.50
0.1	0.034	8	25	0.75
0.1	0.035	6.7	30	0.75

PART NUMBER	OPERATING TEMP (C) MIN	MAX	OUTPUT VOLTAGE (V) MIN	TYP	MAX
LM340L-15	0	70	14.7	15	15.3
LM340T-5	0	70	4.8	5	5.2
LM340T-12	0	70	11.5	12	12.5
LM340T-15	0	70	14.4	15	15.6
LM341-5	0	70	4.8	5	5.2
LM341-12	0	70	11.5	12	12.5
LM341-15	0	70	14.4	15	15.6
LM342-5	0	70	4.8	5	5.2
LM342-12	0	70	11.5	12	12.5
LM342-15	0	70	14.4	15	15.6
LM345-5	0	125	-4.8	-5	-5.2
LM345-5.2	0	125	-5	-5.2	-5.4
LM350K	0	125	1.2	ADJ	33
LM350T	0	125	1.2	ADJ	33
LM376	0	70	5	ADJ	37
LM396	0	125	1.25	ADJ	15
LM723	-55	125	2	ADJ	37

NATIONAL

MAXIMUM OUTPUT CURRENT (A)	MAX REGULATION LINE (%/V)	LOAD (%/A)	MAXIMUM INPUT VOLTAGE (V)	UNIT PRICE ($)
0.1	0.037	6.7	30	0.75
1.5	0.079	0.67	20	0.95
1.5	0.081	0.67	27	0.95
1.5	0.081	0.67	30	0.95
0.5	0.11	4	25	0.75
0.5	0.13	4	30	0.75
0.5	0.16	4	30	0.75
0.25	0.062	4	25	0.68
0.25	0.054	4	30	0.68
0.25	0.054	4	30	0.68
3	0.04	0.67	-20	9.35
3	0.038	0.64	-20	9.35
3	0.03	0.16		5.50
3	0.03	0.16		4.75
0.025	0.03	8	40	0.56
10	0.02	0.1		18.50
0.05	0.033	3	40	3.20

PART NUMBER	OPERATING TEMP (C)		OUTPUT VOLTAGE (V)		
	MIN	MAX	MIN	TYP	MAX
LM723C	0	70	2	ADJ	37
LM2930-5	-40	85	4.5	5	5.5
LM2930-8	-40	85	7.2	8	8.8
LM2931A-5	-40	85	4.75	5	5.25
LM2931T	-40	85	3	ADJ	24
LM7805C	0	70	4.8	5	5.2
LM7812C	0	70	11.5	12	12.5
LM7815C	0	70	14.4	15	15.6
LM78L05AC	0	70	4.8	5	5.2
LM78L12AC	0	70	11.5	12	12.5
LM78L15AC	0	70	14.4	15	15.6
LM78L05C	0	70	4.6	5	5.4
LM78L12C	0	70	11.1	12	12.9
LM78L15C	0	70	13.8	15	16.2
LM78M05C	0	70	4.8	5	5.2
LM78M12C	0	70	11.5	12	12.5
LM78M15C	0	70	14.4	15	15.6

NATIONAL

MAXIMUM OUTPUT CURRENT (A)	MAX REGULATION		MAXIMUM INPUT VOLTAGE (V)	UNIT PRICE ($)
	LINE (%/V)	LOAD (%/A)		
0.05	0.033	4	40	0.64
0.15	0.071	6.7	26	1.15
0.15	0.095	4.2	26	1.15
0.15	0.029	6.7	26	1.15
0.15	0.15	6.7	26	1.75
1.5	0.079	0.67	20	0.95
1.5	0.081	0.67	27	0.95
1.5	0.081	0.67	30	0.95
0.1	0.09	12	20	0.43
0.1	0.083	8.3	27	0.43
0.1	0.093	10	30	0.43
0.1	0.25	12	20	0.43
0.1	0.17	8.3	27	0.43
0.1	0.17	10	30	0.43
0.5	0.11	4	25	0.75
0.5	0.13	4	30	0.75
0.5	0.16	4	30	0.75

PART NUMBER	OPERATING TEMP (C) MIN	MAX	OUTPUT VOLTAGE (V) MIN	TYP	MAX
LM7905C	0	70	-4.8	-5	-5.2
LM7912C	0	70	-11.5	-12	-12.5
LM7915C	0	70	-14.4	-15	-15.6
LM79L05AC	0	70	-4.8	-5	-5.2
LM79L12AC	0	70	-11.5	-12	-12.5
LM79L15AC	0	70	-14.4	-15	-15.6
LM79M05C	0	70	-4.8	-5	-5.2
LM79M12C	0	70	-11.5	-12	-12.5
LM79M15C	0	70	-14.4	-15	-15.6

NATIONAL

MAXIMUM OUTPUT CURRENT (A)	MAX REGULATION		MAXIMUM INPUT VOLTAGE (V)	UNIT PRICE ($)
	LINE (%/V)	LOAD (%/A)		
1.5	0.075	1.3	-25	1.08
1.5	0.042	1.1	-30	1.08
1.5	0.055	0.89	-30	1.08
0.1	0.095	10	-20	1.00
0.1	0.030	8.3	-27	1.00
0.1	0.024	8.3	-30	1.00
0.5	0.06	4	-25	1.19
0.5	0.025	4	-30	1.19
0.5	0.033	3.2	-30	1.19

VOLTAGE REGULATORS:

PART NUMBER	OPERATING TEMP (C)		OUTPUT VOLTAGE (V)		
	MIN	MAX	MIN	TYP	MAX
CA723	-55	125	2	ADJ	37
CA723C	0	70	2	ADJ	37
CA3085	-55	125	1.8	ADJ	26
CA3085A	-55	125	1.7	ADJ	36
CA3085B	-55	125	1.7	ADJ	46

VOLTAGE REGULATORS:

PART NUMBER	OPERATING TEMP (C)		OUTPUT VOLTAGE (V)		
	MIN	MAX	MIN	TYP	MAX
NE5553	0	125	±11.5	±12	±12.5
SA723C	-40	85	2	ADJ	37
SE5553	-55	150	±11.5	±12	±12.5

RCA

MAXIMUM OUTPUT CURRENT (A)	MAX REGULATION		MAXIMUM INPUT VOLTAGE (V)	UNIT PRICE ($)
	LINE (%/V)	LOAD (%/A)		
0.05	0.033	3	40	0.74
0.05	0.033	4	40	0.60
0.012	0.1	8.3	30	0.94
0.1	0.075	1.5	40	1.20
0.1	0.04	1.5	50	6.72

SIGNETICS

MAXIMUM OUTPUT CURRENT (A)	MAX REGULATION		MAXIMUM INPUT VOLTAGE (V)	UNIT PRICE ($)
	LINE (%/V)	LOAD (%/A)		
0.2	0.125	4.2	±30	1.50
0.05	0.033	4	40	0.76
0.2	0.25	8.4	±30	3.20

VOLTAGE REGULATORS:

PART NUMBER	OPERATING TEMP (C)		OUTPUT VOLTAGE (V)		
	MIN	MAX	MIN	TYP	MAX
TL317C	0	125	2	ADJ	37
TL780-05C	0	125	4.95	5	5.05
TL780-12C	0	125	11.88	12	12.12
TL780-15C	0	125	14.85	15	15.15
TL783C	0	125	1.25	ADJ	125

SWITCH-MODE POWER SUPPLY CONTROLLER IC'S:

PART NUMBER	OPERATING TEMP (°C)		NO. OF OUTPUTS	MAXIMUM OUTPUT CURRENT (A)	MAXIMUM INPUT VOLTAGE (V)
	MIN	MAX			
SH1605	0	70	1	5	30
SH1605SM	-55	125	1	5	30
UA494C	0	70	2	0.2	40
UA78S40C	0	70	1	1	40
UA78S40M	-55	125	1	1	40

TEXAS INSTRUMENTS

MAXIMUM OUTPUT CURRENT (A)	MAX REGULATION		MAXIMUM INPUT VOLTAGE (V)	UNIT PRICE ($)
	LINE (%/V)	LOAD (%/A)		
0.1	0.02	5		0.83
1.5	0.025	0.33	25	1.03
1.5	0.017	0.33	30	1.03
1.5	0.017	0.33	30	1.03
0.7	0.01	0.71		2.94

FAIRCHILD

REFERENCE VOLTAGE (V)			OSCILLATOR FREQ (kHz)		UNIT PRICE ($)
MIN	TYP	MAX	MIN	MAX	
	2.5		5	100	15.10
	2.5		5	100	75.00
4.75	5	5.25	1	300	2.19
1.18	1.245	1.31	0.2	500	2.79
1.18	1.245	1.31	0.2	500	8.84

SWITCH-MODE POWER SUPPLY CONTROLLER IC'S:

PART NUMBER	OPERATING TEMP (°C) MIN	MAX	NO. OF OUTPUTS	MAXIMUM OUTPUT CURRENT (A)	MAXIMUM INPUT VOLTAGE (V)
MC33063	-40	85	1	1	40
MC3420	0	70	2	0.04	30
MC34060	0	70	1	0.2	40
MC34063	0	70	1	1	40
MC3520	-55	125	2	0.04	30
MC35060	-55	125	1	0.2	40
MC35063	-55	125	1	1	40

SWITCH-MODE POWER SUPPLY CONTROLLER IC'S:

PART NUMBER	OPERATING TEMP (°C) MIN	MAX	NO. OF OUTPUTS	MAXIMUM OUTPUT CURRENT (A)	MAXIMUM INPUT VOLTAGE (V)
LH1605C	0	70	1	5	30
LM3524	0	70	2	0.05	40

MOTOROLA

REFERENCE VOLTAGE (V)			OSCILLATOR FREQ (kHz)		UNIT PRICE ($)
MIN	TYP	MAX	MIN	MAX	
1.18	1.25	1.32	0.1	100	4.17
7.4	7.8	8.2	5	100	6.79
4.75	5	5.25	1	200	1.50
1.18	1.25	1.32	0.1	100	2.75
7.6	7.8	8	5	100	18.15
4.75	5	5.25	1	200	11.75
1.18	1.25	1.32	0.1	100	9.65

NATIONAL

REFERENCE VOLTAGE (V)			OSCILLATOR FREQ (kHz)		UNIT PRICE ($)
MIN	TYP	MAX	MIN	MAX	
	2.5		5	100	13.60
4.6	5	5.4	1	350	3.20

SWITCH-MODE POWER SUPPLY CONTROLLER IC'S:

PART NUMBER	OPERATING TEMP (°C) MIN	MAX	NO. OF OUTPUTS	MAXIMUM OUTPUT CURRENT (A)	MAXIMUM INPUT VOLTAGE (V)
CA1524	-55	125	2	0.05	40
CA2524	0	70	2	0.05	40
CA3524	0	70	2	0.05	40

SWITCH-MODE POWER SUPPLY CONTROLLER IC'S:

PART NUMBER	OPERATING TEMP (°C) MIN	MAX	NO. OF OUTPUTS	MAXIMUM OUTPUT CURRENT (A)	MAXIMUM INPUT VOLTAGE (V)
NE5560	0	70	1	0.04	30
NE5561	0	70	1	0.02	24
SE5560	-55	125	1	0.04	30
SE5561	-55	125	1	0.02	22

RCA

REFERENCE VOLTAGE (V)			OSCILLATOR FREQ (kHz)		UNIT PRICE
MIN	TYP	MAX	MIN	MAX	($)
4.8	5	5.2	1	300	2.60
4.8	5	5.2	1	300	2.40
4.6	5	4.4	1	300	2.90

SIGNETICS

REFERENCE VOLTAGE (V)			OSCILLATOR FREQ (kHz)		UNIT PRICE
MIN	TYP	MAX	MIN	MAX	($)
3.57	3.72	3.95	0.05	100	2.99
3.57	3.75	3.96	0.05	100	1.20
3.69	3.72	3.81	0.05	100	4.12
3.69	3.75	3.84	0.05	100	2.20

SWITCH-MODE POWER SUPPLY CONTROLLER IC'S:

PART NUMBER	OPERATING TEMP (°C) MIN	MAX	NO. OF OUTPUTS	MAXIMUM OUTPUT CURRENT (A)	MAXIMUM INPUT VOLTAGE (V)
TL493C	0	70	2	0.2	40
TL494C	0	70	2	0.2	40
TL494I	-25	85	2	0.2	40
TL494M	-55	125	2	0.2	40
TL495C	0	70	2	0.2	40
TL497AC	0	70	1	0.5	12
TL497AI	-25	85	1	0.5	12
TL497AM	-55	125	1	0.5	12
TL593C	0	70	2	0.2	40
TL593I	-25	85	2	0.2	40
TL594C	0	70	2	0.2	40
TL594I	-25	85	2	0.2	40
TL594M	-55	125	2	0.2	40
TL595C	0	70	2	0.2	40
TL595I	-25	85	2	0.2	40

TEXAS INSTRUMENTS (TI)

REFERENCE VOLTAGE (V)			OSCILLATOR FREQ (kHz)		UNIT PRICE ($)
MIN	TYP	MAX	MIN	MAX	
4.75	5	5.25	1	300	4.60
4.75	5	5.25	1	300	2.30
4.75	5	5.25	1	300	3.96
4.75	5	5.25	1	300	18.00
4.75	5	5.25	1	300	2.88
1.08	1.2	1.32			2.19
1.14	1.2	1.26			5.18
1.14	1.2	1.26			10.00
4.95	5	5.05	1	300	2.88
4.95	5	5.05	1	300	4.60
4.95	5	5.05	1	300	2.07
4.95	5	5.05	1	300	4.03
4.95	5	5.05	1	300	18.00
4.95	5	5.05	1	300	3.22
4.95	5	5.05	1	300	4.26

Appendix E

Digital-to-Analog Converters

PART NUMBER	OPERATING TEMP (C) MIN	MAX	RESOLUTION (BITS)	OUTPUT TYPE
Am6012DC	0	70	12	CURRENT
Am6012DM	-55	125	12	CURRENT
Am6012ADC	0	70	12	CURRENT

DIGITAL-TO-ANALOG CONVERTERS:

PART NUMBER	OPERATING TEMP (C) MIN	MAX	RESOLUTION (BITS)	OUTPUT TYPE
AD390J	0	70	12 (QUAD)	VOLTAGE
AD390K	0	70	12 (QUAD)	VOLTAGE
AD390S	-55	125	12 (QUAD)	VOLTAGE
AD390T	-55	125	12 (QUAD)	VOLTAGE
AD558J	0	70	8	VOLTAGE
AD558K	0	70	8	VOLTAGE
AD558S	-55	125	8	VOLTAGE
AD558T	-55	125	8	VOLTAGE
AD561J	0	70	10	CURRENT
AD561K	0	70	10	CURRENT
AD561S	-55	125	10	CURRENT
AD561T	-55	125	10	CURRENT
AD562A	-25	85	12	CURRENT
AD562K	0	70	12	CURRENT

ADVANCED MICRO DEVICES (AMD)

MAXIMUM NON-LINEARITY (+/- LSB)	MAXIMUM GAIN TEMPCO (PPM/C)	TYPICAL SETTLING TIME (uS)	UNIT PRICE ($)
2	40	0.25	14.95
2	40	0.25	44.95
2	20	0.25	29.95

ANALOG DEVICES

MAXIMUM NON-LINEARITY (+/- LSB)	MAXIMUM GAIN TEMPCO (PPM/C)	TYPICAL SETTLING TIME (uS)	UNIT PRICE ($)
0.75	40	4	148.00
0.5	20	4	175.00
0.75	40	4	392.00
0.5	20	4	460.00
0.5	60	0.8	8.95
0.25	30	0.8	13.50
0.5	60	0.8	19.50
0.25	30	0.8	26.95
0.5	80 TYP	0.25	17.40
0.25	30	0.25	28.75
0.5	60 TYP	0.25	37.90
0.25	30	0.25	77.60
0.5	5	1.5	96.40
0.5	5	1.5	73.50

PART NUMBER	OPERATING TEMP (C) MIN	MAX	RESOLUTION (BITS)	OUTPUT TYPE
AD562S	-55	125	12	CURRENT
AD563J	0	70	12	CURRENT
AD563K	0	70	12	CURRENT
AD563S	-55	125	12	CURRENT
AD563T	-55	125	12	CURRENT
AD565AJ	0	70	12	CURRENT
AD565AK	0	70	12	CURRENT
AD565AS	-55	125	12	CURRENT
AD565AT	-55	125	12	CURRENT
AD566AJ	0	70	12	CURRENT
AD566AK	0	70	12	CURRENT
AD566AS	-55	125	12	CURRENT
AD566AT	-55	125	12	CURRENT
AD567J	0	70	12	CURRENT
AD567K	0	70	12	CURRENT
AD567S	-55	125	12	CURRENT

ANALOG DEVICES

MAXIMUM NON-LINEARITY (+/- LSB)	MAXIMUM GAIN TEMPCO (PPM/C)	TYPICAL SETTLING TIME (uS)	UNIT PRICE ($)
0.25	5	1.5	209.70
0.5	50	1.5	53.30
0.25	20	1.5	82.10
0.25	30	1.5	230.60
0.25	10	1.5	293.60
0.5	50	0.15	23.95
0.25	20	0.15	37.50
0.5	30	0.15	97.00
0.25	15	0.15	142.00
0.5	10	0.25	19.95
0.25	3	0.25	29.95
0.5	10	0.25	82.00
0.25	3	0.25	129.00
0.5	50	0.4	22.50
0.25	20	0.4	34.50
0.5	30	0.4	89.00

PART NUMBER	OPERATING TEMP (C) MIN	MAX	RESOLUTION (BITS)	OUTPUT TYPE
AD1408-7	0	75	8	CURRENT
AD1408-8	0	75	8	CURRENT
AD1408-9	0	75	8	CURRENT
AD1508-8	-55	125	8	CURRENT
AD1508-9	-55	125	8	CURRENT
AD3860K	0	70	12	VOLTAGE
AD3860S	-55	125	12	VOLTAGE
AD6012N	0	70	12	CURRENT
AD7240A	-25	85	12	VOLTAGE
AD7240B	-25	85	12	VOLTAGE
AD7240J	0	70	12	VOLTAGE
AD7240K	0	70	12	VOLTAGE
AD7240S	-55	125	12	VOLTAGE
AD7240T	-55	125	12	VOLTAGE

ANALOG DEVICES

MAXIMUM NON-LINEARITY (+/- LSB)	MAXIMUM GAIN TEMPCO (PPM/C)	TYPICAL SETTLING TIME (uS)	UNIT PRICE ($)
1	20 TYP	0.25	3.50
0.5	20	0.25	4.30
0.25	20	0.25	9.50
0.5	20	0.25	14.60
0.25	20	0.25	21.30
0.5	10	3	104.00
0.5	10	3	162.75
2	40	0.25	13.95
+0.5/-1.25	6	0.9 MAX	18.50
+0.5/-1	6	0.9 MAX	20.15
+0.5/-1.25	6	0.9 MAX	16.00
+0.5/-1	6	0.9 MAX	17.65
+0.5/-1.25	6	0.9 MAX	65.90
+0.5/-1	6	0.9 MAX	73.90

PART NUMBER	OPERATING TEMP (C) MIN	MAX	RESOLUTION (BITS)	OUTPUT TYPE
AD7520JD	-25	85	10	CURRENT
AD7520JN	0	70	10	CURRENT
AD7520KD	-25	85	10	CURRENT
AD7520KN	0	70	10	CURRENT
AD7520LD	-25	85	10	CURRENT
AD7520LN	0	70	10	CURRENT
AD7520S	-55	125	10	CURRENT
AD7520T	-55	125	10	CURRENT
AD7520U	-55	125	10	CURRENT
AD7521JD	-25	85	12	CURRENT
AD7521JN	0	70	12	CURRENT
AD7521KD	-25	85	12	CURRENT
AD7521KN	0	70	12	CURRENT
AD7521LD	-25	85	12	CURRENT
AD7521LN	0	70	12	CURRENT
AD7521S	-55	125	12	CURRENT

ANALOG DEVICES

MAXIMUM NON-LINEARITY (+/- LSB)	MAXIMUM GAIN TEMPCO (PPM/C)	TYPICAL SETTLING TIME (uS)	UNIT PRICE ($)
2	10	0.5	20.75
2	10	0.5	15.75
1	10	0.5	29.50
1	10	0.5	20.50
0.5	10	0.5	39.50
0.5	10	0.5	27.00
2	10	0.5	42.00
1	10	0.5	70.00
0.5	10	0.5	97.00
4	10	0.5	25.75
4	10	0.5	20.75
2	10	0.5	34.50
2	10	0.5	25.50
1	10	0.5	44.50
1	10	0.5	32.00
4	10	0.5	47.00

DIGITAL-TO-ANALOG CONVERTERS:

PART NUMBER	OPERATING TEMP (C)		RESOLUTION (BITS)	OUTPUT TYPE
	MIN	MAX		
AD7521T	-55	125	12	CURRENT
AD7521U	-55	125	12	CURRENT
AD7522JD	-25	85	10	CURRENT
AD7522JN	0	70	10	CURRENT
AD7522KD	-25	85	10	CURRENT
AD7522KN	0	70	10	CURRENT
AD7522LD	-25	85	10	CURRENT
AD7522LN	0	70	10	CURRENT
AD7522S	-55	125	10	CURRENT
AD7522T	-55	125	10	CURRENT
AD7522U	-55	125	10	CURRENT
AD7523J	0	70	8	CURRENT
AD7523K	0	70	8	CURRENT
AD7523L	0	70	8	CURRENT
AD7524A	-25	85	8	CURRENT
AD7524B	-25	85	8	CURRENT

ANALOG DEVICES

MAXIMUM NON-LINEARITY (+/- LSB)	MAXIMUM GAIN TEMPCO (PPM/C)	TYPICAL SETTLING TIME (uS)	UNIT PRICE ($)
2	10	0.5	75.00
1	10	0.5	102.00
2	10	0.5	26.50
2	10	0.5	15.00
1	10	0.5	29.00
1	10	0.5	17.50
0.5	10	0.5	31.50
0.5	10	0.5	20.00
2	10	0.5	64.00
1	10	0.5	70.25
0.5	10	0.5	76.50
0.5	120	0.1	4.00
0.25	120	0.1	6.00
0.125	120	0.1	8.00
0.5	10	0.1 MAX	13.50
0.25	10	0.1 MAX	16.00

DIGITAL-TO-ANALOG CONVERTERS:

PART NUMBER	OPERATING TEMP (C) MIN	MAX	RESOLUTION (BITS)	OUTPUT TYPE
AD7524C	-25	85	8	CURRENT
AD7524J	0	70	8	CURRENT
AD7524K	0	70	8	CURRENT
AD7524L	0	70	8	CURRENT
AD7524S	-55	125	8	CURRENT
AD7524T	-55	125	8	CURRENT
AD7524U	-55	125	8	CURRENT
AD7525B	-25	85	3.5 DIGITS	CURRENT
AD7525C	-25	85	3.5 DIGITS	CURRENT
AD7525K	0	70	3.5 DIGITS	CURRENT
AD7525L	0	70	3.5 DIGITS	CURRENT
AD7525T	-55	125	3.5 DIGITS	CURRENT
AD7525U	-55	125	3.5 DIGITS	CURRENT
AD7528A	-25	85	8 (DUAL)	CURRENT
AD7528B	-25	85	8 (DUAL)	CURRENT
AD7528C	-25	85	8 (DUAL)	CURRENT

ANALOG DEVICES

MAXIMUM NON-LINEARITY (+/- LSB)	MAXIMUM GAIN TEMPCO (PPM/C)	TYPICAL SETTLING TIME (uS)	UNIT PRICE ($)
0.125	10	0.1 MAX	18.50
0.5	10	0.1 MAX	7.50
0.25	10	0.1 MAX	10.00
0.125	10	0.1 MAX	12.50
0.5	10	0.1 MAX	32.25
0.25	10	0.1 MAX	38.50
0.125	10	0.1 MAX	44.75
1	25	1 MAX	32.25
0.5	25	1 MAX	35.25
1	25	1 MAX	20.00
0.5	25	1 MAX	23.00
1	25	1 MAX	79.00
0.5	25	1 MAX	82.00
1	35	0.18 MAX	16.95
0.5	35	0.18 MAX	19.80
0.5	35	0.18 MAX	23.10

DIGITAL-TO-ANALOG CONVERTERS:

PART NUMBER	OPERATING TEMP (C)		RESOLUTION (BITS)	OUTPUT TYPE
	MIN	MAX		
AD7528J	0	70	8 (DUAL)	CURRENT
AD7528K	0	70	8 (DUAL)	CURRENT
AD7528L	0	70	8 (DUAL)	CURRENT
AD7528S	-55	125	8 (DUAL)	CURRENT
AD7528T	-55	125	8 (DUAL)	CURRENT
AD7528U	-55	125	8 (DUAL)	CURRENT
AD7530JD	-25	85	10	CURRENT
AD7530JN	0	70	10	CURRENT
AD7530KD	-25	85	10	CURRENT
AD7530KN	0	70	10	CURRENT
AD7530LD	-25	85	10	CURRENT
AD7530LN	0	70	10	CURRENT
AD7531JD	-25	85	12	CURRENT
AD7531JN	0	70	12	CURRENT
AD7531KD	-25	85	12	CURRENT
AD7531KN	0	70	12	CURRENT

ANALOG DEVICES

MAXIMUM NON-LINEARITY (+/- LSB)	MAXIMUM GAIN TEMPCO (PPM/C)	TYPICAL SETTLING TIME (uS)	UNIT PRICE ($)
1	35	0.18 MAX	9.80
0.5	35	0.18 MAX	13.05
0.5	35	0.18 MAX	16.40
1	35	0.18 MAX	32.90
0.5	35	0.18 MAX	35.90
0.5	35	0.18 MAX	41.90
2	10	0.5	15.00
2	10	0.5	9.75
1	10	0.5	18.50
1	10	0.5	13.50
0.5	10	0.5	28.50
0.5	10	0.5	18.50
4	10	0.5	15.00
4	10	0.5	9.75
2	10	0.5	18.50
2	10	0.5	13.50

DIGITAL-TO-ANALOG CONVERTERS:

PART NUMBER	OPERATING TEMP (C) MIN	MAX	RESOLUTION (BITS)	OUTPUT TYPE
AD7531LD	-25	85	12	CURRENT
AD7531LN	0	70	12	CURRENT
AD7533A	-25	85	10	CURRENT
AD7533B	-25	85	10	CURRENT
AD7533C	-25	85	10	CURRENT
AD7533J	0	70	10	CURRENT
AD7533K	0	70	10	CURRENT
AD7533L	0	70	10	CURRENT
AD7533S	-55	125	10	CURRENT
AD7533T	-55	125	10	CURRENT
AD7533U	-55	125	10	CURRENT
AD7541A	-25	85	12	CURRENT
AD7541B	-25	85	12	CURRENT
AD7541J	0	70	12	CURRENT
AD7541K	0	70	12	CURRENT
AD7541S	-55	125	12	CURRENT

ANALOG DEVICES

MAXIMUM NON-LINEARITY (+/- LSB)	MAXIMUM GAIN TEMPCO (PPM/C)	TYPICAL SETTLING TIME (uS)	UNIT PRICE ($)
1	10	0.5	39.50
1	10	0.5	29.00
2	10	0.6 MAX	9.00
1	10	0.6 MAX	10.25
0.5	10	0.6 MAX	12.55
2	10	0.6 MAX	5.90
1	10	0.6 MAX	7.35
0.5	10	0.6 MAX	8.85
2	10	0.6 MAX	23.55
1	10	0.6 MAX	28.15
0.5	10	0.6 MAX	32.05
1	10	1 MAX	37.50
0.5	10	1 MAX	40.00
1	10	1 MAX	27.50
0.5	10	1 MAX	30.00
1	10	1 MAX	88.00

DIGITAL-TO-ANALOG CONVERTERS:

PART NUMBER	OPERATING TEMP (C) MIN	MAX	RESOLUTION (BITS)	OUTPUT TYPE
AD7541T	-55	125	12	CURRENT
AD7541AA	-25	85	12	CURRENT
AD7541AB	-25	85	12	CURRENT
AD7541AJ	0	70	12	CURRENT
AD7541AK	0	70	12	CURRENT
AD7541AS	-55	125	12	CURRENT
AD7541AT	-55	125	12	CURRENT
AD7542A	-25	85	12	CURRENT
AD7542B	-25	85	12	CURRENT
AD7542J	0	70	12	CURRENT
AD7542K	0	70	12	CURRENT
AD7542S	-55	125	12	CURRENT
AD7542T	-55	125	12	CURRENT
AD7542GB	-25	85	12	CURRENT
AD7542GK	0	70	12	CURRENT
AD7542GT	-55	125	12	CURRENT

ANALOG DEVICES

MAXIMUM NON-LINEARITY (+/- LSB)	MAXIMUM GAIN TEMPCO (PPM/C)	TYPICAL SETTLING TIME (uS)	UNIT PRICE ($)
0.5	10	1 MAX	98.00
1	5	0.6	14.30
0.5	5	0.6	15.75
1	5	0.6	12.85
0.5	5	0.6	14.30
1	5	0.6	41.85
0.5	5	0.6	48.80
1	5	2 MAX	24.40
0.5	5	2 MAX	26.60
1	5	2 MAX	21.40
0.5	5	2 MAX	23.60
1	5	2 MAX	63.20
0.5	5	2 MAX	71.80
0.5	5	2 MAX	33.50
0.5	5	2 MAX	31.15
0.5	5	2 MAX	96.85

DIGITAL-TO-ANALOG CONVERTERS:

PART NUMBER	OPERATING TEMP (C) MIN	MAX	RESOLUTION (BITS)	OUTPUT TYPE
AD7543A	-25	85	12	CURRENT
AD7543B	-25	85	12	CURRENT
AD7543J	0	70	12	CURRENT
AD7543K	0	70	12	CURRENT
AD7543S	-55	125	12	CURRENT
AD7543T	-55	125	12	CURRENT
AD7543GB	-25	85	12	CURRENT
AD7543GK	0	70	12	CURRENT
AD7543GT	-55	125	12	CURRENT
AD7545A	-25	85	12	CURRENT
AD7545B	-25	85	12	CURRENT
AD7545C	-25	85	12	CURRENT
AD7545J	0	70	12	CURRENT
AD7545K	0	70	12	CURRENT
AD7545L	0	70	12	CURRENT
AD7545S	-55	125	12	CURRENT

ANALOG DEVICES

MAXIMUM NON- LINEARITY (+/- LSB)	MAXIMUM GAIN TEMPCO (PPM/C)	TYPICAL SETTLING TIME (uS)	UNIT PRICE ($)
1	5	2 MAX	24.40
0.5	5	2 MAX	26.60
1	5	2 MAX	21.40
0.5	5	2 MAX	23.60
1	5	2 MAX	63.20
0.5	5	2 MAX	71.80
0.5	5	2 MAX	33.50
0.5	5	2 MAX	31.15
0.5	5	2 MAX	96.85
2	5	2 MAX	15.70
1	5	2 MAX	19.70
0.5	5	2 MAX	21.35
2	5	2 MAX	13.20
1	5	2 MAX	17.25
0.5	5	2 MAX	18.90
2	5	2 MAX	44.55

PART NUMBER	OPERATING TEMP (C) MIN	MAX	RESOLUTION (BITS)	OUTPUT TYPE
AD7545T	-55	125	12	CURRENT
AD7545U	-55	125	12	CURRENT
AD7545GC	-25	85	12	CURRENT
AD7545GL	0	70	12	CURRENT
AD7545GU	-55	125	12	CURRENT
AD7546A	-25	85	16	VOLTAGE
AD7546B	-25	85	16	VOLTAGE
AD7546J	0	70	16	VOLTAGE
AD7546K	0	70	16	VOLTAGE
AD7548A	-25	85	12	CURRENT
AD7548B	-25	85	12	CURRENT
AD7548J	0	70	12	CURRENT
AD7548K	0	70	12	CURRENT
AD7548S	-55	125	12	CURRENT
AD7548T	-55	125	12	CURRENT
AD9768S	-30	115	8	CURRENT

ANALOG DEVICES

MAXIMUM NON-LINEARITY (+/- LSB)	MAXIMUM GAIN TEMPCO (PPM/C)	TYPICAL SETTLING TIME (uS)	UNIT PRICE ($)
1	5	2 MAX	59.15
0.5	5	2 MAX	64.10
0.5	5	2 MAX	31.70
0.5	5	2 MAX	27.65
0.5	5	2 MAX	95.04
32	2	4	39.35
8	2	4	55.10
32	2	4	28.90
8	2	4	44.60
1	5	1.5	18.64
0.5	5	1.5	21.78
1	5	1.5	15.64
0.5	5	1.5	18.97
1	5	1.5	56.10
0.5	5	1.5	63.77
0.5	70	0.005	34.60

DIGITAL-TO-ANALOG CONVERTERS:

PART NUMBER	OPERATING TEMP (C) MIN	MAX	RESOLUTION (BITS)	OUTPUT TYPE
DAC10HT	-55	200	12	CURRENT
DAC10HT-1	-55	200	12	CURRENT
DAC60-10	0	70	12	CURRENT
DAC60-12	0	70	12	CURRENT
DAC63BG	-25	85	12	CURRENT
DAC63CG	-25	85	12	CURRENT
DAC70-CCD-I	-25	85	4 DIGITS	CURRENT
DAC70-COB-I	-25	85	16	CURRENT
DAC70-CSB-I	-25	85	16	CURRENT
DAC70C-CCD-I	0	70	4 DIGITS	CURRENT
DAC70C-COB-I	0	70	16	CURRENT
DAC70C-CSB-I	0	70	16	CURRENT
DAC71-CCD-I	0	70	4 DIGITS	CURRENT
DAC71-CCD-V	0	70	4 DIGITS	VOLTAGE
DAC71-COB-I	0	70	16	CURRENT
DAC71-COB-V	0	70	16	VOLTAGE

BURR BROWN

MAXIMUM NON-LINEARITY (+/- LSB)	MAXIMUM GAIN TEMPCO (PPM/C)	TYPICAL SETTLING TIME (uS)	UNIT PRICE ($)
0.5	10	0.2	295.00
1	25	0.2	273.00
1	30	0.04	158.00
0.5	30	0.04	172.00
0.5	40	0.045	108.00
0.5	30	0.045	119.00
2	7	50	177.50
2	7	50	177.50
2	7	50	177.50
3.3	14	50	124.25
3.3	14	50	124.25
3.3	14	50	124.25
3.3	45	1 MAX	55.00
3.3	15	5	58.00
2	45	1 MAX	55.00
2	15	5	58.00

DIGITAL-TO-ANALOG CONVERTERS:

PART NUMBER	OPERATING TEMP (C) MIN	MAX	RESOLUTION (BITS)	OUTPUT TYPE
DAC71-CSB-I	0	70	16	CURRENT
DAC71-CSB-V	0	70	16	VOLTAGE
DAC72-CCD-V	-25	85	4 DIGITS	VOLTAGE
DAC72-COB-I	-25	85	16	CURRENT
DAC72-COB-V	-25	85	16	VOLTAGE
DAC72-CSB-I	-25	85	16	CURRENT
DAC72-CSB-V	-25	85	16	VOLTAGE
DAC72C-CCD-V	0	70	4 DIGITS	VOLTAGE
DAC72C-COB-I	0	70	16	CURRENT
DAC72C-COB-V	0	70	16	VOLTAGE
DAC72C-CSB-I	0	70	16	CURRENT
DAC72C-CSB-V	0	70	16	VOLTAGE
DAC73J	0	70	16	SELECTABLE
DAC73K	0	70	16	SELECTABLE
DAC74	0	70	16	VOLTAGE
DAC80-CBI-I	0	70	12	CURRENT

BURR BROWN

MAXIMUM NON-LINEARITY (+/- LSB)	MAXIMUM GAIN TEMPCO (PPM/C)	TYPICAL SETTLING TIME (uS)	UNIT PRICE ($)
2	45	1 MAX	55.00
2	15	5	58.00
3.3	15	5	91.00
2	35	1 MAX	79.00
2	15	5	91.00
2	35	1 MAX	79.00
2	15	5	91.00
3.3	15	5	77.00
2	45	1 MAX	69.00
2	15	5	77.00
2	45	1 MAX	69.00
2	15	5	77.00
1	10	50 MAX	242.00
0.5	10	50 MAX	286.00
0.5	5	50 MAX	1495.00
0.5	30	0.3	34.50

DIGITAL-TO-ANALOG CONVERTERS:

PART NUMBER	OPERATING TEMP (C) MIN	MAX	RESOLUTION (BITS)	OUTPUT TYPE
DAC80-CBI-V	0	70	12	VOLTAGE
DAC80-CCD-I	0	70	3 DIGITS	CURRENT
DAC80-CCD-V	0	70	3 DIGITS	VOLTAGE
DAC82KG	-25	85	8	SELECTABLE
DAC85-CBI-I	-25	85	12	CURRENT
DAC85-CBI-V	-25	85	12	VOLTAGE
DAC85C-CBI-I	0	70	12	CURRENT
DAC85C-CBI-V	0	70	12	VOLTAGE
DAC85LD-CBI-V	-25	85	12	VOLTAGE
DAC87-CBI-V	-55	125	12	VOLTAGE
DAC87U-CBI-V	-55	125	12	VOLTAGE
DAC90BG	-25	85	8	CURRENT
DAC90SG	-55	125	8	CURRENT
DAC701BH	-25	85	16	VOLTAGE
DAC701KH	0	70	16	VOLTAGE
DAC703BH	-25	85	16	VOLTAGE

BURR-BROWN

MAXIMUM NON-LINEARITY (+/- LSB)	MAXIMUM GAIN TEMPCO (PPM/C)	TYPICAL SETTLING TIME (uS)	UNIT PRICE ($)
0.5	30	3	36.50
0.25	30	0.3	34.50
0.25	30	3	36.50
0.5	50	2	33.90
0.5	20	0.3	104.00
0.5	20	3	107.00
0.5	20	0.3	77.00
0.5	20	3	79.00
0.5	10	0.3	142.75
0.5	20	3	120.00
0.5	60	3	95.00
0.5	50	0.2	19.40
0.5	50	0.2	26.50
2	15	4	58.00
2	25	4	48.00
2	15	4	58.00

DIGITAL-TO-ANALOG CONVERTERS:

PART NUMBER	OPERATING TEMP (C) MIN	MAX	RESOLUTION (BITS)	OUTPUT TYPE
DAC703KH	0	70	16	VOLTAGE
DAC736J	0	70	16	SELECTABLE
DAC736K	0	70	16	SELECTABLE
DAC800-CBI-I	0	70	12	CURRENT
DAC800-CBI-V	0	70	12	VOLTAGE
DAC850-CBI-I	-25	85	12	CURRENT
DAC850-CBI-V	-25	85	12	VOLTAGE
DAC851-CBI-I	-55	125	12	CURRENT
DAC851-CBI-V	-55	125	12	VOLTAGE

BURR-BROWN

MAXIMUM NON-LINEARITY (+/- LSB)	MAXIMUM GAIN TEMPCO (PPM/C)	TYPICAL SETTLING TIME (uS)	UNIT PRICE ($)
2	25	4	48.00
1	10	50 MAX	220.00
0.5	10	50 MAX	260.00
0.5	30	0.3	23.95
0.5	30	2.5	29.95
0.5	20	0.3	39.00
0.5	20	2.5	47.00
0.5	25	0.3	69.00
0.5	25	2.5	69.00

DIGITAL-TO-ANALOG CONVERTERS:

PART NUMBER	OPERATING TEMP (C) MIN	MAX	RESOLUTION (BITS)	OUTPUT TYPE
UA565J	0	70	12	CURRENT
UA565K	0	70	12	CURRENT
UA565S	-55	125	12	CURRENT
UA565T	-55	125	12	CURRENT
UA801	-55	125	8	CURRENT
UA801C	0	70	8	CURRENT
UA801E	0	70	8	CURRENT
UA802	-55	125	8	CURRENT
UA802A	0	70	8	CURRENT
UA802B	0	70	8	CURRENT
UA802C	0	70	8	CURRENT

FAIRCHILD

MAXIMUM NON-LINEARITY (+/- LSB)	MAXIMUM GAIN TEMPCO (PPM/C)	TYPICAL SETTLING TIME (uS)	UNIT PRICE ($)
0.5	30	0.2	28.50
0.25	20	0.2	34.20
0.5	30	0.2	95.00
0.25	20	0.2	142.50
0.5	50	0.135	3.80
1	50	0.15	1.54
0.5	50	0.15	1.52
0.5	20 TYP	0.135	2.85
0.5	20 TYP	0.135	1.50
1	20 TYP	0.135	1.50
2	20 TYP	0.135	1.50

DIGITAL-TO-ANALOG CONVERTERS:

PART NUMBER	OPERATING TEMP (C) MIN	MAX	RESOLUTION (BITS)	OUTPUT TYPE
HI-562A-2	-55	125	12	CURRENT
HI-562A-5	0	75	12	CURRENT
HI-5610-2	-55	125	10	CURRENT
HI-5610-5	0	75	10	CURRENT
HI-5618A-2	-55	125	8	CURRENT
HI-5618A-5	0	75	8	CURRENT
HI-5618B-2	-55	125	8	CURRENT
HI-5618B-5	0	75	8	CURRENT
HI-7541A	-25	85	12	CURRENT
HI-7541B	-25	85	12	CURRENT
HI-7541J	0	75	12	CURRENT
HI-7541K	0	75	12	CURRENT
HI-7541S	-55	125	12	CURRENT
HI-7541T	-55	125	12	CURRENT

HARRIS

MAXIMUM NON-LINEARITY (+/- LSB)	MAXIMUM GAIN TEMPCO (PPM/C)	TYPICAL SETTLING TIME (uS)	UNIT PRICE ($)
0.25	10 TYP	0.3	75.18
0.5	10 TYP	0.3	17.69
0.5	5 TYP	0.085	61.24
0.5	5 TYP	0.085	16.04
0.25		0.065	15.56
0.25		0.065	7.03
0.5		0.065	13.37
0.5		0.065	6.46
1	5	1 MAX	23.87
0.5	5	1 MAX	28.56
1	5	1 MAX	14.97
0.5	5	1 MAX	16.31
1	5	1 MAX	96.00
0.5	5	1 MAX	103.50

PART NUMBER	OPERATING TEMP (C) MIN	MAX	RESOLUTION (BITS)	OUTPUT TYPE
ICL7134BJC	0	70	14	CURRENT
ICL7134BJI	-20	85	14	CURRENT
ICL7134BJM	-55	125	14	CURRENT
ICL7134BKC	0	70	14	CURRENT
ICL7134BKI	-20	85	14	CURRENT
ICL7134BKM	-55	125	14	CURRENT
ICL7134BLC	0	70	14	CURRENT
ICL7134BLI	-20	85	14	CURRENT
ICL7134UJC	0	70	14	CURRENT
ICL7134UJI	-20	85	14	CURRENT
ICL7134UJM	-55	125	14	CURRENT
ICL7134UKC	0	70	14	CURRENT
ICL7134UKI	-20	85	14	CURRENT
ICL7134UKM	-55	125	14	CURRENT
ICL7134ULC	0	70	14	CURRENT
ICL7134ULI	-20	85	14	CURRENT

INTERSIL

MAXIMUM NON-LINEARITY (+/- LSB)	MAXIMUM GAIN TEMPCO (PPM/C)	TYPICAL SETTLING TIME (uS)	UNIT PRICE ($)
2	8	0.9	21.40
2	8	0.9	47.85
2	8	0.9	148.50
1	8	0.9	32.95
1	8	0.9	61.05
1	8	0.9	193.50
0.5	8	0.9	49.45
0.5	8	0.9	74.25
2	8	0.9	21.40
2	8	0.9	47.85
2	8	0.9	148.50
1	8	0.9	32.95
1	8	0.9	61.05
1	8	0.9	193.50
0.5	8	0.9	49.45
0.5	8	0.9	74.25

DIGITAL-TO-ANALOG CONVERTERS:

PART NUMBER	OPERATING TEMP (C) MIN	MAX	RESOLUTION (BITS)	OUTPUT TYPE
ICL7145JC	0	70	16	CURRENT
ICL7145JI	-25	85	16	CURRENT
ICL7145KC	0	70	16	CURRENT
ICL7145KI	-25	85	16	CURRENT

DIGITAL-TO-ANALOG CONVERTERS:

PART NUMBER	OPERATING TEMP (C) MIN	MAX	RESOLUTION (BITS)	OUTPUT TYPE
MC1406L	0	70	6	CURRENT
MC1408P6	0	70	8	CURRENT
MC1408P7	0	70	8	CURRENT
MC1408P8	0	70	8	CURRENT
MC1506L	-55	125	6	CURRENT
MC1508L8	-55	125	8	CURRENT
MC3410	0	70	10	CURRENT
MC3410C	0	70	10	CURRENT
MC3412	0	70	12	CURRENT
MC3510	-55	125	10	CURRENT
MC3512	-55	125	12	CURRENT
MC6890	0	70	8	VOLTAGE
MC6890A	-55	125	8	VOLTAGE

INTERSIL

MAXIMUM NON-LINEARITY (+/- LSB)	MAXIMUM GAIN TEMPCO (PPM/C)	TYPICAL SETTLING TIME (uS)	UNIT PRICE ($)
4	1 TYP	1	43.50
4	1 TYP	1	55.50
2	1 TYP	1	52.50
2	1 TYP	1	67.50

MOTOROLA

MAXIMUM NON-LINEARITY (+/- LSB)	MAXIMUM GAIN TEMPCO (PPM/C)	TYPICAL SETTLING TIME (uS)	UNIT PRICE ($)
0.5	80 TYP	0.15	7.29
2	20 TYP	0.3	1.46
1	20 TYP	0.3	1.72
0.5	20 TYP	0.3	1.89
0.5	80 TYP	0.15	11.05
0.5	20 TYP	0.3	11.46
0.5	60	0.25	14.92
1	60	0.25	10.40
0.5	30	0.2	26.95
0.5	60	0.25	23.85
0.5	30	0.2	95.25
0.5	50	0.2	10.90
0.5	50	0.2	19.45

DIGITAL-TO-ANALOG CONVERTERS:

PART NUMBER	OPERATING TEMP (C) MIN	MAX	RESOLUTION (BITS)	OUTPUT TYPE
MC10318CL6	0	70	8	CURRENT
MC10318CL7	0	70	8	CURRENT
MC10318L	0	70	8	CURRENT
MC10318L9	0	70	8	CURRENT

MOTOROLA

MAXIMUM NON-LINEARITY (+/- LSB)	MAXIMUM GAIN TEMPCO (PPM/C)	TYPICAL SETTLING TIME (uS)	UNIT PRICE ($)
2	10	0.01	19.45
1	10	0.01	28.45
0.5	10	0.01	42.00
0.25	10	0.01	70.60

DIGITAL-TO-ANALOG CONVERTERS:

PART NUMBER	OPERATING TEMP (C) MIN	MAX	RESOLUTION (BITS)	OUTPUT TYPE
DAC0800L	-55	125	8	CURRENT
DAC0800LC	0	70	8	CURRENT
DAC0801LC	0	70	8	CURRENT
DAC0802L	-55	125	8	CURRENT
DAC0802LC	0	70	8	CURRENT
DAC0806LC	0	75	8	CURRENT
DAC0807LC	0	75	8	CURRENT
DAC0808L	-55	125	8	CURRENT
DAC0808LC	0	75	8	CURRENT
DAC0830LCD	-40	85	8	CURRENT
DAC0830LCN	0	70	8	CURRENT
DAC0830LD	-55	125	8	CURRENT
DAC0831LCN	0	75	8	CURRENT
DAC0832LCD	-40	85	8	CURRENT
DAC0832LCN	0	75	8	CURRENT
DAC1000LCD	-40	85	10	CURRENT

NATIONAL

MAXIMUM NON-LINEARITY (+/- LSB)	MAXIMUM GAIN TEMPCO (PPM/C)	TYPICAL SETTLING TIME (uS)	UNIT PRICE ($)
0.5	50	0.1	7.10
0.5	50	0.1	2.80
1	50	0.1	2.30
0.25	50	0.1	8.25
0.25	50	0.1	4.15
2	20 TYP	0.15	1.55
1	20 TYP	0.15	1.75
0.5	20 TYP	0.15	6.00
0.5	20 TYP	0.15	2.40
0.125	6	1	8.60
0.125	6	1	6.35
0.125	6	1	11.25
0.25	6	1	5.63
0.5	6	1	6.90
0.5	6	1	3.75
0.5	10	0.5	16.10

DIGITAL-TO-ANALOG CONVERTERS:

PART NUMBER	OPERATING TEMP (C) MIN	MAX	RESOLUTION (BITS)	OUTPUT TYPE
DAC1000LCN	0	70	10	CURRENT
DAC1000LD	-55	125	10	CURRENT
DAC1001LCD	-40	85	10	CURRENT
DAC1001LCN	0	70	10	CURRENT
DAC1001LD	-55	125	10	CURRENT
DAC1002LCD	-40	85	10	CURRENT
DAC1002LCN	0	70	10	CURRENT
DAC1002LD	-55	125	10	CURRENT
DAC1006LCD	-40	85	10	CURRENT
DAC1006LCN	0	70	10	CURRENT
DAC1006LD	-55	125	10	CURRENT
DAC1007LCD	-40	85	10	CURRENT
DAC1007LCN	0	70	10	CURRENT
DAC1007LD	-55	125	10	CURRENT
DAC1008LCD	-40	85	10	CURRENT
DAC1008LCN	0	70	10	CURRENT

NATIONAL

MAXIMUM NON-LINEARITY (+/- LSB)	MAXIMUM GAIN TEMPCO (PPM/C)	TYPICAL SETTLING TIME (uS)	UNIT PRICE ($)
0.5	10	0.5	10.45
0.5	10	0.5	46.65
1	10	0.5	12.90
1	10	0.5	8.95
1	10	0.5	43.15
2	10	0.5	12.90
2	10	0.5	7.45
2	10	0.5	40.00
0.5	10	0.5	16.10
0.5	10	0.5	10.45
0.5	10	0.5	39.25
1	10	0.5	12.90
1	10	0.5	8.95
1	10	0.5	34.65
2	10	0.5	12.90
2	10	0.5	7.45

DIGITAL-TO-ANALOG CONVERTERS:

PART NUMBER	OPERATING TEMP (C) MIN	MAX	RESOLUTION (BITS)	OUTPUT TYPE
DAC1008LD	-55	125	10	CURRENT
DAC1020LCD	-40	85	10	CURRENT
DAC1020LCN	0	70	10	CURRENT
DAC1020LD	-55	125	10	CURRENT
DAC1021LCD	-40	85	10	CURRENT
DAC1021LCN	0	70	10	CURRENT
DAC1021LD	-55	125	10	CURRENT
DAC1022LCD	-40	85	10	CURRENT
DAC1022LCN	0	70	10	CURRENT
DAC1022LD	-55	125	10	CURRENT
DAC1208LCD	-40	85	12	CURRENT
DAC1209LCD	-40	85	12	CURRENT
DAC1210LCD	-40	85	12	CURRENT
DAC1218LCD	-40	85	12	CURRENT
DAC1219LCD	-40	85	12	CURRENT
DAC1220LCD	-40	85	12	CURRENT

NATIONAL

MAXIMUM NON-LINEARITY (+/- LSB)	MAXIMUM GAIN TEMPCO (PPM/C)	TYPICAL SETTLING TIME (uS)	UNIT PRICE ($)
2	10	0.5	31.15
0.5	10	0.5	10.70
0.5	10	0.5	9.00
0.5	10	0.5	30.00
1	10	0.5	9.15
1	10	0.5	7.50
1	10	0.5	27.15
2	10	0.5	8.00
2	10	0.5	6.00
2	10	0.5	23.30
0.5	6	1	17.15
1	6	1	14.60
2	6	1	11.90
0.5	6	1	14.90
1	6	1	13.40
2	10	0.5	10.80

PART NUMBER	OPERATING TEMP (C) MIN	MAX	RESOLUTION (BITS)	OUTPUT TYPE
DAC1220LCN	0	70	12	CURRENT
DAC1220LD	-55	125	12	CURRENT
DAC1221LCD	-40	85	12	CURRENT
DAC1221LCN	0	70	12	CURRENT
DAC1221LD	-55	125	12	CURRENT
DAC1222LCD	-40	85	12	CURRENT
DAC1222LCN	0	70	12	CURRENT
DAC1222LD	-55	125	12	CURRENT
DAC1230LCD	-40	85	12	CURRENT
DAC1231LCD	-40	85	12	CURRENT
DAC1232LCD	-40	85	12	CURRENT

NATIONAL

MAXIMUM NON-LINEARITY (+/- LSB)	MAXIMUM GAIN TEMPCO (PPM/C)	TYPICAL SETTLING TIME (uS)	UNIT PRICE ($)
2	10	0.5	9.75
2	10	0.5	32.15
4	10	0.5	9.30
4	10	0.5	8.25
4	10	0.5	28.75
8	10	0.5	8.25
8	10	0.5	6.75
8	10	0.5	27.85
0.5	6	1	17.15
1	6	1	14.60
2	6	1	11.90

PART NUMBER	OPERATING TEMP (C) MIN	MAX	RESOLUTION (BITS)	OUTPUT TYPE
DAC01	-55	125	6	VOLTAGE
DAC01A	-55	125	6	VOLTAGE
DAC01B	-55	125	6	VOLTAGE
DAC01C	0	70	6	VOLTAGE
DAC01D	0	70	6	VOLTAGE
DAC01F	-55	125	6	VOLTAGE
DAC01H	0	70	6	VOLTAGE
DAC02A	0	70	10 + SIGN	VOLTAGE
DAC02B	0	70	10 + SIGN	VOLTAGE
DAC02C	0	70	10 + SIGN	VOLTAGE
DAC02D	0	70	10 + SIGN	VOLTAGE
DAC03A	0	70	10	VOLTAGE
DAC03B	0	70	10	VOLTAGE
DAC03C	0	70	10	VOLTAGE
DAC03D	0	70	10	VOLTAGE

* = 25 PIECE PRICE

PRECISION MONOLITHICS INCORPORATED (PMI)

MAXIMUM NON-LINEARITY (+/- LSB)	MAXIMUM GAIN TEMPCO (PPM/C)	TYPICAL SETTLING TIME (uS)	UNIT PRICE ($)
0.25	80	1.5	45.00
0.125	80	1.5	75.00
0.25	120	1.5	37.50
0.25	160	1.5	22.50
0.5	160	1.5	9.75
0.25	80	1.5	30.00
0.25	160	1.5	4.80*
1	60	2	112.50
1	60	2	67.50
2	60	2	45.00
4	150	2	30.00
1	60 TYP	2	27.00
1	60 TYP	2	21.00
2	60 TYP	2	14.93
4	60 TYP	2	11.93

DIGITAL-TO-ANALOG CONVERTERS:

PART NUMBER	OPERATING TEMP (C) MIN	MAX	RESOLUTION (BITS)	OUTPUT TYPE
DAC05A	0	70	10 + SIGN	VOLTAGE
DAC05C	0	70	10 + SIGN	VOLTAGE
DAC05E	0	70	10 + SIGN	VOLTAGE
DAC05G	0	70	10 + SIGN	VOLTAGE
DAC06B	-55	125	10	VOLTAGE
DAC06C	-55	125	10	VOLTAGE
DAC06E	0	70	10	VOLTAGE
DAC06F	0	70	10	VOLTAGE
DAC06G	0	70	10	VOLTAGE
DAC08	-55	125	8	CURRENT
DAC08A	-55	125	8	CURRENT
DAC08C	0	70	8	CURRENT
DAC08E	0	70	8	CURRENT
DAC08H	0	70	8	CURRENT
DAC10B	-55	125	10	CURRENT
DAC10C	-55	125	10	CURRENT

PRECISION MONOLITHICS INCORPORATED (PMI)

MAXIMUM NON-LINEARITY (+/- LSB)	MAXIMUM GAIN TEMPCO (PPM/C)	TYPICAL SETTLING TIME (uS)	UNIT PRICE ($)
2	60	2	202.50
5	120	2	90.00
2	100	2	67.50
5	100	2	22.50
2	90	1.5	127.50
4	120	1.5	90.00
1	100	1.5	67.50
2	100	1.5	33.00
4	100	1.5	22.50
0.5	80	0.15	7.13
0.25	50	0.135	13.50
1	80	0.15	2.33
0.5	50	0.15	2.93
0.25	50	0.135	3.68
0.5	25	0.135	30.00
1	50	0.15	18.00

PART NUMBER	OPERATING TEMP (C) MIN	MAX	RESOLUTION (BITS)	OUTPUT TYPE
DAC10F	0	70	10	CURRENT
DAC10G	0	70	10	CURRENT
DAC20C	0	70	2 DIGIT	CURRENT
DAC100AAQ7	-25	85	10	CURRENT
DAC100AAQ8	-25	85	10	CURRENT
DAC100ABQ7	-25	85	10	CURRENT
DAC100ABQ8	-25	85	10	CURRENT
DAC100ACQ3	0	70	10	CURRENT
DAC100ACQ4	0	70	10	CURRENT
DAC100ACQ5	-55	125	10	CURRENT
DAC100ACQ6	-55	125	10	CURRENT
DAC100ACQ7	-25	85	10	CURRENT
DAC100ACQ8	-25	85	10	CURRENT
DAC100BBQ5	-55	125	10	CURRENT
DAC100BBQ6	-55	125	10	CURRENT
DAC100BBQ7	-25	85	10	CURRENT

PRECISION MONOLITHICS INCORPORATED (PMI)

MAXIMUM NON-LINEARITY (+/- LSB)	MAXIMUM GAIN TEMPCO (PPM/C)	TYPICAL SETTLING TIME (uS)	UNIT PRICE ($)
0.5	25	0.135	12.75
1	50	0.15	8.25
0.5	80	0.15	4.43
0.5	15	0.375	202.50
0.5	15	0.375	243.00
0.5	30	0.375	130.50
0.5	30	0.375	156.00
0.5	60	0.375	46.80
0.5	60	0.375	46.80
0.5	60	0.375	150.00
0.5	60	0.375	180.00
0.5	60	0.375	78.75
0.5	60	0.375	93.00
1	30	0.375	160.80
1	30	0.375	192.90
1	30	0.375	100.50

PART NUMBER	OPERATING TEMP (C) MIN	MAX	RESOLUTION (BITS)	OUTPUT TYPE
DAC100BBQ8	-25	85	10	CURRENT
DAC100BCQ3	0	70	10	CURRENT
DAC100BCQ4	0	70	10	CURRENT
DAC100BCQ5	-55	125	10	CURRENT
DAC100BCQ6	-55	125	10	CURRENT
DAC100BCQ7	-25	85	10	CURRENT
DAC100BCQ8	-25	85	10	CURRENT
DAC100CCQ3	0	70	10	CURRENT
DAC100CCQ4	0	70	10	CURRENT
DAC100CCQ5	-55	125	10	CURRENT
DAC100CCQ6	-55	125	10	CURRENT
DAC100CCQ7	-25	85	10	CURRENT
DAC100CCQ8	-25	85	10	CURRENT
DAC100DDQ3	0	70	10	CURRENT
DAC100DDQ4	0	70	10	CURRENT
DAC100DDQ5	-55	125	10	CURRENT

PRECISION MONOLITHICS INCORPORATED (PMI)

MAXIMUM NON-LINEARITY (+/- LSB)	MAXIMUM GAIN TEMPCO (PPM/C)	TYPICAL SETTLING TIME (uS)	UNIT PRICE ($)
1	30	0.375	120.00
1	60	0.375	30.60
1	60	0.375	30.60
1	60	0.375	97.50
1	60	0.375	117.00
1	60	0.375	65.55
1	60	0.375	76.50
2	60	0.375	21.60
2	60	0.375	21.60
2	60	0.375	67.50
2	60	0.375	90.00
2	60	0.375	45.00
2	60	0.375	54.00
3	120	0.375	17.93
3	120	0.375	17.93
3	120	0.375	52.50

DIGITAL-TO-ANALOG CONVERTERS:

PART NUMBER	OPERATING TEMP (C) MIN	MAX	RESOLUTION (BITS)	OUTPUT TYPE
DAC100DDQ6	-55	125	10	CURRENT
DAC100DDQ7	-25	85	10	CURRENT
DAC100DDQ8	-25	85	10	CURRENT
DAC208A	-55	125	8 + SIGN	VOLTAGE
DAC208B	-55	125	8 + SIGN	VOLTAGE
DAC208E	0	70	8 + SIGN	VOLTAGE
DAC208F	0	70	8 + SIGN	VOLTAGE
DAC210A	-55	125	10 + SIGN	VOLTAGE
DAC210B	-55	125	10 + SIGN	VOLTAGE
DAC210E	0	70	10 + SIGN	VOLTAGE
DAC210F	0	70	10 + SIGN	VOLTAGE
DAC210G	0	70	10 + SIGN	VOLTAGE
DAC312B	-55	125	12	CURRENT
DAC312E	0	70	12	CURRENT
DAC312F	0	70	12	CURRENT
DAC808A	-55	125	8	CURRENT

PRECISION MONOLITHICS INCORPORATED (PMI)

MAXIMUM NON-LINEARITY (+/- LSB)	MAXIMUM GAIN TEMPCO (PPM/C)	TYPICAL SETTLING TIME (uS)	UNIT PRICE ($)
3	120	0.375	70.50
3	120	0.375	30.00
3	120	0.375	36.00
0.5	40	0.75	42.00
1	60	0.75	25.50
0.5	40	0.75	18.00
1	60	0.75	12.75
0.5	40	1.5	82.50
0.5	60	1.5	45.00
0.5	40	1.5	28.50
0.5	60	1.5	21.75
1	30 TYP	1.5	14.25
1	40	0.25	40.43
0.5	30	0.25	26.93
1	40	0.25	13.43
0.25	50	0.3	21.00

DIGITAL-TO-ANALOG CONVERTERS:

PART NUMBER	OPERATING TEMP (C) MIN	MAX	RESOLUTION (BITS)	OUTPUT TYPE
DAC808B	-55	125	8	CURRENT
DAC808E	-25	85	8	CURRENT
DAC808F	-25	85	8	CURRENT
DAC808G	0	70	8	CURRENT
DAC888A	-55	125	8	CURRENT
DAC888B	-55	125	8	CURRENT
DAC888E	-25	85	8	CURRENT
DAC888F	-25	85	8	CURRENT

PRECISION MONOLITHICS INCORPORATED (PMI)

MAXIMUM NON-LINEARITY (+/- LSB)	MAXIMUM GAIN TEMPCO (PPM/C)	TYPICAL SETTLING TIME (uS)	UNIT PRICE ($)
0.5	80	0.3	14.93
0.25	50	0.3	13.50
0.5	80	0.3	9.00
1	80	0.3	7.50
0.25	50	0.3	21.00
0.5	80	0.3	14.93
0.25	50	0.3	13.50
0.5	80	0.3	9.00

DIGITAL-TO-ANALOG CONVERTERS:

PART NUMBER	OPERATING TEMP (C) MIN	MAX	RESOLUTION (BITS)	OUTPUT TYPE
NE5018	0	70	8	VOLTAGE
NE5019	0	70	8	VOLTAGE
NE5020	0	70	10	VOLTAGE
NE5118	0	70	8	CURRENT
NE5119	0	70	8	CURRENT
NE5410	0	70	10	CURRENT
SE5018	-55	125	8	VOLTAGE
SE5019	-55	125	8	VOLTAGE
SE5118	-55	125	8	CURRENT
SE5119	-55	125	8	CURRENT
SE5410	-55	125	10	CURRENT

SIGNETICS

MAXIMUM NON-LINEARITY (+/- LSB)	MAXIMUM GAIN TEMPCO (PPM/C)	TYPICAL SETTLING TIME (uS)	UNIT PRICE ($)
0.5	20 TYP	1.8	7.00
0.25	20 TYP	1.8	11.00
1	20 TYP	5	12.00
0.5	20 TYP	0.2	5.50
0.25	20 TYP	0.2	10.50
0.5	40	0.25	13.45
0.5	20 TYP	1.8	15.00
0.25	20 TYP	1.8	27.00
0.5	20 TYP	0.2	15.50
0.25	20 TYP	0.2	24.00
0.5	40	0.25	27.00

Appendix F
Analog-to-Digital Converters

ANALOG-TO-DIGITAL CONVERTERS:

PART NUMBER	OPERATING TEMP (C) MIN	MAX	RESOLUTION (BITS)	MAXIMUM NON-LINEARITY (+/- LSB)
Am6108	0	70	8	0.5
Am6112	0	70	12	1 TYP
Am6148	0	70	8	0.5

ADVANCED MICRO DEVICES (AMD)

MAXIMUM GAIN TEMPCO (PPM/C)	ANALOG INPUT RANGE (V)		CONVERSION TIME (uS)		UNIT PRICE ($)
	MIN	MAX	TYP	MAX	
	-5	+5	1	2	22.45
	-5	+5	3.3		67.50
	-5	+5	1	2	19.40

PART NUMBER	OPERATING TEMP (C) MIN	MAX	RESOLUTION (BITS)	MAXIMUM NON-LINEARITY (+/- LSB)
AD5201B	-25	85	12	0.5
AD5201T	-55	125	12	0.5
AD5202B	-25	85	12	0.5
AD5202T	-55	125	12	0.5
AD5204B	-25	85	12	0.5
AD5204T	-55	125	12	0.5
AD5205B	-25	85	12	0.5
AD5205T	-55	125	12	0.5
AD5211B	-25	85	12	0.5
AD5211T	-55	125	12	0.5
AD5212B	-25	85	12	0.5
AD5212T	-55	125	12	0.5
AD5214B	-25	85	12	0.5
AD5214T	-55	125	12	0.5
AD5215B	-25	85	12	0.5
AD5215T	-55	125	12	0.5
AD5240B	-25	85	12	0.5

ANALOG DEVICES

MAXIMUM GAIN TEMPCO (PPM/C)	ANALOG INPUT RANGE (V)		CONVERSION TIME (uS)		UNIT PRICE ($)
	MIN	MAX	TYP	MAX	
	-5	+5		50	207.00
	-5	+5		50	300.00
	-10	+10		50	207.00
	-10	+10		50	300.00
	-5	+5		50	207.00
	-5	+5		50	300.00
	-10	+10		50	207.00
	-10	+10		50	300.00
	-5	+5		13	200.00
	-5	+5		13	330.00
	-10	+10		13	200.00
	-10	+10		13	330.00
	-5	+5		13	200.00
	-5	+5		13	330.00
	-10	+10		13	200.00
	-10	+10		13	330.00
25	-10	+10		5	254.00

ANALOG-TO-DIGITAL CONVERTERS:

PART NUMBER	OPERATING TEMP (C) MIN	MAX	RESOLUTION (BITS)	MAXIMUM NON-LINEARITY (+/- LSB)
AD5240K	0	70	12	0.5
AD570J	0	70	8	0.5
AD570S	-55	125	8	0.5
AD571J	0	70	10	1
AD571K	0	70	10	0.5
AD571S	-55	125	10	1
AD572AD	-25	85	12	0.5
AD572BD	-25	85	12	0.5
AD572SD	-55	125	12	0.5
AD573J	0	70	10	1
AD573K	0	70	10	0.5
AD573S	-55	125	10	1
AD574AJ	0	70	12	1
AD574AK	0	70	12	0.5
AD574AL	0	70	12	0.5
AD574AS	-55	125	12	1
AD574AT	-55	125	12	0.5

ANALOG DEVICES

MAXIMUM GAIN TEMPCO (PPM/C)	ANALOG INPUT RANGE (V)		CONVERSION TIME (uS)		UNIT PRICE ($)
	MIN	MAX	TYP	MAX	
30	-10	+10		5	193.00
	-5	+5	25	40	28.60
	-5	+5	25	40	61.30
	-5	+5	25	40	45.30
	-5	+5	25	40	52.20
	-5	+5	25	40	103.40
30	-10	+10		25	160.00
15	-10	+10		25	191.00
25	-10	+10		25	383.00
	-5	+5	20	30	21.70
	-5	+5	20	30	29.20
	-5	+5	20	30	103.40
50	-10	+10	25	35	49.50
27	-10	+10	25	35	65.00
10	-10	+10	25	35	95.00
50	-10	+10	25	35	145.00
25	-10	+10	25	35	195.00

ANALOG-TO-DIGITAL CONVERTERS:

PART NUMBER	OPERATING TEMP (C) MIN	MAX	RESOLUTION (BITS)	MAXIMUM NON-LINEARITY (+/- LSB)
AD574AU	-55	125	12	0.5
AD578J	0	70	12	0.5
AD578K	0	70	12	0.5
AD578L	0	70	12	0.5
AD579J	0	70	10	0.5
AD579K	0	70	10	0.5
AD579T	-55	125	10	0.5
AD673J	0	70	8	0.5
AD673S	-55	125	8	0.5
AD7550B	-25	85	13	1
AD7552K	0	70	12 + SIGN	1
AD7571A	-25	85	10 + SIGN	1
AD7571B	-25	85	10 + SIGN	0.75
AD7571J	0	70	10 + SIGN	1
AD7571K	0	70	10 + SIGN	1
AD7571S	-55	125	10 + SIGN	1
AD7571T	-55	125	10 + SIGN	0.75

ANALOG DEVICES

MAXIMUM GAIN TEMPCO (PPM/C)	ANALOG INPUT RANGE (V) MIN	MAX	CONVERSION TIME (uS) TYP	MAX	UNIT PRICE ($)
12.5	-10	+10	25	35	285.00
30	-10	+10		6	141.75
30	-10	+10		4.5	166.50
30	-10	+10		3	207.25
40	-10	+10		2.2	145.50
40	-10	+10		1.8	177.00
40	-10	+10		1.8	253.50
	-5	+5	20	30	12.35
	-5	+5	20	30	61.30
1 TYP	-2	+2	40000		36.00
	-2	+2	160000		16.10
5 TYP	-10	+10	80		40.45
5 TYP	-10	+10	80		46.10
5 TYP	-10	+10	80		32.20
5 TYP	-10	+10	80		37.90
5 TYP	-10	+10	80		121.30
5 TYP	-10	+10	80		138.35

ANALOG-TO-DIGITAL CONVERTERS:

PART NUMBER	OPERATING TEMP (C) MIN	MAX	RESOLUTION (BITS)	MAXIMUM NON-LINEARITY (+/- LSB)
AD7574A	-25	85	8	0.75
AD7574B	-25	85	8	0.5
AD7574J	0	70	8	0.75
AD7574K	0	70	8	0.5
AD7574S	-55	125	8	0.75
AD7574T	-55	125	8	0.5
AD9000J	0	70	6	0.4 TYP
AD9000S	-55	125	6	0.75

ANALOG DEVICES

MAXIMUM GAIN TEMPCO (PPM/C)	ANALOG INPUT RANGE (V)		CONVERSION TIME (uS)		UNIT PRICE ($)
	MIN	MAX	TYP	MAX	
	-10	+10	15		15.50
	-10	+10	15		18.00
	-10	+10	15		12.50
	-10	+10	15		15.00
	-10	+10	15		30.00
	-10	+10	15		35.00
	-2	+2	0.013		73.00
	-2	+2	0.013		109.00

ANALOG-TO-DIGITAL CONVERTERS:

PART NUMBER	OPERATING TEMP (C) MIN	MAX	RESOLUTION (BITS)	MAXIMUM NON-LINEARITY (+/- LSB)
ADC100-BOB	0	70	16	4
ADC100-SMD	0	70	4 DIG+SGN	1
ADC10HT	-55	200	12	0.5
ADC10HT-1	-55	200	12	2
ADC60-08	0	70	8	0.5
ADC60-10	0	70	10	0.5
ADC60-12	0	70	12	1
ADC71J	0	70	16	4
ADC71K	0	70	16	2
ADC72A	-25	85	16	4
ADC72B	-25	85	16	2
ADC72J	0	70	16	4
ADC72K	0	70	16	2
ADC731J	0	70	16	1
ADC731K	0	70	16	0.5
ADC73J	0	70	16	1
ADC73K	0	70	16	0.5

BURR-BROWN

MAXIMUM GAIN TEMPCO (PPM/C)	ANALOG INPUT RANGE (V)		CONVERSION TIME (uS)		UNIT PRICE ($)
	MIN	MAX	TYP	MAX	
10	-10	+10		200000	387.00
5	-10	+10		30	387.00
35	-10	+10	30	50	485.00
100	-10	+10	30	50	448.00
20	-10	+10		0.88	285.00
20	-10	+10		1.88	316.00
15	-10	+10		1.88	326.00
15	-10	+10		50	129.00
15	-10	+10		50	161.00
15	-10	+10		50	207.00
15	-10	+10		50	258.00
20	-10	+10		50	173.00
20	-10	+10		50	216.00
10	-10	+10	150	170	335.00
10	-10	+10	150	170	395.00
10	-10	+10	150	170	285.00
10	-10	+10	150	170	345.00

ANALOG-TO-DIGITAL CONVERTERS:

PART NUMBER	OPERATING TEMP (C) MIN	MAX	RESOLUTION (BITS)	MAXIMUM NON-LINEARITY (+/- LSB)
ADC76J	0	70	16	4
ADC76K	0	70	16	2
ADC80AG-10	-25	85	10	0.5
ADC80AG-12	-25	85	12	0.5
ADC82	-25	85	8	0.5
ADC84KG-10	0	70	10	0.5
ADC84KG-12	0	70	12	0.5
ADC85-10	-25	85	10	0.5
ADC85-12	-25	85	12	0.5
ADC85C-10	0	70	10	0.5
ADC85C-12	0	70	12	0.5
ADC87	-55	125	12	0.5
ADC87U	-55	125	12	0.5

BURR-BROWN

MAXIMUM GAIN TEMPCO (PPM/C)	ANALOG INPUT RANGE (V) MIN	MAX	CONVERSION TIME (uS) TYP	MAX	UNIT PRICE ($)
15	-10	+10		15	229.00
15	-10	+10		15	265.00
30	-10	+10		21	85.00
30	-10	+10		25	87.00
40	-10	+10		2.8	69.00
	-10	+10		6	105.00
30	-10	+10		10	119.00
20	-10	+10		6	143.00
15	-10	+10		10	172.00
40	-10	+10		6	119.00
25	-10	+10		10	132.00
15	-10	+10	7.5	8	240.00
15	-10	+10	7.5	8	172.00

ANALOG-TO-DIGITAL CONVERTERS:

PART NUMBER	OPERATING TEMP (C) MIN	MAX	RESOLUTION (BITS)	MAXIMUM NON-LINEARITY (+/- LSB)
UA571J	0	70	10	1
UA571K	0	70	10	0.5
UA571S	-55	125	10	0.5

ANALOG-TO-DIGITAL CONVERTERS:

PART NUMBER	OPERATING TEMP (C) MIN	MAX	RESOLUTION (BITS)	MAXIMUM NON-LINEARITY (+/- LSB)
HI-5712-2	-55	125	12	0.5
HI-5712-5	0	75	12	0.5
HI-5712A-2	-55	125	12	0.5
HI-5712A-5	0	75	12	0.5
HI-574AJD-5	0	70	12	1
HI-574AKD-5	0	70	12	0.5
HI-574ASD-2	-55	125	12	1
HI-574ATD-2	-55	125	12	0.5

FAIRCHILD

MAXIMUM GAIN TEMPCO	ANALOG INPUT RANGE (V)		CONVERSION TIME (uS)		UNIT PRICE
(PPM/C)	MIN	MAX	TYP	MAX	($)
88	-5	+5	25	30	42.47
44	-5	+5	25	30	49.40
50	-5	+5	25	30	96.62

HARRIS

MAXIMUM GAIN TEMPCO	ANALOG INPUT RANGE (V)		CONVERSION TIME (uS)		UNIT PRICE
(PPM/C)	MIN	MAX	TYP	MAX	($)
20	-10	+10	9	10	286.50
20	-10	+10	9	10	135.76
10	-10	+10	9	10	436.90
10	-10	+10	9	10	273.00
50	-10	+10	25	35	46.07
27	-10	+10	25	35	59.82
50	-10	+10	25	35	123.76
25	-10	+10	25	35	178.75

ANALOG-TO-DIGITAL CONVERTERS:

PART NUMBER	OPERATING TEMP (C) MIN	MAX	RESOLUTION (BITS)	MAXIMUM NON-LINEARITY (+/- LSB)
ICL7106	0	70	3.5 DIGIT	1 COUNT
ICL7107	0	70	3.5 DIGIT	1 COUNT
ICL7109C	0	70	12 + SIGN	1 COUNT
ICL7109I	-20	85	12 + SIGN	1 COUNT
ICL7109M	-55	125	12 + SIGN	1 COUNT
ICL7115JC	0	70	14	2
ICL7115JI	-25	85	14	2
ICL7115KC	0	70	14	1
ICL7115KI	-25	85	14	1
ICL7115LC	0	70	14	0.5
ICL7115LI	-25	85	14	0.5
ICL7116	0	70	3.5 DIGIT	1 COUNT
ICL7117	0	70	3.5 DIGIT	1 COUNT
ICL7126	0	70	3.5 DIGIT	1 COUNT
ICL7129	0	70	4.5 DIGIT	1 COUNT
ICL7135	0	70	4.5 DIGIT	1 COUNT
ICL7136	0	70	3.5 DIGIT	1 COUNT
ICL7137	0	70	3.5 DIGIT	1 COUNT

INTERSIL

MAXIMUM GAIN TEMPCO (PPM/C)	ANALOG INPUT RANGE (V) MIN	MAX	CONVERSION TIME (uS) TYP	MAX	UNIT PRICE ($)
5	−2	+2	66667		12.00
5	−2	+2	66667		12.00
5	−4.1	+4.1	33333		15.00
5	−4.1	+4.1	33333		29.70
5	−4.1	+4.1	33333		61.15
4	0	+5		40	58.50
4	0	+5		40	88.50
4	0	+5		40	102.00
4	0	+5		40	132.00
4	0	+5		40	148.50
4	0	+5		40	178.50
5	−2	+2	66667		12.00
5	−2	+2	66667		12.00
5	−2	+2	333333		12.00
5	−2	+2	166667		19.70
5	−2	+2	66667		15.00
5	−2	+2	333333		12.00
5	−2	+2	333333		12.00

ANALOG-TO-DIGITAL CONVERTERS:

PART NUMBER	OPERATING TEMP (C) MIN	MAX	RESOLUTION (BITS)	MAXIMUM NON-LINEARITY (+/- LSB)
MC10315L	0	70	7	0.2 TYP
MC10317L	0	70	7	0.2 TYP

MOTOROLA

MAXIMUM GAIN TEMPCO (PPM/C)	ANALOG INPUT RANGE (V)		CONVERSION TIME (uS)		UNIT PRICE ($)
	MIN	MAX	TYP	MAX	
	-2	+2	0.07		73.00
	-2	+2	0.07		73.00

ANALOG-TO-DIGITAL CONVERTERS:

PART NUMBER	OPERATING TEMP (C) MIN	MAX	RESOLUTION (BITS)	MAXIMUM NON-LINEARITY (+/- LSB)
ADC0800P	-55	125	8	1
ADC0800PC	0	70	8	1
ADC0801L	-55	125	8	0.25
ADC0801LC	-40	85	8	0.25
ADC0802L	-55	125	8	0.5
ADC0802LC	-40	85	8	0.5
ADC0803LC	-40	85	8	0.5
ADC0804LC	0	70	8	1
ADC0805LC	-40	85	8	1
ADC0808C	-55	125	8	0.5
ADC0808CC	-40	85	8	0.5
ADC0809CC	-40	85	8	1
ADC0816C	-55	125	8	0.5
ADC0816CC	-40	85	8	0.5
ADC0817CC	-40	85	8	1
ADC0833BCJ	-40	85	8	0.5
ADC0833BCN	0	70	8	0.5

NATIONAL

MAXIMUM GAIN TEMPCO (PPM/C)	ANALOG INPUT RANGE (V) MIN	MAX	CONVERSION TIME (uS) TYP	MAX	UNIT PRICE ($)
100	-5	+5	50		35.05
100	-5	+5	50		16.60
	0	5	110		50.25
	0	5	110		17.00
	0	5	110		41.60
	0	5	110		9.25
	0	5	110		6.20
	0	5	110		4.05
	0	5	110		5.15
	0	5	100		62.00
	0	5	100		10.12
	0	5	100		5.37
	0	5	100		74.00
	0	5	100		14.92
	0	5	100		8.92
	0	+5	80		21.75
	0	+5	80		9.75

ANALOG-TO-DIGITAL CONVERTERS:

PART NUMBER	OPERATING TEMP (C) MIN	MAX	RESOLUTION (BITS)	MAXIMUM NON-LINEARITY (+/- LSB)
ADC0833CCJ	-40	85	8	1
ADC0833CCN	0	70	8	1
ADC1021CCD	-40	85	10	1
ADC1210H	-55	125	12	0.5
ADC1210HC	-25	85	12	0.5
ADC1211H	-55	125	12	2
ADC1211HC	-25	85	12	2

ANALOG-TO-DIGITAL CONVERTERS:

PART NUMBER	OPERATING TEMP (C) MIN	MAX	RESOLUTION (BITS)	MAXIMUM NON-LINEARITY (+/- LSB)
CA3162	0	75	3 DIGIT	1 COUNT
CA3300	-40	85	6	0.8

NATIONAL

MAXIMUM GAIN TEMPCO (PPM/C)	ANALOG INPUT RANGE (V)		CONVERSION TIME (uS)		UNIT PRICE ($)
	MIN	MAX	TYP	MAX	
	0	+5	80		12.75
	0	+5	80		4.80
	0	+5	200		31.40
	0	+10	100	200	125.45
	0	+10	100	200	54.65
	0	+10	100	200	90.00
	0	+10	100	200	45.00

RCA

MAXIMUM GAIN TEMPCO (PPM/C)	ANALOG INPUT RANGE (V)		CONVERSION TIME (uS)		UNIT PRICE ($)
	MIN	MAX	TYP	MAX	
50 TYP	0	+1	10400		6.02
250 TYP	0	+5	0.07		21.25

ANALOG-TO-DIGITAL CONVERTERS:

PART NUMBER	OPERATING TEMP (C) MIN	MAX	RESOLUTION (BITS)	MAXIMUM NON-LINEARITY (+/- LSB)
NE5034	0	70	8	0.5
NE5036	0	70	6	0.5
NE5037	0	70	6	0.5

ANALOG-TO-DIGITAL CONVERTERS:

PART NUMBER	OPERATING TEMP (C) MIN	MAX	RESOLUTION (BITS)	MAXIMUM NON-LINEARITY (+/- LSB)
TL520	-40	85	8	0.25 TYP
TL521	-40	85	8	0.5 TYP
TL522	-40	85	8	0.25 TYP
TL530	-40	85	8	0.5
TL531	-40	85	8	0.5 TYP
TL532	-40	85	8	0.5
TL533	-40	85	8	0.5 TYP
TLC532AI	-40	85	8	0.5
TLC533AI	-40	85	8	0.5
TLC540I	-40	85	8	0.5
TLC541I	-40	85	8	0.5

SIGNETICS

MAXIMUM GAIN TEMPCO (PPM/C)	ANALOG INPUT RANGE (V)		CONVERSION TIME (uS)		UNIT PRICE ($)
	MIN	MAX	TYP	MAX	
	-10	+10	17		10.00
	0	+2	23		2.20
	0	+2	9		2.40

TEXAS INSTRUMENTS

MAXIMUM GAIN TEMPCO (PPM/C)	ANALOG INPUT RANGE (V)		CONVERSION TIME (uS)		UNIT PRICE ($)
	MIN	MAX	TYP	MAX	
	-5	+5	70 MIN		10.81
	-5	+5	100 MIN		6.10
	-5	+5	200 MIN		6.39
	0	+5	300		12.65
	0	+5	300		6.79
	0	+5	300		10.35
	0	+5	300		6.10
	0	+5	15		11.20
	0	+5	30		8.03
	0	+5	10		5.08
	0	+5	19		3.86

Appendix G
Sample-and-Hold Amplifiers

SAMPLE-AND-HOLD AMPLIFIERS:

PART NUMBER	TEMP GRADE	TYPICAL ACQUIS. TIME (uS)	TYPICAL APERTURE TIME (nS)	TYPICAL SETTLING TIME (uS)
AD346J	COM	1.6	30	0.5
AD346S	MIL	1.6	30	0.5
AD582K	COM	6	200	0.5
AD582S	MIL	6	200	0.5
AD583K	COM	5	50	
ADSHC-85	COM	5 MAX	25	0.5 MAX
ADSHC-85ET	MIL	5 MAX	25	0.5 MAX
HTC-0300	COM	0.17	6	0.1
HTC-0300A	COM	0.15	6	0.1
HTC-0300AM	MIL	0.15	6	0.1
HTC-0300M	MIL	0.17	6	0.1
HTC-0500A	IND	0.85	30	0.46
HTC-0500S	MIL	0.85	30	0.46
HTS-0010K	COM	0.016	-2	0.005
HTS-0010S	MIL	0.016	-2	0.005
HTS-0025	COM	0.03	5	0.02
HTS-0025M	MIL	0.03	5	0.02

ANALOG DEVICES

TYPICAL DROOP RATE (mV/mS)	TYPICAL S/H OFFSET (mV)	INPUT VOLTAGE RANGE (±V)	MAXIMUM INPUT OFFSET (mV)	UNIT PRICE ($)
0.1	2	10	3	69.00
0.1	2	10	3	108.00
0.1 MAX	0.5	10	6	12.50
0.1 MAX	0.5	10	6	34.80
0.005		10	6	23.90
0.2	1	10	6	91.00
0.2	1	10	6	138.00
5	5	10	20	223.00
1	5	10	20	191.00
1	5	10	20	259.00
5	5	10	20	305.00
0.5	5	10	5	106.00
0.5	5	10	5	144.00
0.1 MAX	2	1	5	369.00
0.1 MAX	2	1	5	438.00
0.2	5	1	20	288.00
0.2	5	1	20	365.00

SAMPLE-AND-HOLD AMPLIFIERS:

PART NUMBER	TEMP GRADE	TYPICAL ACQUIS. TIME (uS)	TYPICAL APERTURE TIME (nS)	TYPICAL SETTLING TIME (uS)
SHC298AM	IND	6	30	1
SHC80KP	COM	10 MAX	40	1
SHC85	COM	4.5 MAX	30	0.5
SHC85ET	MIL	4.5 MAX	30	0.5
SHM60	COM	0.8	12	0.2 MAX

SAMPLE-AND-HOLD AMPLIFIERS:

PART NUMBER	TEMP GRADE	TYPICAL ACQUIS. TIME (uS)	TYPICAL APERTURE TIME (nS)	TYPICAL SETTLING TIME (uS)
UA198	MIL	20	125	0.8
UA298	IND	20	125	0.8
UA398	COM	20	125	0.8

BURR-BROWN

TYPICAL DROOP RATE (mV/mS)	TYPICAL S/H OFFSET (mV)	INPUT VOLTAGE RANGE (±V)	MAXIMUM INPUT OFFSET (mV)	UNIT PRICE ($)
0.025	15	10	7	6.95
0.5	2 MAX	10	2	51.00
0.125	2 MAX	10	2	95.00
0.125	2 MAX	10	2	129.00
1		10	3 TYP	154.00

FAIRCHILD

TYPICAL DROOP RATE (mV/mS)	TYPICAL S/H OFFSET (mV)	INPUT VOLTAGE RANGE (±V)	MAXIMUM INPUT OFFSET (mV)	UNIT PRICE ($)
0.003	0.5	10	3	24.93
0.003	0.5	10	3	10.91
0.003	1	10	7	3.33

SAMPLE-AND-HOLD AMPLIFIERS:

PART NUMBER	TEMP GRADE	TYPICAL ACQUIS. TIME (uS)	TYPICAL APERTURE TIME (nS)	TYPICAL SETTLING TIME (uS)
HA-2420-2	MIL	5	30	
HA-2425-5	COM	5	30	
HA-5320-2	MIL	1.5	25	
HA-5320-5	COM	1.5	25	

HARRIS

TYPICAL DROOP RATE (mV/mS)	TYPICAL S/H OFFSET (mV)	INPUT VOLTAGE RANGE (±V)	MAXIMUM INPUT OFFSET (mV)	UNIT PRICE ($)
0.005	0.005	10	4	44.66
0.005	0.005	10	6	9.77
0.1	2.5			61.00
0.1	2.5			13.38

SAMPLE-AND-HOLD AMPLIFIERS:

PART NUMBER	TEMP GRADE	TYPICAL ACQUIS. TIME (uS)	TYPICAL APERTURE TIME (nS)	TYPICAL SETTLING TIME (uS)
IH5110I	IND	4	120	
IH5110M	MIL	4	120	
IH5111I	IND	4	120	
IH5111M	MIL	4	120	
IH5112I	IND	4	120	
IH5112M	MIL	4	120	
IH5113I	IND	4	120	
IH5113M	MIL	4	120	
IH5114I	IND	4	120	
IH5114M	MIL	4	120	
IH5115I	IND	4	120	
IH5115M	MIL	4	120	

INTERSIL

TYPICAL DROOP RATE (mV/mS)	TYPICAL S/H OFFSET (mV)	INPUT VOLTAGE RANGE (±V)	MAXIMUM INPUT OFFSET (mV)	UNIT PRICE ($)
0.0005	5 MAX	7.5	40	8.95
0.0005	5 MAX	7.5	40	13.50
0.0005	5 MAX	10	40	10.45
0.0005	5 MAX	10	40	16.80
0.0005	5 MAX	7.5	10	11.25
0.0005	5 MAX	7.5	10	16.50
0.0005	5 MAX	10	10	12.40
0.0005	5 MAX	10	10	17.65
0.0005	5 MAX	7.5	5	12.75
0.0005	5 MAX	7.5	5	18.00
0.0005	5 MAX	10	5	13.90
0.0005	5 MAX	10	5	19.15

SAMPLE-AND-HOLD AMPLIFIERS:

PART NUMBER	TEMP GRADE	TYPICAL ACQUIS. TIME (uS)	TYPICAL APERTURE TIME (nS)	TYPICAL SETTLING TIME (uS)
LF198	MIL	20	200	0.8
LF198A	MIL	20	200	0.8
LF298	IND	20	200	0.8
LF398	COM	20	200	0.8
LF398A	COM	20	200	0.8
LH0023	MIL	50	150	
LH0023C	IND	50	150	
LH0043	MIL	30	20	
LH0043C	IND	30	20	
LH0053	MIL	5	10	
LH0053C	IND	8	10	

NATIONAL

TYPICAL DROOP RATE (mV/mS)	TYPICAL S/H OFFSET (mV)	INPUT VOLTAGE RANGE (±V)	MAXIMUM INPUT OFFSET (mV)	UNIT PRICE ($)
0.003	0.5	10	3	34.60
0.003	0.5	10	1	51.20
0.003	0.5	10	3	16.50
0.003	1	10	7	4.25
0.003	1	10	2	6.50
0.001		10	20	34.20
0.002		10	20	27.60
0.001		10	40	38.50
0.002		10	40	22.25
0.006		10	7	60.00
0.01		10	10	42.15

SAMPLE-AND-HOLD AMPLIFIERS:

PART NUMBER	TEMP GRADE	TYPICAL ACQUIS. TIME (uS)	TYPICAL APERTURE TIME (nS)	TYPICAL SETTLING TIME (uS)
SMP-10A	MIL	5	50	7
SMP-10B	MIL	5	50	7
SMP-10E	COM	5	50	7
SMP-10F	COM	5	50	7
SMP-11A	MIL	5	50	1.5
SMP-11B	MIL	5	50	1.5
SMP-11E	COM	5	50	1.5
SMP-11F	COM	5	50	1.5
SMP-11G	COM	5	50	1.5
SMP-81E	COM	3.5	50	1.5
SMP-81F	COM	3.5	50	1.5

P M I

TYPICAL DROOP RATE (mV/mS)	TYPICAL S/H OFFSET (mV)	INPUT VOLTAGE RANGE (±V)	MAXIMUM INPUT OFFSET (mV)	UNIT PRICE ($)
0.005	1.5	10	1.5	52.50
0.005	1.5	10	3	33.53
0.005	1.5	10	1.5	26.25
0.005	1.5	10	3	14.25
0.06		10	1.5	52.50
0.07		10	3	33.53
0.06		10	1.5	26.25
0.07		10	3	14.25
0.08		10	7	4.13
0.1		10	1.6	16.13
0.1		10	3.5	8.25

SAMPLE-AND-HOLD AMPLIFIERS:

PART NUMBER	TEMP GRADE	TYPICAL ACQUIS. TIME (uS)	TYPICAL APERTURE TIME (nS)	TYPICAL SETTLING TIME (uS)
NE5537	COM	20	125	0.8
SE5537	MIL	20	125	0.8

SIGNETICS

TYPICAL DROOP RATE (mV/mS)	TYPICAL S/H OFFSET (mV)	INPUT VOLTAGE RANGE (±V)	MAXIMUM INPUT OFFSET (mV)	UNIT PRICE ($)
0.002	1	10	7	2.70
0.002	0.5	10	3	11.40

Index

Edited by Roland S. Phelps